A History of Ecology and Environmentalism in Spanish American Literature

A History of Ecology and Environmentalism in Spanish American Literature

Scott M. DeVries

Lewisburg
Bucknell University Press

Published by Bucknell University Press
Co-published with Rowman & Littlefield
4501 Forbes Boulevard, Suite 200, Lanham, Maryland 20706
www.rowman.com

10 Thornbury Road, Plymouth PL6 7PP, United Kingdom

British Library Cataloguing in Publication Information Available

Library of Congress Cataloging-in-Publication Data

DeVries, Scott M., 1973–
A history of ecology and environmentalism in Spanish American literature / Scott M. DeVries.
pages cm
Includes bibliographical references and index.
ISBN 978-1-61148-515-8 (cloth : alk. paper) — ISBN 978-1-61148-516-5 (electronic) 1. Spanish American literature—History and criticism. 2. Ecology in literature. 3. Environmentalism in literature. I. Title.
PQ7081.D478 2013
860.9'3556--dc23
2013020602

∞™ The paper used in this publication meets the minimum requirements of American National Standard for Information Sciences Permanence of Paper for Printed Library Materials, ANSI/NISO Z39.48-1992.

Printed in the United States of America

Para Jorge, que me dio la idea.

Contents

Acknowledgments

As with any book, what is bound up between these covers is not the product of the efforts of a single person. I had help from various quarters and the generous support of many people and institutions. Except for their encouragement early in my career, I do not think I would have chosen this path: the kind commentary from Phyllis and Christine so many years ago is never far from my mind as I work. As this project developed, I enjoyed financial support and release from teaching responsibilities from the trustees and administration at Bethel College; specifically, I should thank Jim, Alesha, Dennis, Bradley, Tom, and the Lilly Foundation for the resources that were required to complete research and the time it took to write what is in the pages that follow. Colleagues and students at Bethel are also to be thanked for their many contributions: Gerardo for research assistance, Cristian and Tim for comments on early drafts, Mark and Kevin for securing hard to find titles, and Roxanne for her help in so many other ways. Jennifer French read a draft of chapter 4 and provided very generous input before a version of that chapter was originally published in *Hispania*. The manuscript has benefitted from her ecumenical suggestions as well as from those by the anonymous reader at Bucknell University Press.

For his inspiration, advice, instruction, and friendship, I dedicate this book to Jorge Marcone. Over coffee, tapas, barbecue, picnics, and various other meals in New Jersey, Salamanca, Kentucky, North Carolina, and wherever else our paths have crossed before and after I completed the graduate program at Rutgers University, Jorge has thoughtfully given his time and support to my work. It was he who introduced me to ecocriticism and to the idea of using it as an approach to Spanish American literature. We worked together on several pages from part 3 in another context and in an earlier form. The entire scholarly community focused on ecology and Spanish American literature owes him a great debt of gratitude, none more than me.

Finally, I cannot have completed this work without the support, patience, and encouragement from my family. I am exceedingly grateful to my parents, John and Susan, whose enthusiasm (and financial provision) for education has been foundational to my formation as a student and scholar. To my children, Joaquín, Ana, and Sofia: gracias niños por aguantar tanto tiempo con Papi en la computadora, trabajando en su libro. And to my wife María, thank you for your help, your encouragement, and the time you invested in making each and every page possible.

Introduction

Ecology and environmentalism have had a long if unfamiliar history in Spanish American[1] literature. In 1989, when the musician Sting founded the Rainforest Foundation and spent quite a bit of time in the South American rainforest, Chilean Luis Sepúlveda released his explicitly environmentalist novel *El viejo que leía novelas de amor* (*The Old Man who Read Love Stories*), a novel that has become the third most popular Spanish-language book in print after the *Quijote* and Gabriel García Márquez's masterpiece *Cien años de soledad* (*One Hundred Years of Solitude*) (1967).[2] Famed Mexican monarch butterfly activist, author, and intellectual Homero Aridjis had just published two apocalyptic novels about the future of Mexico City. Gioconda Belli, sometime director of State Communications for the Sandinista regime in Nicaragua, had left that position by the 1990s, and published *Waslala* (1996), a work of environmentalist science fiction. Elsewhere in Central America, Costa Rica was cultivating a reputation for eco-tourism and authors Anacristina Rossi and Fernando Contreras Castro wrote novels that advocated for the conservation of national parks and exposed the environmental fallout from trash disposal practices around the capital San José. And back in the Southern Cone, Mempo Giardinelli, noted Argentine columnist and winner of the prestigious Venezuelan Rómulo Gallegos literary prize, published the ecologically themed *Imposible equilibrio* (*An Impossible Balance*) (1995), followed by a pseudo-sequel *Final de novela en Patagonia* (Concluding a novel in Patagonia) in 2000. This batch of explicitly environmentalist writing seemed to assume the contours of a sub-genre within the ill-defined contemporary literary period after the "Boom." When I came across these novels in the 1990s, I assumed other scholars would think the ecological tendencies of recent texts an invitation to undertake an ecocritical reevaluation of earlier literary periods: the "foundational fictions" of the nineteenth century, the "Boom," and other quote-worthy currents of the literature. Indeed, as I write these words, the interest in recent Spanish American environmentalist fictions is booming, yet an attempt at the construction of a longer eco-literary history has yet to be made.

An earlier, shorter version of just such an effort may be found in an article from 2000 by Jorge Paredes and Benjamin McLean: "Hacia una tipología de la literatura ecológica en español" (Toward a typology of ecological literature in Spanish). In that article, the authors are careful to denominate the object of their "typology" with the term "literatura ecológica" (ecological literature) and to define it as works that transmit:

> un claro mensaje acerca de la necesidad de reexaminar las relaciones entre el ser humano y los demás elementos de la Naturaleza, lo que deberá redundar en una toma de conciencia sobre la necesidad de respetar y defender nuestro entorno . . . donde el ser humano no tenga privilegios con relación a los demás seres vivos, ni derechos dictatoriales sobre la suerte del planeta.[3]
>
> [a clear message about the necessity to reexamine relationships between humankind and other elements of nature that should have the effect of a dawning awareness about the necessity to respect and defend the environment, . . . an awareness that human beings do not live a privileged existence with regard to other living things nor possess a kind of dictatorial right of dominion over the planet.]

The definition of "ecological literature" as writing with a kind of eco-activist subtext leads the authors to suggest that in the context of Spanish America, it did not happen until the publication in 1983 of both *Dolor de patria* (Suffering homeland) by Salvadoran José Rutilio Quesada and the neo-*indigenista* testimonial text *Me llamo Rigoberta Menchú* (*I, Rigoberta Menchu*, compiled by Elizabeth Burgos-Debray). While each of these titles clearly fits within the parameters of their definition, the fixture of such a late date for texts that qualify as "ecological" seems shortsighted. Indeed, José Manuel Marrero Henríquez argues that

> limitar el estudio de la ecocrítica a la literatura producida a partir de los años ochenta vinculada a las culturas prehispánicas . . . abandona la posibilidad de hallar contenidos y formas ecologistas tanto en otros tiempos como en ámbitos de la literatura hispanoamericana enraizados en tradiciones europeas.[4]
>
> [to limit ecocritical approaches only to literature produced in the [nineteen] eighties in the pre-Hispanic tradition . . . abandons the effort to find ecological content and forms from other times or in Spanish American literature rooted in the European tradition.]

Aware of these internal debates within the ecocritical community of scholars, readers, and students of the region's literature, I believe the time has come to more exhaustively plumb the depths and even go beyond traditional or seemingly established boundaries of the canons of poetry, novels, stories, and other texts from the whole of Spanish American literary history. What I lay out in the following pages and chapters demonstrates that, contrary to the date fixed at 1983 by Paredes and McLean, the tradition of an ecological literature from Mexico to Patagonia and from Puerto Rico to Easter Island has a history that goes much farther back. But before I turn to that effort, it will be a useful point of departure to survey what has been done so far.

Jorge Marcone was among the first to author scholarly analyses that featured ecocritical approaches to Spanish American literature in a series of articles about novels set in the Amazon River basin.[5] By now, it would take a great deal of space to enumerate all of the other scholars that have published journal articles with ecocritical approaches to Spanish American literature and poetry, but Laura Barbas-Rhoden; Adrian Taylor Kane; Carmen Flys Junquera, José Manuel Marrero Henríquez, and Julia Barella Vigal; Beatriz Rivera-Barnes and Jerry Hoeg; and Germán Bula and Ronald Bermúdez have all recently published volumes of collected essays on ecology and Latin American literature, particularly focused on the nineteenth century.[6] As Barbas-Rhoden puts it, these articles and collections are "an examination of selected fiction about different historical moments, . . . [and are] one way to tell [Latin American] environmental history."[7] The quantity of scholarly work referenced above demonstrates the expanding interest in ecocritical approaches to recent works of fiction and poetry written in Spanish. I take this as the point of departure for the work in the following chapters, but with a view from the present concerns of ecology and environmentalism looking back. With regard to ecocriticism as a form of textual analysis, the frame may be placed to forty years ago or more;[8] but for the articulation of a broad history of literary ecology in Spanish America, the starting point must go much farther back. So in the following pages, I range beyond an analysis confined to selected novels, period genres, or individual authors to trace the record starting from the very first examples of fiction, poetry, and essay written in the New World.

What I have found is that most of the issues of interest to present environmentalists—conservationism, environmental justice, deforestation, the exhaustion of natural resources, climate change, species extinction, ecological contamination, alternative models of consumption, representations of sublime nature, etc.—have been present for quite some time in the region's literature. And not only that, but these environmental concerns have been expressed rather distinctively in the Spanish American literary tradition. In the nineteenth century, the articulation of competing political ecologies occurs almost exclusively in fictional forms; at the dawn of the new century, the representation of sublime nature is uniquely evoked through the highly aesthetic conventions of *modernismo*; the nineteenth century regional novels achieve a fictionalization of wilderness, rainforest, landscape, and topography that complicates simplistic representations typical of North American and English writers;[9] the unique framework of the *indigenista* text takes up questions of environmental justice from the perspective of native peoples; and, as inheritors of the legacy of the literary innovations of the "Boom" of the 1960s and 1970s, more recent novels tend to feature environmentalist ideas that are put to the service of literary innovation. But to appreciate the distinctiveness of these environmental emphases in Spanish American literature on the whole will require a comprehensive accounting, one that begins with the nineteenth century authors from the newly independent nations of the Spanish-speaking Americas and concludes with its most recent fiction.

After an extension of temporal bounds, the next significant challenge for such a project is the question of the importance of novels, stories, myths, legends, poetry, and other fictional texts when the degradation of the natural world is so acute, particularly in a region like Latin America with its unsafe mining practices, petroleum contamination, deforestation, violation of the land rights of indigenous groups, and other pressing ecological concerns. In *Going Away to Think*, Scott Slovic considers Charles Wilkinson's proposal to re-connote language to more accurately reflect environmental reality. Wilkinson had challenged a forest ranger's claim about "forest productivity" by referencing the perspective of a vole; Slovic observes that such a rearrangement of language and values can only come in the form of literature: "the process of transforming our official legal, economic, and governmental language from that of bloodless contractual arrangements to empathetic stories that might enable us to imagine the issue of forest productivity from the perspective of a vole (a small, mouse-like mammal) will require monumental ambition and verbal acrobatics."[10] But that is what Spanish American literature has been doing all along, no acrobatics required. Uruguayan Horacio Quiroga, for example, has written several short stories where the narrative proceeds directly from the perspective of animals: hunting dogs, venomous snakes, tigers, and horses. Furthermore, many other of the region's most well-known writers were also high-ranking politicians who used literature to communicate ideas, many of them ecological in tenor. In fact, several of the foundational principles and philosophical approaches of environmentalism, ecologism, and ecocriticism involve many of the same issues that were of concern to the authors of the earliest examples of the literature: civilization and barbarism, the representation of nature, the place of Latin American nature mythology, political ecology, geographic determinism, the representation of indigenous peoples and phenomena. Later authors took up issues concerning morality and nature, environmental degradation, utopian dreams, recovery from guerilla war, warnings about ecological apocalypse. All the while, these ideas rarely take the form of the non-fiction treatise,[11] but are expressed through fictional literature or poetry. In chapters 1 and 2 for example, I articulate certain thematic tendencies in important texts from the nineteenth century; subsequent chapters examine the way in which different literary periods or genres take up the issues enumerated above. Some of these are exclusive to regional contexts, others are common to ecocriticism as a whole; all of them have an extensive history in Spanish American literature.

ECOLOGY AND ENVIRONMENTALISM

Terms like "environmentalism" and "ecocriticism" and the imprecise adjective "ecological" are conspicuously ill-defined and I want to clear up any doubts before going further. The most elemental of the concepts—"ecology"—has attained a dual meaning; first, it can mean either the set of relations between flora, fauna, land, and other elements of the biosphere in an ecosystem, but it also commonly signifies efforts undertaken to protect or conserve the conditions in which those relations occur. One of its earliest occurrences dates from the 1870s in translations of German zoologist Ernst Haeckel's texts who used it in the first sense as "the study of the relations of plants and animals with each other and with their habitat."[12] When humans are introduced into the set of relations, ecology evokes the term environment in the sense that humans are the center around which these relations can be observed.

In *Ecology in the 20th Century*, Anna Bramwell notes that the etymology of "ecology"—"used originally as coterminous with . . . *oekonomie*, the concept of 'economical' household management"—indicates that ecology invokes a moral imperative and thus the involvement of humans. She indicates that the "normative sense of the word has come to mean the belief that severe or drastic change within [a] system, or indeed any change which can damage any specie within it, or that disturbs the system, is seen as wrong."[13] Bramwell observes that presently, ecology is widely used in the normative sense as evidenced by the "eco-" adjunct for a prefix in an array of conservationist, protectionist, and/or activist-oriented terminology. From a Latin American perspective, Brazilian liberation theologian Leonardo Boff makes a further distinction concerning ecology, one that spreads its meaning to principles of ecological justice: it "is concerned not only with animals, plants, and the purity of the atmosphere, but with the joint and global relations of human beings and nature. The authentic notion of ecology is always holistic and maintains an alliance of solidarity with nature."[14] Ecological injustice has likewise become as important an issue for the concerns of environmentalism as sustainable development and conservation;[15] to wit, the link between ecology and justice means the effort to protect the environment, not just for the sake of nature itself or for future generations, but for current populations of people who are most directly victimized by ecological contamination. Predominantly, these populations tend to be poor, uneducated, politically disenfranchised, female, and non-white; the ecological or environmental justice movement seeks to protect the rights of the members of these groups to an uncontaminated environment.

The inception of the "deep ecology" movement, articulated by Arne Naess at the World Future Research conference in Bucharest (1972), further contributed to the normative sense of ecology. Naess proposed a program of radical societal transformation in response to failures of what he called the "shallow" efforts at conservation: faith in technology to curb pollution, corporate-sponsored resource management, and "mild changes in lifestyle (such as recycling)."[16] Naess's ideas were based upon the assumption that individuals ought to "feel inclined to act benevolently towards nature [and that] identification with nature and natural things is the process through which this inclination towards benevolent action is thought to develop. . . . Anyone who identifies with natural things, who sees them as part of themselves, is therefore likely to feel inclined to protect them."[17] The idea of environmentalism and environmentalist are similarly related to the normative sense of ecology: both terms connote dual, sometimes contradictory, sometimes synonymous meanings while their derivatives—ecocriticism or environmental criticism, ecological justice or environmental justice, eco-activist or environmental activist or simply environmentalist—are terms that connote the normative concerns of ecology.

Likewise, environmentalism is a multi-faceted social phenomenon comprised of various tendencies, foci, and emphases. Anthropologist Kay Milton defines it as the expression of "concern that the environment should be protected, particularly from the harmful effects of human activities [expressed] through government policies, [or] through demands for changes in land use"; it can also involve "protest against a commercial ethos," or the emulation of non-industrial societies, perceived as "models for a sustainable or conserver society" that maintain "spiritual ties"[18] to their land. She references Jonathan Bate to affirm that the romantics (and subsequently, the environmentalists) idealized such societies because, as Bate put it, their members "live in rhythm with nature."[19] Milton's anthropological analysis of nature protectionists identifies a similar spiritual or emotional commitment, a profound identification with and even love for nature: the "deep ecology" mentioned in the previous paragraph.

ECOCRITICISM

The reflections in the previous few paragraphs have mostly to do with a description of the natural world and activism or advocacy to conserve, protect, and/or identify with it. All of these are somewhat different from ecocriticism which has more to do with the texts—fiction and non-fiction, prose and poetry, on paper and on screen—that represent such things. And here, precision is foiled by the same binary definition that characterized the idea of ecology: a subtle merging of descriptive and normative attributes is central to the critical debate concerning a definition for "ecocriticism." In the introduction to his anthology of "Green Studies," Laurence Coupe states that ecocriticism is "in need of clarification and organisation. I was tempted to say 'definition' but that would be fatal. . . . Green connections really do stop

Canon

With regard to the corpus of ecological texts that "should" be read, Murphy suggests that the promotion of ecological awareness can be effected through decisions by faculty members in university departments of literature to expand the canon to include texts "that are devoted to nature and environmental topics."[27] More specifically, John Tallmadge describes the political aspects of canon reevaluation as an inevitable product of ecocriticism which he characterizes as having been used "thematically, to denote ordinary analysis of writing concerned with environmental issues, and politically, to revisit classics or valorize neglected works by reading them ecocentrically. This sort of criticism has already produced impressive results by opening the canon, revitalizing the curriculum, and deepening our understanding of many great writers."[28] These comments are particularly relevant to the history I aim to construct in chapters 1–7 with an eye toward both uncovering the ecology in texts where it may not have been seen before as well as discovering texts that have much to do with ecology and environment but which have been neglected or forgotten by scholars, students, and readers of the literature. In chapter 4, for example, I assert the ecocritical importance of Nicaraguan Hernán Robleto's *Una mujer en la selva* (A woman in the jungle) (1936), a title that merits the obligatory mention in a series of introductions and companions to Spanish American fiction, but which has not appeared as the subject of a monograph or scholarly article in any context at all.

The story is set on an *hacienda* that borders undeveloped jungle into which a woman is kidnapped by a large monkey with whom she eventually identifies very closely and finally grows to love. The novel, in all likelihood, took advantage of the similar story and international success of Hollywood's *King Kong* (released just a few years before) as a hook for publication, but was eventually relegated to just one or two imprints and soon forgotten. But *Una mujer en la selva* deals with many issues that are of current ecological concern including the beauty of pristine nature and the unfortunate advancement of development to such places; the identity and value of nonhuman nature; animal rights; the human capacity to get back to nature; etc. It is only through an extensive reexamination of the corpus (not the canon) of the literature that such titles can be uncovered.[29] The method can be quixotic (I stumbled upon quite a few texts simply browsing the PQ stacks in the Hesburgh Library at Notre Dame) or methodical (I read my eyes blurry going through several companions to the literature[30] and blurrier still reading the titles on various lists I composed as a result). The process involves a return to literary periods when ecological ideas were not necessarily in vogue and supposes that the expression of such ideas is often discovered implicitly and in retrospect. There are often incomplete thoughts, asides, or throw-away lines, undeveloped, infrequent, and without elaboration; perhaps the beginnings of an argument are made, but because the didactic, artistic, or literary emphases are elsewhere the ecological implications are never pursued to their logical conclusion. Thus, a re-reading of the literature along broad lines of inquiry and over vast periods and diverse regions with consideration for neglected titles alongside what has come to be thought of as canonical texts can uncover ecocritical sentiments in the literature that have not heretofore been thoroughly explored.

Even when texts only contain one or two moments of environmental awareness, the compendium of proto-environmentalist material amassed through a broader reading of the corpus evidences the anticipation of several key ecological ideas. I discovered that with the exception of the more explicitly environmentalist novels considered in chapters 8, 9, and 10, the development of a sense of the ecological tenor in Spanish American literature required a wide-ranging search. It is part of the work of an ecocritic to undertake such searches, bring candidate texts to light, and restore them (or include them for the first time) to reading lists commonly recognized by scholars, students, and the general reading public.

Political Ecology

The other normative aspect of ecocriticism that is particularly relevant here is the way in which it advocates for political change, a point on which scholars seem unequivocal. For example, Laurence Coupe elaborates an approximation of what he calls "green studies" specifically with regard to behavioral changes: "It sees planetary life as being in a 'critical' condition; and it is to this sense of 'crisis' that it offers a response. If green studies does not have an effect on this way of thinking, does not change behaviour, does not encourage resistance to planetary pollution and degradation, it cannot be called fully 'ecocritical.'"[31] Andrew Husband seems to echo the same idea when he affirms the political aspects of theoretical approaches: "Any lacking of awareness of the politically conscious and unconscious motivations that fuel ecocriticism will nullify the validity and efficiency of its development."[32] Both Coupe and Husband are highly skeptical of an ecocritical agenda that does not advocate for real action as is evident in almost all of the definitions, characterizations, and approximations to ecocriticism that they elaborate. But there have also been some doubts among ecocritics and eventually the (inevitable) manifestation of critical perspectives on ecocriticism: the very lack of a definable project that identifies what ecocritics are doing makes for a theoretical approach that seems a little wishy-washy. John Tallmadge goes so far as to say that "the 'eco' refers more often to the content of the work or the purpose of the critic than it does to the critical method itself."[33] Still, when it comes to Spanish American literature, the persistent appearance of political ecology from the nineteenth century onward makes ecocritical approaches that foreground praxis an inevitability of the literary analysis.

In the introduction to *Liberation Ecology*, Richard Peet and Michael Watts offer a summary of the various characteristics, tendencies, and possibilities of political ecology. I have distilled their analysis to this definition: the concept can be defined as analysis (approaching advocacy) of the way in which certain communities engage with structures of political power for access to natural resources and interactions with the environment. A concise and narrow definition of political ecology is attractive for its unambiguous categorical finality, but the term is much like its root "ecology" in that it is both descriptive and prescriptive. That is, political ecology attempts to describe negotiations for access to resources, but it also simultaneously advocates for certain arrangements of access, usually in favor of local entities and opposed to globalized capitalism. Raymond L. Bryant and Sinéad Bailey describe political ecology as a contest mainly among a variety of actors to whom costs and benefits of economic development are distributed unequally. They find that wealth inequality creates differential power relations among factions within a society or region where degrees of power translate into the ability of actors to control their own and others' access to resources. Bryant and Bailey describe the way such arrangements may function differently for the state, for transnational corporations, for non-governmental organizations, and for grassroots organizations in (what they continue to refer to as) the "Third World." In the end, they find that "political ecology is a research field that seeks to explain the topography of a politicized environment, and the role that diverse actors play in the 'moulding' of that environment, so as to better assist those actors in society who are fighting for social justice and environmental conservation."[34] Thus, political ecology is not only a negotiation for access to resources, it is also descriptive of that effort, and a form of advocacy that prescribes and proscribes the boundaries within which these negotiations take place. Bryant and Bailey make reference to aspects of politics, anthropology, sociology, and geography in their discussion of political ecology, but Andrew Dobson takes a still another tack and construes political ecology as nature protectionism conceived as a public good and a matter of active citizenship, political virtue, and a prerequisite for the enjoyment of other formal rights.[35] And Alleta Biersack emphasizes the idea (after anthropologist Arturo Escobar)[36] that political ecology can also be an exercise in deconstruction: that reality is produced discursively (through language) and that "nature" now goes in quotes in the sense that it is secondary, socially produced, and must be concerned with the impact of human actions upon nature: "Second nature is nature as it has been transformed though human activity . . . a nature that is *humanly* produced (through conceptualization as well as activity) and that therefore partakes, but without being entirely, of the human."[37] As with many terms, "political ecology" seems to have passed through various iterations and become difficult to manage. Therefore, I will use the term exclusively to refer to the following distillation of the distillation of the definition that I reference above: analysis and/or advocacy concerning access to natural resources. In cases where there are contests among, rights construed to, or conceptualizations promulgated by any of the actors involved in these negotiations, I will be explicit and specific rather than generalize all of these as "political ecology."

Peet and Watts suggest several directions for political ecology which make "the causal connections between the logics and dynamics of capitalist growth and specific environmental outcomes rigorous and explicit," explore how agents of a state mediate conflicts concerning "nature, labor power, and communal conditions of production," consider the influence of "the institutions of civil society" such as environmental movements or local practices, and tackle issues concerning "the plurality of perceptions and definitions of environmental and resource problems."[38] In the pages that follow, I find that political ecology in the literature of Spanish America invokes each of the directions mapped out by Peet and Watts as well as by the other scholars cited above: sometimes with an agenda that decries environmental contamination or one that condemns the way in which agents of the state neglect the victimization of marginalized groups or one that defends local and indigenous institutions against foreign and external incursions or one that articulates the effects of environmental disruptions, biological scarcity, and poor stewardship of resources.

When in *Facundo* (1845) Argentine Domingo Faustino Sarmiento argues that the waterways that branch out from the River Platte ought to be exploited as a means to develop the country's hinterlands, his novel/biography/essay was engaged in the analyses and advocacies of political ecology. In this case, Sarmiento is concerned with the way in which a land that previously produced barbarism can be altered so that it will produce civilization. The political ecology of *Facundo* envisions cities as the ideal conductors for the transmission of civilization because "la pampa es un malísimo conductor para llevarla y distribuirla en las provincias" ["the pampa or Argentine plains are an awful conductor of civilization to the interior"]. Civilization—"las leyes, las ideas de progreso, los medios de instrucción, alguna organización municipal, el gobierno regular" ["laws, ideas about progress, methods of instruction, municipal organization, normal government"][39] —proceeds from urban centers and ought to go forth with the assistance of science and industry to tame wild, barbarous nature. From the moment that Sarmiento's proposal for a political ecology of possession and industrialization of the land rang out from the pages of *Facundo*, a capitalist mode of land exploitation becomes the standard model for the development of Latin America. But most often, the political ecologies of Spanish American literature challenge Sarmentine assumptions and take ideological positions that have arisen to challenge the many troubling aspects of development and modernization as they have been applied in Latin America. As a participant in the global economy, the region finds itself in the precarious position of supplying markets, products, and resources to an economy dominated by the United States, Europe, and now China. While the inclusion of Brazil in the Group of Eight developing countries may counterbalance U.S./European influence, the continuing power of the latter (as well as of the growing Chinese economy) still inordinately affects the way that Latin American economies and markets operate. The larger states leverage their considerable financial power to influence the import and export of products and natural resources, the kinds of government that ought to exist, as well as the environmental policies to be pursued. In the context of economies dominated by North American, European, and Chinese interests, what parts of "nature" ought to be conserved and what parts developed and utilized becomes a question of neo-imperial power relations. The role of power in the use of nature is a particularly important issue in Latin America which, as a post-colonized region, object of neo-imperialism, and inheritor of the discourse of development, must still wrestle with questions of hegemony in terms of land use and control of natural resources.[40]

Development

It was as early as the appearance in 1891 of the essay "Nuestra América" ("Our America"), by Cuban literary hero José Martí, that we find one of the earliest challenges to the influence of development discourse: the references to stolen lands and to the value of local knowledge; the rejection of traditional forms of governance, education, and class ideology; the call for a change of intellectual spirit and an embrace of the natural man. All of these ideas challenge the central suppositions upon which the imported concept of development was founded. Although the meaning of "development" can vary depending on the field in which it is applied, Raymond Williams observes that the modern usage of the term "relates to certain ideas of the nature of economic change."[41] Development theory hinges upon the notion that regions should, can, and will experience economic growth through a series of stages usually culminating in globalized trade, but when Martí describes his America—"Nuestra América"—as a region marked by the Río Bravo and the Magellan Strait, he proposes geographies that reject globalized economies in favor of localized and individuated alternatives.

In Latin America, the diverse processes of modernization and development have resulted in a kind of politics of dependency based upon what Chilean Fernando Mires describes as an "economics of the absurd":

> Entre los "economistas del crecimiento" existe una suerte de pacto tácito, que va más allá de toda convicción política. Ese pacto se hace presente en el desprecio absoluto

que manifiestan frente a toda noción que altere en lo más mínimo el sentido puramente cuantitativo de la economía que ellos representan.

Los "economistas del crecimiento", en sus versiones de derecha e izquierda, al no integrar los valores no cuantificables de la naturaleza en las teorías del cálculo económico, tampoco pueden integrar al ser humano, dado que éste no puede ser únicamente entendido a partir de simples relaciones de cuantificación.[42]

[Among "growth economists" there exists a kind of tacit pact that goes far beyond all political convictions. This pact is made evident in the absolute disregard that they show toward any notion that deviates to even the smallest degree from the purely quantitative foundations of the economy that they represent.

"Growth economists," in their conservative and liberal manifestations, because they do not take into account the unquantifiable value of nature in their theories of economic calculation, also cannot properly account for humanity as humans simply cannot be understood in quantifiable terms.]

Although it puts a price on the development of nature, an "economics of the absurd" does not, indeed cannot, correctly value the ultimate costs of doing so. The price of the transformation of an oil field into usable energy is determined under the capitalist system based only upon what is involved in the exploration and drilling for, then pumping, refining, and transportation of the oil. But the permanent loss of the resource, the destruction of the land upon which the drills and refineries are built, the contamination of air by the by-products of combustion, and any other destructive environmental side effects are not included in the calculation. The "absurdity" of the system lies in its qualitative failures: Mires's criticism of the failures of capitalist economic calculus has been central to the political ecologies of Spanish American literature as far back as the nineteenth century.

A collection of novels and short stories (called alternately the regional novels, the *novela de la tierra,* or the *novela de la selva* according to the generalities or specificities required by the critic)[43] contains marked criticism of globalized development that features ideas central to the current environmentalist critique. The regional novels often portray social injustice not just in terms of suffering endured by workers but also in terms of the damage to nature occasioned by modernized industry. Conservationism, the environmentalist response to the "economy of the absurd," emerges as an important part of the political ecology of these novels as do aspects of deep ecology. Similarly, Spanish American *indigenista* literature represents the harsh social and ecological by-products of capitalist economics as these are experienced in indigenous communities. Inasmuch as the literary traditions of *indigenismo* make connections between the natural and human worlds (or rather, make no distinction between the two), they offer an argument against development when it comes at the expense of human suffering, as it so often has in indigenous communities.

Ecofeminism

Ecological feminism represents a critique of oppressive institutions or social structures that result in the domination and destruction of the natural world. While the specific concerns of ecofeminism are rather diverse, one unifying feature is that "the oppressive conceptual frameworks that have justified the domination of women and nonhuman nature have been patriarchal."[44] In Karen J. Warren's account of the philosophy of ecofeminism, she also articulates such elements as the historical pervasiveness of patriarchal domination; the raw data of the way in which poor environmental stewardship adversely affects women; the importance of earth-based, feminine symbols such as nature goddesses, Gaia, Mother Nature, etc.; the importance of communion with nature; and the value of flourishing or well-being for individuals, communities, and species. Greg Garrard notes that "ecofeminism emphasizes environmental justice to a far greater degree than deep ecology. The logic of domination is implicated in discrimination and oppression on grounds of race, sexual orientation and class as well as species and gender."[45] And with regard to the analysis of literature from that perspective, Karla Armbruster holds that "in order to make a significant impact on literary criticism and theory, ecofeminist literary criticism must offer a perspective that complicates cultural conceptions of human identity and human relationships with nonhuman nature."[46] Literary works as manifestations of political ecology, environmentalism, and other nature-protectionist postures can be particularly effective to complicate such "cultural conceptions of human identity."

From the feminist context, Warren comes to a similar conclusion as did Scott Slovic in his attempt to imagine the perspective of a vole; that is, the use of novels, poetry, and other texts "can be more than just a helpful literary device for describing ineffable experience or a mythology for documenting personal and social history; it can be an invaluable vehicle for revealing what is ethically significant in human interactions with humans and the nonhuman world."[47] Narrative expression as a means to express ecological principles of particular concern to feminists may be manifest in four distinct ways: it can "give voice to a felt sensitivity, an emotional disposition, or . . . a sensitivity to conceiving of oneself as fundamentally being in relationships with others including the nonhuman world"; it "gives expression to a variety of ethical attitudes and behaviors overlooked or underplayed in mainstream Western ethics"; it can "express felt differences between conquering and caring attitudes in relation to the nonhuman natural environment"; and it "has argumentative significance by suggesting *what counts* as an appropriate conclusion to an ethical situation."[48] When literary criticism approaches ecology and environmental justice with an emphasis upon relational identification and caring attitudes, rejecting the domination of one gender over another, it displays uniquely ecofeminist attributes. And whereas ecofeminist literary criticism affirms the value of finding such elements without regard to textual genre, canonical status, or fictional posture, it becomes a highly useful analytical tool for the work of the following chapters: the imperative to include literary representations of appropriate relational behaviors and caring attitudes encourages the expansion of the canon of Spanish American literature to include texts with these elements.

LITERATURE, ETHICS, AND THE "ECOSUBLIME"

To contextualize the connections among ethics, ecocriticism, and literary texts as they are relevant to the history that is the focus here, it will be helpful to return to mid-nineteenth century Spanish American literature: the sentimental novel *María* (1867) by Colombian Jorge Isaacs and other Spanish American novels followed the literary forms of European romanticism which were in vogue on the continent, particularly in France. In these novels, the representation of nature follows the romantic tendency to idealize and anthropomorphize nature; characters experience nature as a kind of transcendent encounter with the sublime. And the highly spiritual or mystical human/nature connection has also had long-standing precedent among ecologically minded authors. The idea of the sublime finds expression in analysis by Jonathan Bate who observes a contrast between those that exhibit a sense of belonging in nature as opposed to the experience of "modern men": "For the old woodlanders, there is no division between human intercourse and local environment. The presence of memory means that the countryside is inhabited rather than viewed aesthetically. The condition of the modern man, with his mobility and his displaced knowledge, is never to be able to share this sense of belonging. He will always be an outsider; his return to nature will always be partial, touristic, and semi-detached."[49] While Bate's characterization of the ecological condition of modern individuals seems to preclude anything but a "partial, touristic, and semi-detached" experience of nature, Lee Rozelle's conception of the "ecosublime" redeems the value of such mediated experiences: "There is no affective difference between the natural sublime and the rhetorical ecosublime; both have the power to bring the viewer, reader, or player to heightened awareness of real natural environments. Both can promote advocacy."[50] Part of the ecocritical approach here will be analysis of the way that Spanish American texts can either articulate and/or criticize political ecologies, ideas about development, and the concerns of ecofeminism, but it will also consider how literary phenomena such as the "ecosublime" can arouse environmentalist reactions in readers. That is, just as for political ecologists, the representation of inequitable negotiations for access to resources can help to redress that injustice or for ecofeminists, the representation of a caring relationship between a human and his or her landscape suggests an ethic of responsible stewardship for the whole of the natural world, I follow Rozelle in the belief that transcendent experiences of nature mediated by literature suggest protectionist rather than consumerist or developmentalist attitudes toward the environment.

For the authors themselves, Slovic has noted that the power of environmental literature derives from the fact that writers "understand their work as the effort to achieve not only beautiful lyrical language, but an understanding of human society's relationship to the actualities of the planet."[51] On the part of readers, Lawrence Buell has enumerated four ways in which environmentalist literature can engage readers with the physical world: "They may connect readers vicariously with others' experience, suffering, pain: that of nonhumans as well as humans. They may reconnect readers with places that they have been and send them where they would otherwise never physically go. They may direct thought toward alternative futures. And they may affect one's caring for the physical world: make it feel more or less precious or endangered or disposable."[52] The articulation of these four engagements suggest that there is an ethics of what Buell calls the "environmental imagination," an ethics of writing about the environment that can direct readers toward a consideration of how to act and think with regard to the natural world. Buell's connection between reading and action is particularly apt here as ethics and Spanish American writing have long been closely related: the explicitly didactic aims of nineteenth century and *indigenista* literature were none too subtle. But an ecocritical approach to Spanish American literature can also account for its highly literary moments as these too can effectively articulate ecological principles. For example, critic Aníbal González affirms that the *modernistas* were the ones to inaugurate "el principio en Hispanoamérica de una verdadera ética de la escritura, es decir, de una tendencia a preguntarse no sólo cómo utilizar la escritura como instrumento crítico o herramienta del poder, sino también por qué y para qué se escribe" ["the start in Hispanic America of a true ethics of writing, that is, the tendency to ask oneself not only how to utilize writing as a critical instrument or tool of power, but also why and for what purpose one writes"].[53] González's invocation of Spanish American *modernismo*, with its highly formalized aesthetics and insistent sense of pure literary expression, strongly suggests that he would recognize that even the most aesthetic literature of the region may also express ethical positions. Obviously, the ethical aspects of literature are at the very core of an ecocritical analysis because the way that reading fiction or poetry causes people to *think* about the environment can have an impact on their actions when they come into contact with it in acts of consumption or conservation. Images of nature in film, poetry, essay, and literature must play a fundamental role in the cultural formation of ecological values and in social decisions about environmental behavior.

Nevertheless, the texts to be studied in the following chapters do more than just getting readers to think and act in certain ways. Argentine Jorge Luis Borges's analysis of the North American writer Nathaniel Hawthorne in a series of lectures on "*las letras americanas*" (American letters) warns against the closure of literary interpretation. In a talk entitled "Nathaniel Hawthorne" he cited the author's parable *Earth's Holocaust* (1844) lamenting upon the moral that Hawthorne had inserted at the close of the tale:

> Hawthorne, aquí, se ha dejado arrastrar por la doctrina cristiana, y específicamente calvinista, de la depravación ingénita de los hombres y no parece haber notado que su parábola de una ilusoria destrucción de todas las cosas es capaz de un sentido filosófico y no sólo moral.[54]

> [Here Hawthorne has allowed himself to be overly influenced by the Christian and specifically Calvinist doctrine of original sin and he does not seem to have noted that his parable about the illusory destruction of all things has a philosophical and not just moral significance.]

Borges criticized Hawthorne for inserting the moral because it limits the exegetical framework of the text to the Calvinist doctrine of original sin. I risk imposing the same closure of interpretation upon the works to be considered here if I offer an ethical analysis only. My idea, then, is to integrate historical and ecocritical approaches with ethical analyses of political ecology and with the rhetorical strategies of the ecosublime for several kinds of environmental literary representation. That is, I will be concerned to consider how the textual appearance of the mysterious Amazon, the majestic Andes, the limitless Pampa—now celebrated, now contaminated—can affect both the ethical *and* the formal aspects of literature. The second part of what I propose seems to satisfy Tallmadge's dissatisfaction with ecocriticism as a literary theory that "refers more often to the content of the work or the purpose of the critic than it does to the critical method itself." To the contrary, I will also focus on the literary elements of environmental literature, particularly in the more recent work. Texts may function at an ethical level if they can inspire readers to act or at least think in certain ways; but when it comes to aspects of the literary forms themselves, that will be the focus of part 3.

In chapters 8, 9, and 10, I consider the literary innovations that result from the textual incorporation of activist advocacy, political objectives, and environmentalist principles developed as part and parcel of literary narrative. Ecocriticism therefore, according to the suggested definitions I outline above, is as responsible to identify and endorse an ideologically activist perspective (political ecology and the ecosublime) as it is to identify and analyze new forms of literary expression. For Spanish American literature, the first aspect of ecocriticism is a natural fit: the perspective of the aggrieved is the central characteristic of Spanish American literature in such genres as *indigenismo*, testimonial literature, and most regional novels. But despite their rather long history, ecocritical approaches to these literary traditions has only just been categorized as a "Third Wave" of analysis.

THE THREE WAVES OF ECOCRITICISM

Lawrence Buell was the first to classify ecocriticism using the "wave" metaphor in *The Future of Environmental Criticism*; Joni Adamson and Scott Slovic summarize Buell's characterization of "First wave environmental criticism" as analysis that "concerns itself with conventional nature writing and conservation-oriented environmentalism, which traces its origins to the work of Emerson, Muir, and Thoreau."[55] In his book, Buell distinguishes between "first" and "second wave" ecocriticism by expanding the first wave's "traditional commitment to the nature protection ethic . . . to accommodate the claims of environmental justice."[56] For Adamson and Slovic, this was an important turn in ecocriticism, but one that may have been limited by a "community of ecocritics [that] has been relatively non-diverse and also has been constrained by a perhaps overly narrow construing of 'white' and 'non-white' as the primary categories of ethnicity."[57] As a response to the perceived limitation in the ecocritical community, Adamson and Slovic edited a special issue of *MELUS* (the journal of The Society for the Study of the Multi-Ethnic Literature of the United States) containing a collection of essays from a diverse group of writers, not constrained by racial, national, or gendered categories. The result, as they write in thc introduction to the issue, is "what seems to be a new *third* wave of ecocriticism, which recognizes ethnic and national particularities and yet transcends ethnic and national boundaries; this third wave explores all facets of human experience from an environmental viewpoint" (6–7). Because Spanish American literature transcends these boundaries in several ways, particularly in the novels and other fictions of the twentieth century, the analysis of such texts may well qualify as part of a third wave. And in an analysis of literature written in Spanish, it would be neglectful, as Adamson and Slovic have pointed out, not to include some of the ecocritical theorists who are themselves nationals of the region from which the literature under consideration has been produced. I have already referenced Fernando Mires's idea about the "economics of the absurd," Arturo Escobar's work on political ecology, and the essays by Eduardo Galeano whose main criticism concerns environmental contamination resulting from economic exploitation of the region by multinational corporations, the governments of industrialized nations, and powerful domestic elites. Raúl de la Rosa is another ecocritic whose perspectives on local and global ecology was informed by his position as founder and former president of the Association for Geobiological Study, an NGO with branches throughout the Spanish-speaking world.

Like Galeano, de la Rosa's analysis of corporate and domestic elites' manipulation of the natural world for profit and power leads him to call for "*La revolución ecológica*" (The ecological revolution), in his book by the same name. According to de la Rosa, the basis of such a revolution should be founded upon principles that recognize the global nature of ecological degradation while maintaining respect for local identities and traditions:

> Por una parte respetar la diversidad cultural, las tradiciones, lenguas y modos éticos de vivir de los diferentes pueblos, es decir, la diversidad en su totalidad. Por otra parte . . . una política global para erradicar la pobreza y el impacto ecológico, es decir, la totalidad en la diversidad. Dos aspectos que convergen en la solidaridad con la naturaleza, entre comunidades y entre personas.[58]
>
> [On the one hand, respect for cultural diversity, for the traditions, languages and values of all communities; that is, total diversity. On the other . . . a global politics whose goal is the eradication of poverty and ecological degradation; that is, totality through diversity. These two aspects must come together in solidarity with nature, with communities, and with individual persons.]

Also like Galeano, de la Rosa emphasizes the bilateral aspects of the current ecological crisis by dividing the world into poor and rich countries for the purposes of demonstrating the way in which ecological concerns, environmentalism, conservation, sustainable development, and other related concepts perpetuate a system of exploitative economic profits for the rich at the expense of the environmental well-being of the poor. The following critique of sustainable development is illustrative of de la Rosa's critique on the whole:

> Los países desarrollados implantan políticas medioambientales para proteger sus entornos y reducir la destrucción de su medio, pero se siguen nutriendo de otras zonas, esquilmándolas y devastándolas. Así pues no se trata de ecología sino de nuclearización de la riqueza, pero no perciben—o no quieren advertir—que esta política finalmente se volverá contra ellos ya que no existe daño ecológico que no tenga proporciones planetarias, nadie escapa, tarde o temprano, a ello. (de la Rosa, *Revolución ecológica*, 164)
>
> [Developed nations impose environmental policies in order to protect their own surroundings and reduce the destruction to their own environments, but they do this at the expense of other regions, devouring and destroying them. It's not a matter of ecology but the concentration of wealth; and what's worse, these nations do not (or do not want to) admit that such a policy is a double-

> edged sword: ecological degradation has planetary proportions; no one can escape this fact in the long run.]

The passage cited above opens with a criticism of sustainable development that does not envision itself in global terms and concludes with observations about global interconnectedness that anticipate what Buell (and other ecocritics) would later describe as the ecological effects of translocality: "the increasing sense that regions remain permeable to shock waves potentially extending worldwide. That is why the bioregional horizon must extend beyond a merely local horizon: the locale cannot shut itself off from translocal forces even if it wanted to."[59] Ursula K. Heise has also recently argued for a more globalized environmentalist perspective when she argues that in "a context of rapidly increasing connections around the globe, what is crucial for ecological awareness and environmental ethics is arguably not so much a sense of place as a sense of planet—a sense of how political, economic, technological, social, cultural, and ecological networks shape daily routines."[60] What is noteworthy about de la Rosa's commentary is not that it anticipates Buell and Heise by several years or that it marries the insights of the environmental justice movement with a critique of sustainable development, but that it is articulated from a perspective physically located at many of the actual sites of the ecological devastation denounced in his book: de la Rosa's association (the GEA) is international in scope with branches in Mexico, Colombia, Argentina, and Cuba.

SPACE AND PLACE

One final ecocritical concept that I will pursue in the chapters to come arises from theoretical approaches in geography, anthropology, sociology, literary studies, cultural studies, and other academic disciplines to notions of space and place. For my purposes here, the most succinct and effective definition of the terms comes from Alexandra Kogl's *Strange Places*: "space + meaning = place."[61] The elegantly concise equation frames the debate in the most general of terms, but its mathematical precision is useful because its inverse, "Space = place – meaning," effectively expresses the central criticism of development and modernization in much of Spanish American literature. *Indigenista* literature, for example, affirms the cultural identity maintained by, advocates for the human rights construed to, and exposes the ecological predicaments of indigenous communities largely through the condemnation of political ecologies that seek to discount the meaning and value of place constructed by these peoples. In *Justice, Nature and the Geography of Difference*, David Harvey indicates that:

> material practices transform the spaces of experience from which all knowledge of spatiality is derived. These transformative material practices in part accord with discursive maps and plans (and are therefore expressing of both social relations and power) but they are also manifestations of symbolic meanings, mythologies, desires. The spatialities produced through material practices (be they frameworks for living, for communication, for work, for symbolic activities and rituals, for enjoyment) also constitute the material framework within which social relations, power structures, and discursive practices unfold.[62]

The modifications implied by these material practices presuppose a material reality (space) molded by human desire and will (the creation of a place). When an indigenous community lacks the power to maintain its will for the use of certain sites then those sites are inevitably transformed by external agents.

Following the converse I derived from Kogl's definition—"space = place – meaning"—the erasure of previous notions of place allows non-indigenous actors to categorize traditional lands as nothing more than meaningless space only to transform them anew. *Indigenista* literature is primarily concerned to restore notions of place that have been lost by native peoples. But to further complicate the state of affairs, an environmentalist might argue that nature is a space, existent and valued independent from human desire, especially where competing notions of place and space may be at play and where one may be more benevolent to the natural world. Given the complexities of these real world conflicts, an ecocritical account of Spanish American literature must consider the issues surrounding the representation of nature and the power of those who define the parameters of place. In addition, it must also be concerned with the way in which material space and human interaction with that space influences its discursive articulation; in each of the works selected for study in the chapters to follow, the literary properties of the texts themselves are influenced by the dynamics of the space/place problematic. Many feature conflicts that arise from difficulties generated by the carving out of a place from space, or resistance to the ways in which place-making may occur, or the proposal of alternative places not founded upon the logic of development or modernization.

The ecocritical approach continues to evolve at a fascinating rate and from a vast array of critical perspectives: literary scholars, historians, anthropologists, economists, environmental scientists, feminists, Marxists, activists, and others. The approaches sometimes overlap and sometimes contradict. Disciplinary boundaries are blurred and ignored. Ecocriticism can both be intensely specific and widely generalized. Because of the complex and heterogeneous condition of ecological approaches, I will limit the focus of my analysis to the issues mentioned above: the ecological implications of the cultural understanding of the categories civilization and barbarity enshrined in Sarmiento's *Facundo*; the legacy of the ecosublime inherited from the romantic tendencies of nineteenth century texts; the frequent incorporation of political ecology; the challenges to globalization, development, and modernization; the meaning of nature in the Spanish American literary context; and the contested categories of place and space. These ecological concerns are those suggested by the texts themselves to be considered according to the following organization in the subsequent chapters.

OUTLINES

In chapters 1 and 2, I consider the ecological foundations of Spanish American literature, beginning with an examination of the way in which Sarmiento's categories of "civilization" and "barbarity" from *Facundo* are systematically challenged and revised in the rest of the century. In particular, I emphasize the way in which Sarmiento's proposal for the development of Argentina through the exploitation of its natural waterways and seemingly inexhaustible resources inaugurates a literature of political ecology that influences Spanish American ecological literature up to the very present. But I also find that Sarmiento's association of nature, indigenous elements, and *gauchos* (a kind of Argentine or Uruguayan cowboy) with barbarity is problematized through representations of sublime nature that are as numerous in the literature as the expression of political ecology. While the advent of environmentalism as an established movement was still more than one hundred years in the future, I document those elements of current ecological concern that found clear expression in the fictions and poetry of the nineteenth century. These include expressions of love for nature, conservationist nature protectionism, environmental injustice, and deep ecology; articulations of place; resistance to commercialization and industrialization; emulation of non-modernized communities; and even an early expression of animal rights advocacy. All of these represent a dialogue with Sarmiento's ideas of civilization and barbarity that expand, refine, and challenge his categorizations and point the way to the development of an ever-more-explicitly ecological Spanish American literature in the twentieth century.

In chapter 3, I consider essays, poetry, novels, and short stories in the tradition of *modernismo* which was considered the first truly Spanish American literature; the *modernistas*' highly aesthetic, formal literary innovation was unlike anything that had come before in the region's literary work, nor indeed in anything produced by European authors. Critical attention tends to focus on the period's highly original prose and poetry, the authors' unsettling rejection of progress and modernization, and the resistance to what they considered a vulgar bourgeois culture in the region enabled by the steady economic development of Latin America. But often, the *modernista* critical posture adopts an ecological flavor through the representation of an idealized nature, through disturbing imagery of contaminated urban environments, and through nightmarish visions of a future cultural, economic, and environmental apocalypse. My argument in chapter 3 is that a vital part of the cultural critique and stylistic, formal, and aesthetic innovation of Spanish American *modernismo* involves an ecological challenge to the uncritical imposition of boundless development modeled after what transpired in Europe and in the emerging power from the North, the United States. Authors like Cuban José Martí, Uruguayan José Enrique Rodó, Nicaraguan Rubén Darío, and others who are less well-known represent the alternative of *their* America; from an ecological perspective, these texts offer a conception of the natural world that eventually represents a return to an idea for which the genre is known: as art is produced for its own sake, so the conception of nature is based upon its valuation not for economic profit or the price that can be got for exploiting its resources, but for its own sake.

In part 2, I begin chapter 4 with the identification of a moment of intertextuality between *Un viejo que leía novelas de amor* by Luis Sepúlveda and *La vorágine* (*The Vortex*) (1924) by Colombian José Eustasio Rivera as an analytical motif for a reevaluation of the environmentalism and political ecologies in the Spanish American *novela de la selva* tradition. I find that many of the well-established titles from the genre feature a variety of political ecology characterized by appeals to agents of the state. I then propose a countertradition to the canonical *novela de la selva* titles and, in chapter 5, I do the same for the *novela de la tierra* genre. I discuss numerous neglected titles that express aspects of environmentalism such as the principles of "deep ecology," the role of emotion in nature protectionism, conservationism, the rights of nonhuman nature, etc. These works are precursors to the literary environmentalism of Sepúlveda's novel and to other recent environmentalist novels and thus deserve a place in the canon among what has come to be called the "regional authors." But more than that, they anticipate and inform specific connections between Spanish American literature and environmentalism in general, particularly if that movement wishes to include local perspectives on such globally important sources of natural resources as the Amazonian jungle, the Lake Maracaibo oil fields, the Argentine *pampa*, the mountains of the Andean cordillera, and the fruit and sugar plantations of the Caribbean.

I begin chapter 6 by framing the corpus of *indigenista* literature with analysis of *Los caminos de Paxil* (The ways of Paxil) (1990) by Guatemalan Arturo Arias; the ecological aspects of that novel include references to the contaminating effects of oil extraction, contrasting conceptions of the rightful authority to determine land use, and invocations of the Mayan cosmovision as valid positions from which to articulate a political ecology. These elements, as they are represented in *Los caminos de Paxil*, tend toward a more conservationist view of traditional Mayan lands as opposed to one that embraces development. And that tendency seems to invite a reexamination of previous examples of the genre to see if conservationism or other ecological ideas should be included as defining ideologies of *indigenismo* in the tradition as a whole. Through a survey of titles that begins with one of the very earliest written in and about the Americas—Spanish Dominican friar Bartolomé de las Casas's *Brevísima relación de la destrucción de las Indias* (*A Brief Account of the Destruction of the West Indies*) (1542, published 1552)—and concludes with Arias's novel, I trace the development of *indigenista* ecological thought. After las Casas, I examine nineteenth century examples from the genre including Dominican Manuel de Jesús Galván's historical novel *Enriquillo* (1882) and *Tabaré* (1888) by Uruguayan Juan Zorilla de San Martín, continue with the corpus from the mid-twentieth century when the form was at its numerical height, and conclude with more recent texts including Arias's novel. The focus in these two chapters emphasizes what turns out to be the most common ideological position in *indigenista* literature: a critique of political ecologies enshrined in Spanish American law, decree, or tradition that were nothing more than outright robbery of traditional indigenous lands. I conduct my analysis through an examination of the articulation of perspectives on place and space and discover that many *indigenista* authors invoke the ecology of generational attachments to, spiritual connections with, and violent resistance as a valid action for a defense of the land. As I follow the development of critical ideologies in early twentieth century texts and in later neo-*indigenista* works, the emphasis shifts to environmental injustice suffered by indigenous communities as a result of industrialized agriculture, mining, oil drilling, the exploitation of timber, and the construction of dams.

Ultimately, I come full circle in part 3 with a return to authors writing in the 1990s and the first decade of the twenty-first century and identify explicit environmentalism as a distinct tendency in recent, post-"boom" Spanish American fiction. There are several titles that could have gone in these chapters, but I limit myself because ecology and environmentalism, green studies and ecocriticism, conservationism, nature protectionism, and other manifestations of environmentalist principles have now become quite common in the literature. It is no longer remarkable to discover such ideas in recent works. But what I explore in part 3 is that the articulation of these ideas—the expression of ecological advocacy, even the outright expression of environmentalist ideals—can actually impact the formal elements of literature in unique ways. I consider textual conventions such as language and imagery, narrative sequence and register, metafiction, intertextuality, the use of humor, satire, and parody, focalization, and the incorporation of utopian scenarios such that these elements represent innovations of form that proceed directly from the ethical aspects of environmentalism. I conclude chapter 10 with a glimpse into the future of Spanish American ecological literature through the consideration of a still-to-be completed trilogy of science fiction by Chilean Fernando Raga about a kind of totalitarian ecologism that takes over the United States, Europe, and other allied nations, 200 years in the future. The trilogy does not have much in the way of formal literary innovation, but represents the first example of an explicitly critical take on the politics, dogma, and sacred tenets of environmentalist doctrines in recent Spanish American fiction. As such, it represents what I perceive to be the clear markings of a mature expression of ecological ideas in the literature of the region: Raga's trilogy begins the process of a literary dialogue within the environmentalist texts of recent Spanish American fiction. The corpus has become heterogeneous, multi-faceted, diverse, skeptical, and richer as a result.

I conclude the analysis as a whole with comments concerning what I believe must follow as the result of what I have discovered in chapters 1-10 about the ecology of Spanish American literature. There were several texts that (based mainly on their titles) seemed promising, but did not have much along the lines of my interests here. And in certain areas, particularly contemporary poetry, a good deal of work has already been done. I conclude with the proposal of two specific projects as one way forward: the revision of canon and in-depth studies of certain issues that I raise in the pages that follow but did not pursue for a more global picture of ecology and Spanish American literature. What comes between the end of this introduction and my conclusions after chapter 10 should not be considered only a survey of texts that contain representations of nature, ecology, and environmentalism; while such a project is worthy for the sheer volume of nature representation in the literature, the term "survey" can imply summary and little else. Rather, I attempt to develop a sense of the way in which ecological ideas have developed over time, particularly the way in which most of the Spanish American texts considered in the pages that follow anticipate many of the most central elements of "green" studies and the environmentalist project; of political ecology, conservationism, and environmental justice; and of several other convictions that have become a predominant aspect of global culture, a constant in current political ideologies, and pivotal for the ecological future of humankind.

NOTES

1. Although fully aware of the troublesome nature of transnational identity terminology, I will employ the adjectives Spanish American and Latin American in slightly different senses. While cultural, social, geographical, and political phenomena in the region as a whole are generally referred to as Latin American, I prefer a linguistic descriptor—"Spanish American"—to designate the literature under consideration here: literature written by authors whose national identities can be associated with countries where Castilian Spanish is the

primary language. The distinction is important because literary studies tend toward disciplinary organization based upon language while cultural studies tend toward organization according to regional peculiarities.

2. Paché Merayo, "Un relato de geografía circular," *El Comercio*, September 4, 2008.

3. Jorge Paredes and Benjamin McLean, "Hacia una tipología de la literatura ecológica," *Ixquic: Revista Hispánica Internacional de Análisis y Creación* 2 (August 2000): 6–7. Translations are by the author unless otherwise indicated.

4. José Manuel Marrero Henríquez, "Ecocrítica y hispanismo," in *Ecocríticas: Literatura y medio ambiente*, ed. Carmen Flys Junquera, José Manuel Marrero Henríquez, and Julia Barella Vigal (Orlando, FL: Iberoamericana Vervuert Publishing Company, 2010), 204.

5. As early as 1993, Jorge Marcone published his ecocritical analysis "*El hablador* de Mario Vargas Llosa y la imagen de la Amazonía en el Perú contemporáneo," in *LA CHISPA '93: Selected Proceedings*, ed. Gilbert Paolini (New Orleans, LA: Tulane University, 1993), 134–40. Since then his work has included: "Cultural Criticism and Sustainable Development in Amazonia: A Reading from the Spanish American Romance of the Jungle," *Hispanic Journal* 19, no. 2 (Fall 1998): 281–94; "Del retorno a lo natural: *La serpiente de oro*, la 'novela de la selva' y la crítica ecológica," *Hispania: A Journal Devoted to the Teaching of Spanish and Portuguese* 81, no. 2 (May 1998): 299–308; "Jungle Fever: Primitivism in Environmentalism: Rómulo Gallegos's *Canaima* and the Romance of the Jungle," in *Primitivism and Identity in Latin America: Essays on Art, Literature, and Culture*, ed. Erik Camayd-Freixas and José Eduardo González (Tucson: University of Arizona Press, 2000), 157–72; "A Painful Pastoral: Migration and Ecology in Chicana/o Literature," *Pacific Coast Philology* 46, no. 2 (2011): 194–209; and with Priscilla Solis Ibarra, "Mexican and Chicana/o Environmental Writing: Unearthing and Inhabiting," in *Teaching North American Environmental Writing*, ed. Laird Christensen, Mark C. Long, and Fred Waage (New York: Modern Language Association of America, 2008), 93–111.

6. Laura Barbas-Rhoden, *Ecological Imaginations in Latin American Fiction* (Gainesville: University Press of Florida, 2011); Adrian Taylor Kane, ed., *The Natural World in Latin American Literatures: Ecocritical Essays on Twentieth Century Writings* (Jefferson, NC: McFarland, 2010); Carmen Flys Junquera, José Manuel Marrero Henríquez, and Julia Barella Vigal, eds., *Ecocríticas: Literatura y medio ambiente* (Orlando, FL: Iberoamericana Vervuert Publishing Company, 2010); Beatriz Rivera-Barnes and Jerry Hoeg, eds., *Reading and Writing the Latin American Landscape* (New York: Palgrave, 2009); and Germán Bula and Ronald Bermúdez, *Alteridad y pertenencia* (Bogotá: Universidad de La Salle, 2009).

7. Barbas-Rhoden, *Ecological Imaginations*, 16.

8. Laurence Coupe, ed., *The Green Studies Reader: From Romanticism to Ecocriticism* (New York: Routledge, 2000). In his early anthology, Coupe traces ecocritical approaches to Raymond Williams's *The Country and the City* (1973), but the collection of texts contains excerpts from works as old as William Blake's letter to the Reverend Dr. Trusler, titled "Nature as Imagination," (1799), 16.

9. Among the most well-known in English, Jules Vernes's *Eight Hundred Leagues on the Amazon* (1881) and Arthur Conan Doyle's *The Lost World* (1912) are entertaining enough, but neither author ever set foot in South America. *Green Mansions* (1904) by William Henry Hudson and *The Sea and the Jungle* (1912) by Henry Major Tomlinson rehearse the romantic conventions of the jungle love story and the English travelogue respectively, conventions that had been explored several decades earlier by Spanish American authors such as Ecuadoran Juan León Mera and Colombian Jorge Isaacs.

10. Scott Slovic, *Going Away to Think* (Reno: University of Nevada Press, 2008), 135.

11. Eduardo Galeano, *Las venas abiertas de América Latina* (Mexico City: Siglo XXI, 1971); and *Úselo y tírelo*, (Buenos Aires: Planeta, 1994). The Uruguayan Galeano's work is a notable exception; because the long-form prose essay is a literary genre unto itself, it does not easily fit under categories that attempt to distinguish between fiction and non-fiction. See Susana Rotker's seminal anthology: *Ensayistas de nuestra América*, 2 vols. (Buenos Aires: Editorial Losada, 1995).

12. Raymond Williams, *Keywords: A Vocabulary of Culture and Society* (New York: Oxford University Press, 1983), 111.

13. Anna Bramwell, *Ecology in the 20th Century* (New Haven, CT: Yale University Press, 1989), 14–15, 4.

14. Leonardo Boff, *Ecology & Liberation: A New Paradigm* (Maryknoll, NY: Orbis, 1995), 14.

15. See David E. Camacho, ed., "The Environmental Justice Movement: A Political Framework," in *Environmental Injustices, Political Struggles: Race, Class, and the Environment* (Durham, NC: Duke University Press, 1998), 11–30; Giovanna Di Chiro, "Nature as Community: The Convergence of Environment and Social Justice," in *Uncommon Ground: Rethinking the Human Place in Nature*, ed. William Cronon (New York: Norton, 1995), 298–320; Teresa Vicente Giménez, *Justicia ecológica y protección del medio ambiente* (Madrid: Trotta, 2002), 13–67; David Harvey, *Justice, Nature, and the Geography of Difference* (Cambridge, MA: Blackwell, 1996), 329–438; and the essays collected in Joni Adamson, Mei Mei Evans, and Rachel Stein, eds., *The Environmental Justice Reader: Politics, Poetics, Pedagogy* (Tucson: University of Arizona Press, 2002).

16. Alan Drengson and Yuichi Inoue, eds., *The Deep Ecology Movement: An Introductory Anthology* (Berkeley, CA: North Atlantic Books, 1995), xix.

17. Kay Milton, *Loving Nature* (New York: Routledge, 2002), 74–75.

18. Kay Milton, *Environmentalism and Cultural Theory: Exploring the Role of Anthropology in Environmental Discourse* (New York: Routledge, 1996), 27, 28, 29.

19. Jonathan Bate, "Culture and Environment: From Austen to Hardy," *New Literary History* 30, no. 3 (Summer 1999): 542.

20. Coupe, *Green Studies Reader*, xi.

21. Scott Slovic, "Ecocriticism: Containing Multitudes, Practising Doctrine," in Coupe, *Green Studies Reader*, 160.

22. Patrick D. Murphy, *Ecocritical Explorations in Literary and Cultural Studies: Fences, Boundaries, and Fields* (Lanham, MD: Lexington Books, 2009), 7.

23. Slovic, *Going Away*, 27.

24. Murphy, *Ecocritical Explorations*, 6.

25. Robert Kern, "Ecocriticism: What is it Good for?" in *The ISLE Reader: Ecocriticism, 1993–2003*, ed. Michael P. Branch and Scott Slovic (Athens: University of Georgia Press, 2003), 260.

26. In Coupe's *Green Studies Reader*, Coupe himself as well as William Howarth and Jhan Hochman explicitly attempt definitions and there are countless others, many articulating a version of Slovic's two-tiered definition as the explicit study of nature in a text or the study of implicit ecological ideologies even if nature seems absent from a text.

27. Murphy, *Ecocritical Explorations*, 4.

28. John Tallmadge, "Toward a Natural History of Reading," in *The ISLE Reader*, ed. Branch and Slovic, 282.

29. I was lucky enough to be researching an article on the representation of the jungle in Spanish American literature and Robleto's novel came up in searches because its title contains the Spanish word for jungle, *selva*.

30. I used the histories by E. Anderson Imbert, *Historia de la literatura hispanoamericana*, 2 vols. (Mexico City: Fondo de cultura económica, 1997); Jean Franco, *An Introduction to Spanish American Literature* (New York: Cambridge University Press, 1994); and Kessel Schwartz, *A New History of Spanish American Fiction*, 2 vols. (Coral Gables, FL: University of Miami Press, 1971).

31. Coupe, *Green Studies Reader*, 4.

32. Andrew Husband, "Environmental Affects," *ISLE: Interdisciplinary Studies in Literature and Environment* 17, no. 4 (Autumn 2010): 783.

33. Tallmadge, "Toward a Natural History of Reading," 282.

34. Raymond L. Bryant and Sinéad Bailey, *Third World Political Ecology* (New York: Routledge, 1997), 195.

35. Andrew Dobson, "Citizenship," in *Political Theory and the Ecological Challenge*, ed. Andrew Dobson and Robin Eckersley (New York: Cambridge University Press, 2006), 216–31.

36. See Arturo Escobar, "Constructing Nature: Elements for a Poststructuralist Political Ecology," in *Liberation Ecologies: Environment, Development, Social Movements*, ed. Richard Peet and Michael Watts (London: Routledge, 1996), 46–68.

37. Aletta Biersack, "Reimagining Political Ecology: Culture/Power/History/Nature," in *Reimagining Political Ecology*, ed. Aletta Biersack and James B. Greenberg (Durham, NC: Duke University Press, 2006), 14.

38. Richard Peet and Michael Watts, "Liberation Ecology: Development, sustainability, and environment in an age of market triumphalism," in *Liberation Ecologies*, ed. Peet and Watts, 9, 11.

39. Domingo Faustino Sarmiento, *Facundo* (Madrid: Ediciones escolares, 2001), 31, 36.

40. Jean Grugel and Pía Riggirozzi, eds., *Governance after Neo-Liberalism in Latin America* (New York: Palgrave Macmillan, 2009). Most of these essays discuss the influence of resource exploitation in post-neo-liberal Latin America, particularly oil in Hugo Chavez's Venezuela and natural gas in Evo Morales's Bolivia.

41. Williams, *Keywords*, 103

42. Fernando Mires, *El discurso de la naturaleza: Ecología y política en América Latina* (San José, Costa Rica: DEI, 1990), 67.

43. See Carlos J. Alonso, *The Spanish American Regional Novel* (New York: Cambridge University Press, 1990); Jennifer L. French, *Nature, Neo-Colonialism, and the Spanish American Regional Writers* (Hanover, NH: Dartmouth University Press, 2005); Carlos Fuentes, *La nueva novela hispanoamericana* (Mexico City: Cuadernos Joaquín Mortiz, 1969); and Marcone, "Retorno." The works in this tradition are known by various names. French uses the terminology "regional novel" to describe texts with a particular affinity to geographic regions. Alonso uses the same term in the title of his book, but employs the Spanish *novela de la tierra* to mean the same thing. Marcone uses *novela de la selva* to refer to fictions with settings geographically specific or similar to the Amazonian jungle; I follow that usage here and by *novela de la tierra*, I mean novels that primarily take place in any non-urban and non-jungle setting: the mountains, plains, and coasts of Latin America.

44. Karen J. Warren, *Ecofeminist Philosophy: A Western Perspective on What It Is and Why It Matters* (Lanham, MD: Rowman & Littlefield Publishers Inc., 2000), 50.

45. Greg Garrard, *Ecocriticism* (New York: Routledge, 2004), 26.

46. Karla Armbruster, "A Poststructuralist Approach to Ecofeminist Criticism," in *Green Studies Reader*, ed. Coupe, 199.

47. Warren, *Ecofeminist Philosophy*, 102.

48. Ibid., 103–4.

49. Bate, "Culture and Environment," 554.

50. Lee Rozelle, *Ecosublime: Environmental Awe and Terror from New World to Oddworld* (Tuscaloosa: University of Alabama Press, 2006), 3.

51. Slovic, *Going Away*, 153.

52. Lawrence Buell, *Writing for an Endangered World* (Cambridge, MA: Harvard University Press, 2001), 2.

53. Aníbal González, *Abusos y admoniciones: Ética y escritura en la narrativa hispanoamericana moderna* (México City: Siglo Veintiuno Editores, 2001), 20.

54. Jorge Luis Borges, "Nathaniel Hawthorne," in *Prosa completa* (Barcelona: Editorial Bruguera, 1980), 187.

55. Joni Adamson and Scott Slovic, "The Shoulders We Stand On: An Introduction to Ethnicity and Ecocriticism," *Melus* 34, no. 2 (Summer 2009): 6.

56. Lawrence Buell, *The Future of Environmental Criticism: Environmental Crisis and Literary Imagination* (Malden, MA: Blackwell, 2005), 22.

57. Adamson and Slovic, "The Shoulders We Stand On," 6.

58. Raúl de la Rosa, *La revolución ecológica* (Barcelona: Icaria, 2001), 236.

59. Buell, *Environmental Criticism*, 88.

60. Ursula K. Heise, *Sense of Place and Sense of Planet: The Environmental Imagination of the Global* (New York: Oxford University Press, 2008), 55.

61. Alexandra Kogl, *Strange Places: The Political Potentials and Perils of Everyday Spaces* (Lanham, MD: Lexington Books, 2008), 15.

62. David Harvey, *Justice, Nature, and the Geography of Difference* (Cambridge, MA: Blackwell, 1996), 112.

Part One

Foundations, Aesthetics, Ecology

Chapter One

Foundations of Environment: Literary Political Ecologies of Nineteenth Century Southern Cone Literature

In 1826, three years after the decisive Battle of Lake Maracaibo that marked the definitive independence of Venezuela from Spain, Venezuelan poet and historian Andrés Bello wrote "La agricultura de la zona tórrida" (The Agriculture of the Torrid Zone). The poem features one of the earliest representations of Latin American nature in a Spanish American text: the horrors of war can be healed by the balm of nature, the distraction of the city can be relieved by the delights of the country. In Bello's poem, the celebration of the natural world became a model followed by later poets in countless other representations of the sublime wonders of Latin American nature. Just for this, it would be worthwhile including "Agricultura" in any history of the region's nineteenth century literature, but what makes Bello's poem all the more relevant is its seminal expression of a key tenet of current environmentalist thought, the preference for the country over the city:

> ¿Amáis la libertad? El campo habita,
> no allá donde el magnate
> entre armados satélites se mueve,
> . . .
> Del obstruido estanque y del molino
> recuerden ya las aguas y el camino:
> el intricado bosque el hacha rompa,
> consuma el fuego: abrid en luengas calles
> la oscuridad de su infructuosa pompa.[1]

[Love you liberty? Live in the country,
Not where the magnate roves
His armed cronies descend in droves,
. . .
Lo' the windmill and the obstructed pail
And remember now water and the trail:
O'er intricate forest the fire sailed,
And hatchet consumed: Opened lengthy roads
For pomp without circumstance that failed.]

The images from these lines describe a country where freedom reigns and where buildings, hatchets, and roads are seen as intrusions from the unnatural city. Dissatisfaction with urban living, condemnation of deforestation and over-development, and yearning for the country are sentiments that inform much of the current environmentalist project: conservationism, nature protectionism, deep ecology, sustainable development. And in the poem, the valuation of nature occurs along two figurative axes such that it is perceived as precious to human existence, *but also* intrinsically valuable in its own right. The poet expresses joy about treasured places—plains, rivers, mountains, forests, gardens, farms, ancestral hunting grounds, a homeland—and articulates the spiritual feeling of connection to such places. But there are also expressions of despair about endangered species of plants and animals with abstractions and idealizations of the natural world that evoke its intrinsic value. Bello's poem asserts the superiority of the country over the city with both instrumental and non-instrumental imagery: where trees have not been felled and where pompous roads are not the norm is where humans encounter liberty. In many other texts from the nineteenth century, these sentiments are proclaimed and echoed; enough of them in fact, to confirm the project proposed in the introduction: that an ecocritical reevaluation of both canonical and neglected or forgotten novels, essays, and poems can provide new insights into the way in which Spanish American literature expresses ecological and even environmentalist ideas.

While the poetry of Andrés Bello ranks among the most canonical of the literature from the period and region,[2] on the other end of the canonical spectrum resides *La familia de Sconner* (*Sconner's family*) (1858) by Argentine Miguel Cané (father of Miguel Cané of the generation of 1880).[3] Cané's novella deals specifically with the promise of the nascent wool industry in Argentina and dramatizes legal efforts to restore inherited lands to the dispossessed orphans of a wealthy landowner, a story that implicitly advocates for legal title to lands and for an industry that could be exploited for Argentine development. In the early chapters of the story, the orphans' father is praised for his epiphany that "la mejora de la hacienda lanar podría aumentar la riqueza del país" ["the improvement of the wool industry could increase the nation's wealth"]. It was to be "una mina no esplotada todavía" ["an as yet unexploited source of wealth"] and despite the ridicule from a class of conservative landowners, "después de dos ó tres años de trabajo inteligente y perseverante . . . D. Pedro era hombre rico, el primer creador de ovejas finas en el estado de Buenos Aires, el único que esportaba lanas por su cuenta, y el único que podía vender padres para la mejora de la raza lanar indígena" ["after two or three years of intelligent work and perseverance . . . Mr. Pedro was a rich man, the first to create fine fleece in the province of Buenos Aires, the only independent wool exporter, and the only seller of stud rams for the improvement of the domestic wool-producing ovine population"].[4] While the development of the industry does not necessarily represent a more environmentally sensitive alternative to any of the other agriculturally based enterprises that eventually took hold in Argentina, I cite this brief moment from Cané's obscure novella to highlight the way in which the discourse of political ecology was an omnipresent element in Spanish American literature of the nineteenth century. As I mention in the introduction, political ecology was the central concern in Argentine Domingo Faustino Sarmiento's *Facundo* (1845), one of the most foundational and influential of all nineteenth century texts; and it is present, implicitly or explicitly, in almost all of the other titles that I consider in chapters 1 and 2. But I am introducing the first part of this analysis with poetry from Bello and with ideas from the all-but-forgotten *La familia de Sconner* as illustrative of the proposed scope of genres and titles for the book as a whole. I make no distinction between well-known and neglected work even if canonical titles almost surely were more influential in their own time. Instead, because of the current interest in green literary studies and environmentalist ideology, part of the work here is to identify ways in which ecological insights from both familiar and relatively unknown titles may change perceptions about the canon if such insights feature relevant tenets of nature protectionism, political ecology, and other environmentally oriented beliefs. However, before I forge ahead on that account, some literary and historical context will be helpful for what is to follow.

FOUNDATIONS

In *Foundational Fictions*, her thematic study of the connection between romance and ideology in Latin American literature of the nineteenth century, Doris Sommer established the theoretical parameters by which students and scholars have come to understand the texts of that region and century. She employs the metaphor of "erotics as politics"[5] both in analysis of the content of the novels chosen for her study and to describe the extraliterary effect of national identity construction that was the result of reading those novels, poems, and other works of fiction. The point of departure for her analysis considers several love stories from nineteenth century Latin American novels, short stories, and poetry as the ideologically foundational texts for newly independent American nations. Sommer notes the long list of authors from that time who also ascended to the presidencies or to other important political positions in their respective nations and she argues that "there could be no clear epistemological distinction between science and art, narrative and fact, and consequently between ideal projections and real projects." The texts by such politician/authors adopt and co-opt the ideological positions espoused by the actual political individuals who wrote them through a kind of allegorical connection that Sommer describes as "a narrative structure in which one line is a trace of the other, in which each helps to write the other."[6] For Sommer, the intricacies of the love-story genre in these national romances communicate the idealized identities that nineteenth century authors envisioned for their new nations. But my analysis here is more concerned with the ideological effects that such allegories had on the political ecologies which were eventually espoused by the state apparatuses of the new Spanish American states. The nearly invisible line between literary and political ideologies in nineteenth century Spanish American "foundational fictions" allows for a more straight-forward analysis. The ecocritical reconsideration I undertake in the following pages will focus on the politics of nature each texts espouses; or the dynamics of space and place that they evidence; or their philosophies of development, consumption, and conservation. And for those that include representations of indigenous inhabitants, I consider the implications of each one's implicit ecological ideology for questions of environmental justice.

Sommer's work has remained an important theoretical piece, the standard secondary source for nineteenth century Latin American literary studies and her use of the term "romance" suggests that the aesthetic and political legacies of European romanticism were fundamental influences; indeed Jennifer L. French observes, "romanticism . . . was central to the discursive formulations of Spanish America's cultural independence: civic minded writers of the nineteenth century freely appropriated romantic discourses by the likes of Bernardin de Saint Pierre, Chateaubriand, Lord Byron, and Shelley and adapted them to their own purposes."[7] I take the association between romanticism and nineteenth century Spanish American literature as fundamental to the argument that I develop here and in chapter 2. That is, the influence of the romantic ethos upon these authors makes ecocritical approaches to their representations of nature particularly apt. First, Jonathan Bate inaugurates a scholarly appreciation for "romantic ecology": "In Romantic poetics, poetry is to be found not only in language but in nature; it is not only a means of verbal expression, it is also a means of emotional communication between man and the natural world."[8] William Blake himself affirms that "Some see Nature all Ridicule & Deformity. . . . But to the Eyes of the Man of Imagination, Nature is Imagination itself"[9] and John Ruskin goes so far as to identify the love of nature with moral virtue: "[though] the absence of the love of nature is not an assured condemnation, its presence is an invariable sign of goodness of heart and justness of moral *perception*."[10] As evidenced by the quotations from Blake and Ruskin, the articulation, appreciation, and idealization of sublime nature has an indisputable place in the characterization of the work of romantic authors and poets. However, it is something of a quandary for an ecocritical characterization of work by nineteenth century Spanish American authors because they were influenced by European romanticism but they also undertook the writing of national identities in the face of sometimes brutal natural landscapes.

One problem, as observed by Stephen C. Behrendt, was the tendency for texts to conclude on a grim note, something he calls the "romantic dilemma": "The fact that however much we may formulate our private, imaginative Edens, we all live nevertheless in temporal, public Wildernesses."[11] For Behrendt as well as other German, English, and French romantic writers, the notion of "public wilderness" had to do with the war of American Independence, the horrors of the French Revolution, and the continental implications of French imperial designs. In the Spanish American context, the "wildernesses" were rather more literal and the dilemma revolved around different axes. Certainly, the resolutions of novels and poetry were often uncommonly grim, with heroes, lovers, mothers, daughters, revolutionaries, and even idealized indigenous tribes all meeting untimely ends. But the romantic dilemma in the literature of Central and South America often came down to a tension between competing perceptions of nature. For some, it was an awesome and terrible threat to be subdued and transformed while for others, it could lead to personal improvement in the form of the acquisition of wisdom. I discuss these ambivalences and their ecological implications at greater length in chapter 4 in the context of the *novela de la selva*,[12] but for all of nineteenth century Spanish American literature—and especially for Argentine, Chilean, and Uruguayan authors—nature was not just the European romantic sublime but could also represent a kind of barbarity that was contrary to civilization.

A second problem has to do with the reconsideration of the parameters of canon after its inadvertent cementing by Sommer through her use of an analytical characterization of the literature as "national romances." The expression of romance (in the erotic, passionate sense) as a form of literary expression proceeds from an aspect of romanticism that Michael Lowy and Robert Sayre characterize as the tendency to "rediscover paradise in present reality [by] transforming one's immediate environment and one's own life [which] can take the form of dandyism or aestheticism . . . , [or through] the creation of a community of like-minded individuals . . . , [or] a utopian experiment . . . , or simply falling in love. This last element brings us to the popular meaning that is commonly given to Romanticism today."[13] Sommer's notion of "national romances" is key to the corpus of texts that make up her analysis, the selection of which was based upon certain factors that she lays out in the preface and first chapter of her book: those texts that were "frequently required in the nations' secondary schools as a source of local history and literary pride"; those produced through the unique marriage between high-level Spanish American politics and the production of literature; those that encouraged re-population efforts after indigenous demographic decline and modeled the possibilities for cross-class, multi-cultural, and even inter-racial family arrangements; but particularly, those texts whose "coherence comes from their common project to build through reconciliations and amalgamations of national constituencies cast as lovers destined to desire each other."[14] At the level of textuality, the role of nature in these projects was at best implicit. But the advent of ecocriticism requires an update of Sommer's analysis of the foundational nature of nineteenth century literature[15] and I hold that the influence of romanticism (and its simultaneously terrified/awed perceptions of sublime nature) suggests that there must have been a hidden ecological current running through the whole of nineteenth century Spanish American literature.

Finally, and before I turn to a consideration of the Argentine, Chilean, and Uruguayan nineteenth century texts themselves, it will be useful to document some historio-cultural landscapes of the Southern Cone from that century. A short excursus should provide some context for the ecological currents that I identify in the literature from that time and place; I will begin most of the subsequent chapters with something similar: an exploration of the historical, cultural, and ecological realities that informed the production of Spanish American texts from the times of Columbus to the present. In this case, we should begin with the Napoleonic invasion of Spain in 1808 and the subsequent French occupation of the Iberian peninsula until 1814 during which the territories now known as Argentina, Chile, and Uruguay declared their independence from the Spanish empire in 1810 (as did México and Venezuela in the same year). With the natural separation by the Andean mountain range of Chile from what in the early nineteenth century was called the United Provinces of the Río de la Plata, the boundaries of the former as distinct from what came to be known as Argentina were established fairly early. However, on the eastern side of the Andes, the better part of a century was spent in conflict concerning the political organization of the provinces as either a unified nation with strong central governance located in Buenos Aires or as a weak confederacy of territories with sovereign power determined by provincial rather than national borders. That arrangement did not allow for Argentina's definitive territorial organization until 1880, although Uruguay was established as an independent nation in 1828 after the treaty of Montevideo which ended a nearly two-year war between the provinces of Argentina and the newly independent empire of Brazil. With regard to economic history, the extensive plains or pampas that dominate the Argentine landscape were early perceived as particularly apt for animal husbandry and livestock agriculture became the engine of economic development as well as the most impactful agent of ecological change.

Introduced by Spanish colonizers, cattle brought with them (in their gut) new types of vegetation including "most prominently the cardoon thistle which in the pampas' deep soils grew more than 3 meters tall."[16] While it made things difficult for horse and rider, the cattle loved it and with no natural predators, their numbers increased exponentially. It was not until the late eighteenth century, and especially after independence in the nineteenth, that hides, beef, and other bovine products became profitable, limited as the industry was by colonial restraints. Once these restrictions were eased by the Spanish empire in the 1780s and access to international markets finally made available without restriction by the governments of the newly independent nations of the Southern Cone, the commerce in products derived from livestock increased dramatically. Yet the relationship between a taxing metropolis and its hinterland, which in the 1700s was a relationship between empire and colony (Spain and its viceroyalties), was only slightly adjusted in the context of an independent Argentina. The privileged position occupied by the city of Buenos Aires—on the banks of the Río de la Plata estuary just as it begins to widen to the sea—reproduced the colonial management of commerce through control of a body of water that was the principal point of entry for the rivers that served as various routes to the interior. Such elements of the topography—extensive plains, suitably navigable rivers, and access to them through an immense but navally controllable estuary—accounted for the geo-political relationship between Buenos Aires and the interior. The geographically-strategic location of the city explains the torturous path to a fully constituted nation-state for Argentina some seventy years after the first declarations of independence, but these political and physical landscapes also represented the central literary focus for Southern Cone authors from the first part of the nineteenth century.

In the latter part of the century, another issue came to the fore: mass migrations to Montevideo and Buenos Aires and unforeseen growth in these, the two main cities of the eastern Southern Cone. Demographic and physical expansions were accompanied by an increase in trade with European markets and subsequent exposures to imported products available to newly arrived migrants in the city. Luis Vitale notes that from the resultant ecological imbalance between country and city "se estableció una forma de dependencia o colonialismo interno entre la capital y las provincias" ["a form of dependency or internal colonialism was established between the capital and the provinces"]. As a consequence, Vitale finds that "el subsistema urbano agravó significativamente el deterioro ambiental. La expoliación de los ecosistemas estuvo en función de las ciudades y puertos por donde salía y se procesaba la economía agrominera exportadora" ["the urban subsystem worsened environmental decline significantly. The pillaging of ecosystems was a consequence of the mining and agro-industries whose channels of distribution were so intimately connected to cities and ports"].[17] The Argentine and Uruguayan writers of the second half of the nineteenth century express the ecological consequences of migration to and growth of the cities in a variety of ways including characteristic expressions of nostalgia for a disappeared and more "natural" past. As I propose in the analysis below, the literary expression of an ideology of development is both proclaimed and challenged by a first generation of authors, but quickly altered by those writing after 1880. Faced with an ever-increasingly acute environmental decline in Buenos Aires and Montevideo, the next generation expresses nostalgia for the country in terms that would be familiar to nature-protectionists and ecological activists of the present.

FACUNDO AND POLITICAL ECOLOGY

In the annals of nineteenth century Spanish American literature, there is probably no more canonical a text than *Facundo* by Sarmiento for its inauguration of the dialectic between civilization and barbarism. The novel/essay advocates the legitimization of power and authority over non-urban land on the basis of the connection between the peculiarities of topography and the cultures and types of individuals that such land will yield. In particular, Sarmiento is concerned with the way in which physical territories that previously produced barbarism can be altered so as to yield civilization. The highly politicized content in *Facundo* owes its stridency to the specific historical moment in which it was written: three and a half decades of conflict over the shape of governance for the region to be followed by three and a half more before the Argentine state as it is currently known would take final shape. Sarmiento's purpose was to attack the political abuses perpetrated by then-president/dictator Juan Manuel Rosas, a figure associated in the text with the Argentine outback: gauchos and indigenous peoples, the use of violence, and ubiquitous savage barbarity. The policies and machinations of power attributed to Rosas are combined wholesale with those elements of Argentine culture that Sarmiento associates with the dictator and individuals like him. If Rosas is from the pampa, then Sarmiento proposes a political ecology where the city functions as conductor for the transmission of civilization because the Argentine plain was notoriously bad at this. The conflict is defined by efforts to spread civilization to a non-urban, savage land and its inhabitants, but Sarmiento offers no argument for why the land should be civilized, just that the authority to do so proceeds from the self-evident value of civilization and that it cannot be done "sin la posesión permanente del suelo, sin la ciudad, que es la que desenvuelve la capacidad industrial del hombre y le permite extender sus adquisiciones" ["without permanent possession of the land and without the city which is what cultivates man's industrial capacity and permits the spread of his domain"].[18]

In *Facundo*, the image of nature that emerges from the text is as rhetorical as the foundational political ideals that condemn the federalism of the Rosas regime (embodied in the *caudillo* [regional strong-man] Facundo Quiroga) in order to elevate the political vision of the *unitario* manner of thinking.[19] Sarmiento's narrative of nature was "dibujado a partir de recuerdos, reminiscencias sin haber visitado los lugares de la escena o haber interrogado el suelo" ["rendered through the use of memories and reminiscences without having visited or set foot in the places described in the book"].[20] Under these circumstances, it is beside the point to question the accuracy of nature representation in *Facundo*; rather, such representations can be understood in their function as premises that form part of the argument for Sarmiento's vision of Argentine development. As Lelia Area points out, the representation of nature in *Facundo* "funda una *geo-grafía posible*, paradigma fuerte de una argumentación política, fundada en una figuración literaria" ["creates potential *geographies* as a paradigm for political persuasion that emerges from literary representations"].[21] The close relationship between nature, politics, and literature is duly recognized by Area; from an ecocritical perspective, such closeness demonstrates the long history of ecology in the region's "foundational fictions." The opposition between civilization and barbarism was employed as a sweeping dialectic in Sarmiento's analysis of Argentine cultural phenomena. Sarmiento conceived of civilization as a process: a society's acquisition of the values and structures that characterize cities, science, industry, and European culture represents progress in the form of a widening separation from undeveloped natural states.

Sarmiento envisioned the spread of civilization over the American continent as a positive event, but for Western critical dialogue in general and for certain Spanish American thinkers in particular, the notion of civilization could also be criticized because it "had produced not only wealth, order, and refinement, but as part of the same process poverty, disorder, and degradation."[22] Because the actualization of civilization in Latin America involved the processes of industry and modernization, environmental contamination was a by-product and consequently, civilization produced its contrary: barbarism. Throughout the text, Sarmiento despairs the wide open spaces that seem to dominate the Argentine landscape: "El mal que aqueja a la República Argentina es la extensión: el desierto la rodea por todas partes y se le insinúa en las entrañas" ["What ails Argentina is its wide-open spaces: the wilderness goes to the very guts of the country"]. Isolated inhabitants are terrified by the remoteness and by the impenetrable fields of recently endemic cardoon thistle: "¿Qué impresiones ha de dejar en el habitante de la República Argentina el simple acto de clavar los ojos en el horizonte, y ver . . . no ver nada?" ["What happens when the typical inhabitant of Argentina stares at the horizon and sees . . . absolutely nothing?"]. Sarmiento seems to legitimately fear that nature will take over the cities if they are abandoned to the Federalist educational politics of Rosas: "¡Es la *ciudad,* que trata de salvarse de no ser convertida en pampa si abandona la educación que la liga al mundo civilizado!" ["If they want to be civilized, the cities must forget about the plains and embrace education; this represents the cities' real connection to the civilized world!"][23] Such fears are exactly opposite of the ideals of isolation expressed by Sarmiento's North American contemporary Henry David Thoreau who famously wrote about the healing powers of wilderness in *Walden.* When Thoreau wrote that he "would rather sit on a pumpkin and have it all to myself than be crowded on a velvet cushion,"[24] I can almost envision Sarmiento rolling his eyes. Thoreau's praise of wilderness has become dogma in the doctrines of orthodox environmentalism while Sarmiento's yearning for development seems quaint. The man who would eventually become president of the Argentine republic seems almost giddy thinking on the riches that might result from the development of the interior. Here, he personifies the rivers, "pidiendo a gritos que se rompan las cataratas oficiales que les estorban ser navegados" ["begging for an end to the flood of bureaucratic obstacles that have left them unnavigable"]. Sarmiento seems to long for the industrializing proficiency of European powers that can "penetrar hasta el corazón de la América a sacar las riquezas que encierra y que nuestra inhabilidad desperdicia" ["exploit the riches of the very heart of the Americas, riches that our incompetence just lets go to waste"]; he dreams of a process of development and industrialization with "todas las márgenes de los ríos cubiertas de ciudades" ["every riverbank covered with cities"].[25] If these are Sarmiento's dreams, it is no wonder that he has not been beatified an environmental saint like Thoreau. However, it would not be wise to so quickly dismiss the importance of Sarmiento's text as foundational to the ecology of Spanish American literature. His repeated insistence on the importance of natural landscapes to the identity and culture of the nation actually represents what might now be thought of as ecocritical analysis at a time where environmentalism was the farthest thing from his mind.

Sarmiento's insistence on the role of nature in his political analysis represents the inauguration of what has evolved into a heritage of Spanish American literature as a whole: preoccupation with the role of the land, water, and air in the development of culture and identity. In the pages of *Facundo*, it takes the form of a developmentalist political ecology with Sarmiento repeatedly condemning the lack of a politics of nature from Rosas, his political rival:

> Los llanos de La Rioja están hoy desiertos; la población ha emigrado a San Juan; los aljibes que daban de beber a millares de rebaños se han secado. En esos llanos, donde ahora veinte años pacían tantos millares de rebaños, vaga tranquilo el tigre, que ha reconquistado su dominio; algunas familias de pordioseros recogen algarroba para mantenerse. Así han pagado los llanos los males que extendieron sobre la Republica. (Sarmiento, *Facundo*, 112)
>
> [Presently, the plains of the Rioja region are deserted; the populace has emigrated to San Juan; the wells that provided water to thousands of herds have gone dry. Twenty years ago those plains were host to many thousands of herds; but now, the tiger lopes in his reconquered territory and some beggar families still collect mesquite to eke out a living. This is how these plains have paid for the bad government that seems to have taken over the republic.]

The commentary on the ecological processes of an abandoned space includes concern about drought blamed on the evil policies of Rosas and the emphasis throughout *Facundo* takes the form of a political ecology that advocates for the role of a strong centralized government to direct the strategic exploitation of untapped resources in Argentina's vast interior.

While the call for ever-more-extensive development has been steadily tempered in recent decades for a variety of reasons, Sarmiento's absolute categories of civilization and barbarity have also long been challenged. Most current ecocritical perspectives would neither accept Sarmiento's vision for riverbanks covered with cities nor his reductionist division of education and urbanization as "civilization" with nature and indigenous peoples as "barbarity." But there was also an extensive, if underappreciated nineteenth century literary counter-tradition comprised of works by authors writing at the same time as or just after Sarmiento who disputed his categories and how they were defined. In light of the received wisdom of nineteenth century literature as foundational, any hint of a counter-tradition must be taken seriously. So that is what will occupy the analysis here and in chapter 2: an ecocritical re-evaluation of the canon and content of the Spanish American literary corpus from that century. In the following pages, the approach will proceed in two ways: first, a consideration of how contrary definitions of civilization and barbarity unfolded as a function of political ecology in representations of nature; and second, an examination of the role that these representations played in the development of peculiarly Spanish American manifestations of what, in the parlance of contemporary environmentalism, would be called environmental justice, conservationism, and sustainability. While the ideological slant of political ecology shifts radically in what would follow these initial texts, the role of nature as central to literature is inaugurated here in *Facundo* and foreshadows the place that ecological principles will occupy in the region's literature.

POETRY, PARADISE, AND THE PAMPA

A national (and political) compatriot to Sarmiento, Argentine Esteban Echeverría was most well-known for the short story "El matadero" (The Slaughterhouse) (1838, published in 1871) and for the long poem *La cautiva* (*The Captive*) (1837). The former was a political screed critical of the federalist party and the dictatorial Juan Manuel de Rosas; the events in the story develop almost entirely within the confines of an open-air slaughterhouse as a rhetorical strategy to associate federalism with the grotesque processes of cattle butchery. The tale concludes with the senseless lynching of a young *unitario*, the opposition political party from the Argentine political scene of the nineteenth century. While there is not much in the way of nature writing in "El matadero," the explicit political ideology of the narrative associates Echeverría with Sarmiento's politics and illuminates the representation of the pampa in Echeverría's other well-known work, *La cautiva*. Sarmiento's categories of civilization and barbarity included the idea that certain concrete phenomena should be associated with one category or the other. Indigenous peoples, nature, federalism, and gauchos were associated with barbarism; the city, European culture, education, and the model for a strong central government based in Buenos Aires were thought of as manifestations of civilization. The close association between Sarmiento's political philosophy and the narrative in "El matadero" invites a reading of *La cautiva* from that perspective. Thus, in a story about the escape of the captives Brian and María from their indigenous captors by flight through the limitless pampa, it does not surprise that the representation of nature is rather negative:

> Aquí se vía
> Al voraz cuervo, tragando
> Lo más asqueroso y vil;
> . . .
> Más allá se ve al carancho,
> Que jamás presa desdeña,
> Con pico en forma de gancho
> De la expirante alimaña
> Zajar la fétida entraña: –
> Y en aquel páramo yerto,
> Donde a buscar como a puerto
> Refrigerio, van errantes
> Brián y María anhelantes,
> Sólo divisan sus ojos,
> Feos, inmundos despojos
> De la muerte.[26]
>
> [Here crows are greedy;
> They ravenously swallow
> What's gross and seedy.

. . .
And vultures feast
With beaks like hooks
That pull carrion from nooks,
No meat too fell to taste:
Putrid guts, bloody waste.
And lost in the fields
They seek a place that yields
To their search for aid;
But Brian and María find only pain.
What meets their eyes:
The ugly styes
Of Death.]

While these references to crows and buzzards swallowing disgusting entrails and the images of Brian and María surrounded by the filthy remains of rotting flesh and death are shocking, they go well with Sarmiento's views on nature and the pampa as poorly conducive to civilization. As Fernando Operé observes: "la naturaleza en la cultura del continente y en particular en la Argentina, ha sido concebida como una fuerza superior y destructiva cuyas secuelas arrastran al individuo y los grupos humanos a un estado de degeneración animal de irreparables consecuencias" ["nature, in the culture of the continent and particularly in that of Argentina, had been conceived as some kind of superior and destructive force, a force which leaves individuals and communities in an animal state that leads to irreparable harm"].[27] This view is simply and graphically reinforced in Echeverría's texts. But just as the federalist party represented an ideological opposition to the political ideas of the *unitarios*, there was literary resistance as well. The attempt to redeem Argentine nature from its reputation as a well of barbarism was replaced by a view of the pampa and mountains as places of sublime beauty and liberty and by a political ecology highly critical of Sarmiento's optimistic developmentalist vision.

Gauchos and Ecology

The most familiar counter-tradition to the civilization/barbarity dialectic in Sarmiento and Echeverría was the development of a literature of the gaucho, maligned in *Facundo*, but canonically revindicated in Argentine José Hernández's epic *Martín Fierro* ("El gaucho Martín Fierro," 1872 and "La vuelta de Martín Fierro" (The return of Martin Fierro), 1879). The figure of a kind of Southern Cone cowboy had appeared in earlier precedents such as *Caramurú* (1850) by Uruguayan Alejandro Magariños Cervantes in which a gaucho of indigenous heritage is marked by his honor, education, and patriotic sense of nation: characteristics contrary to Sarmiento's disparaging representations of "Indians" and gauchos as ignorant, brutal, nationless savages. Similarly inconsistent with Sarmiento's pessimistic view of nature, Magariños Cervantes makes a point of convincing readers (who were most likely Spanish as the first edition was published in Madrid), that if they could "trasladarse con el pensamiento á las floridas riberas del *Uruguay*, sin duda les encantaría el bellísimo paisaje que presenta el lugar donde comienza nuestra historia" ["transport themselves in their minds' eye to the bountiful banks of the Uruguay River, without a doubt they would be enchanted by the beautiful landscape wherein our story begins"]. Later, he praises "ráfagas impregnadas de los más suaves perfumes. Emanación purísima de las selvas vírgenes del Nuevo Mundo" ["breezes laden with the sweetest of perfumes. Pure emanations from the virginal jungles of the New World"].[28] These sublime images of beautiful landscapes and aromatic jungles are hardly characteristic of Sarmiento's fearsome barbarity-conducting pampa.

The more familiar *Martín Fierro* most famously represents the Argentine, nature-representation counter-tradition to *Facundo's* stark stereotypes. Hernández was a political opponent to Sarmiento with extensive first-hand experience of life on the pampa and the customs and culture of gaucho life. One aspect of the political ecology in *Martín Fierro* criticizes the conscription of gauchos for the way that forced soldiering meant loss of access to property and personal investment in the land as a result of involuntary absence:

> Después me contó un vecino
> Que el campo se lo pidieron,
> La hacienda se la vendieron
> En pago de arrendamientos,
> Y qué sé yo cuántos cuentos;
> Pero todo lo fundieron.[29]
>
> [Later a neighbor told me they wanted his land
> They asked for his property
> And sold it secretly
> To pay accounts
> Of unknown amounts,
> But it was all a travesty.]

And these gaucho conscripts also fall victim to the failure of ill-conceived, large-scale development:

> Hablaban de hacerse ricos
> Con campos en las fronteras;
> . . .
> Todos se güelven proyetos
> De colonias y carriles—
> Y tirar la plata a miles
> En los gringos enganchaos,

Mientras al pobre soldao
Le pelan la chaucha—¡ah, viles!—
Pero si siguen las cosas
Como van hasta el presente
Puede ser que de repente
Veamos el campo desierto,
Y blanqueando solamente
Los güesos de los que han muerto. (Hernández, *Martín Fierro*, 27)

[They spoke of getting rich
On country near the border
. . .
A huge project
Of settlements and works
Wasted money, bribery, and perks
For *gringos* and the rich,
While poor soldiers in the ditch
Beg for pennies. Those jerks!
But if things go on like this
It could be
Before the count of three
Piles of whitened bones instead,
In a desert and empty country
Filled with nothing but the dead.]

The criticism is striking for its vision of these kinds of projects, characterized by the courting of industrializing, commercial multinational entities ("los gringos enganchaos"), the abuse of labor, the eventual abandonment of the project, the land left deserted or converted into a desert (the Spanish is ambiguous). The protest against foreign investment—in the context of *Martín Fierro*, by French and especially English commercial concerns—could just as easily be lines from an activist website or petition advocating local control and resisting multinational corporate development in any locale of the present.

Hernández's epic is most commonly understood as the text that redeems the figure of the gaucho and turns him from pariah into national myth. But as Sommer notes, it was also written with a sense of nostalgia for a way of life that had been lost: "Hernández wrote it when the gauchos had all but disappeared . . . confined to narrowing spaces between modernized, privatized ranches on a once vast pampa."[30] The nostalgia in the text comes from the profound changes in the land and its husbandry as the nineteenth century wore on, changes that led to an economic agrarian culture exactly opposite to what the poem articulates as the central characteristic of gaucho life: "Mi gloria es vivir tan libre / Como el pájaro del cielo" ["My glory is to live / as free as the birds of the sky"].[31] Later, in the introduction to the second part of the poem, Hernández praises the gaucho for being able to deduce, solely from interactions with nature, proverbs and principles similar or identical to the greatest wisdom literature from all history. The representation of liberty and life on the pampa was more mythical than factual, but *Martín Fierro* nevertheless offers another counterpoint to the civilization/barbarism dialectic from Sarmiento: nature was rather a good conductor of wisdom and provider of liberty. What is at the heart of the criticism of Sarmiento's ideas in gauchesque literature represents what is by now a foregone ecological conclusion: the superiority of nature over industrialized urbanization. In Sarmiento, the pampa and other manifestations of the natural world are dismissed as the opposite of civilization; but, keeping in mind the impact of the romantic ethos on the Spanish American writers of the nineteenth century that I discuss above, nature can also be viewed as a source for refreshment and recreation, as a place for sublime contemplation, and as inspiration for poets and artists.

The Ecology of Captivity and Captivating Nature

The gauchesque tradition analyzed above represents a critical dialogue with Sarmiento's categories in the decades that followed the publication of *Facundo*. But the articulation of a counter-tradition was not limited to *Martín Fierro* (and *Caramarú*); there are numerous examples of narrative texts that challenge the negative representation of nature in Sarmiento and Echeverría from several other sub-genres of Southern Cone nineteenth century literature. One is the novelization of the captive "white" woman among "savage Indians," in the tradition of Esteban Echeverría's *La cautiva*. Susana Rotker writes that "there is a rape at the very origin of Argentina. Not the ritual rape of the subjugated indigenous woman by the Spaniard, but the reverse: the violation of a married, Catholic, Spanish lady by a chief of the pampa. . . . The legend is known by the victim's name: Lucía Miranda."[32] While there were several novels of the same title by Argentine writers including Eduarda Mansilla de García (1862), Miguel Ortega (1864), and Celestina Funes (1883), the earlier *Lucía Miranda* (1860) by Rosa Guerra represents a marked divergence from that nature conceived as "crows, buzzards, entrails, and death" in Echevería's *La cautiva.* While in Guerra's novel the indigenous Timbue and their tribal leader Mangoré are regularly described as savages controlled by passion rather than reason or religious devotion, the novel finds a way to redeem them. Rotker observes that Guerra's Lucía "is not locked into the female stereotypes of romantic literature [but] that she, the woman, is the mediator between cultures. It is she who evangelizes, performs works of charity, and manages to save Mangoré's soul before he dies."[33] Guerra's evocation of the Timbue is more sympathetic than in other versions of the "Lucía Miranda" story from Argentine literature and I find that her representation of nature to which the indigenous community is inevitably attached is similarly sympathetic, almost Edenic:

> ¡Qué bellas y espumosas son las plácidas corrientes que bañan las riberas de este soberbio y caudaloso río! Majestosos los bosques que sombrean sus márgenes, magníficos los árboles que bordan su costa. . . .
>
> . . . La margen derecha del Paraná donde estaba levantada la fortaleza, era un verdadero edén. . . .
>
> . . . El Paraná es el más rico jardín del mundo.[34]
>
> [How beautiful and foamy are the peaceful currents that lap against the banks of this proud and voluminous river! And majestic are the forests that shade its edges, magnificent the trees that border its coast. . . .
>
> . . . The right bank of the Paraná where the fort had been built was a veritable Garden of Eden. . . .
>
> . . . The Paraná is the most bountiful garden in the world.]

With images that include "beautiful currents," "magnificent trees," and "bountiful gardens," Guerra's version of the female captive story occurs on a pampa that does not resemble the horrors from Echeverría's epic. The representation of nature as sublimely inspirational challenges Sarmiento's condemnation and demonstrates that the categories articulated in *Facundo* represent an ideology not widely reproduced in other texts. Instead, nature is most commonly described, as it is in Guerra's *Lucía Miranda*, for its beauty and idealized qualities rather than as a conductor of savagery and barbarism.

Like many of the other exceptions to the negative evocation of nature in Sarmiento, another white, female captive story—*Huincahual* (1888) by Chilean Alberto del Solar—also contains several descriptions of sublime nature. And these surpass almost all others from the rest of nineteenth century Spanish American literature:

> ¡Cuánta profusión de detalles, cuánta novedad de forma, cuánta luz, cuánta armonía y riqueza de tintas perdidas en el fondo de aquella escenografía ignota y oculta, como el oro encerrado en inmenso caudal dentro de las entrañas de sus montes y en el seno de sus selvas impenetrables! . . .
>
> ¡Cuánta magnificencia en aquellos paisajes grandiosos que la mirada de ningún hombre civilizado ha contemplado aun![35]
>
> [What array of details, what originality of form, such light, what rich harmony of color lost amid an unknowable and forgotten scenery, like huge deposits of gold hidden under the mountain or in the heart of impenetrable jungles! . . .
>
> What magnificence in those splendid landscapes, landscapes that no civilized human eye has yet beheld!]

While other texts from the nineteenth century have challenged Sarmiento's negative vision of nature, none utilize an extended imagery such as this. The representation of "magnificent harmonies of color" is matched in other parts of the novel where del Solar employs a language of opulence, paradise, and savage majesty to describe nature. That tendency alone would be enough for inclusion here, but *Huincahual* includes other representations of nature that come in for ecocritical analysis of a different sort.

The novel deals with the kidnapping of María by Huincahual, *cacique* (tribal chief) of a Chilean Araucan community; after her capture, she is forced to marry the chief and quickly becomes his favorite from among the rest of numerous other wives. But rather than a narrative solely focalized through the experiences of captives, *Huincahual* attempts to communicate indigenous perspectives on the recent arrival of Spanish conquistadors and colonists. Thus, del Solar also offers a sympathetic representation of the injustice that resulted from the theft of land where the Araucan ancestors of the characters in the novel previously dwelled. In this way, the novel displays characteristics of an *indigenista* text by foregrounding the loss of land and the results of these losses upon subsequent generations and makes gestures toward recognizing the validity of the ideology that is at the core of that genre of Spanish American literature. The connections among access to land, other ecological concerns, and *indigenista* literature are a central focus of chapter 6 where I will more fully analyze these elements and their occurrence in del Solar's novel.

Other Foundations of Argentine Literary Ecology

United by their opposition to the Rosas regime in Argentina, Sarmiento and Echeverría were among the prominent members of the *Asociación de mayo*[36] to which several other Argentine authors belonged including Juan Bautista Alberdi and José Mármol. The lasting conflict between philosophies of federal versus centralized governmental organization divided the Argentine political class as well as its most notable authors. Sarmiento's strident assertions about political identity and the ideological currents that supported or opposed it made for a fertile literary atmosphere in the nineteenth century Southern Cone. But the crusading political ecology of *Facundo* also generated a literature of nature in the region that was similarly diverse and politically committed: competing representations of the natural world in the captive woman genre in Echeverría versus *Lucía Miranda* and *Huincahual*; the vindication of the ruralized gaucho in *Caramarú* and *Martín Fierro*; and, as shall be seen in the rest of this chapter, a challenge to Sarmiento's categories such that nature is no longer a contributing factor to the spread of barbarity but a necessary and ennobling element of civilization.

Argentine Vicente F. López was a contemporary to the *Asociación de mayo* movement, but his *La novia del hereje* (The heretic's lover) (1854) represents an early manifestation of the historical novel, the first among the association's authors to write in that genre. It also represents one of several, lesser-known nineteenth century Spanish American texts that I gather here for the different ecological principles that they express. The bulk of the story deals with the frustrated love between an English pirate—Henderson—in the crew of Sir Francis Drake and a young and wealthy *criolla* (individual born in Spanish America of Spanish-born parents) from Lima— María—who had been briefly captured by the pirates. Subsequently, several members of the Roman Catholic clergy in Lima attempt to use the power of the Inquisition to steal María's wealth by condemning her as a heretic, but the pirates' ingenious resourcefulness liberates her from such a fate. Like others of the anti-Rosas *unitarios*, López takes pains to emulate non-Spanish European models; so English pirates are not criticized for pillaging, but praised for their savvy. In fact, Nina Gerassi-Navarro observes that in *La novia del hereje*, "destruction begins with the arrival of the Spanish. They are the ones who condemned and devoured the land with their despotism."[37] As a contrast to that destruction, López evokes praise for sublime natural beauty as in the following description of the mountains that border the sea along the coasts of Perú: "parecían tender una mirada majestuosa sobre las llanuras movedizas de la mar, desde el vasto pedestal que les servía de trono allá, en el interior remoto de la tierra americana" ["they seemed to render a majestic view of the plain-like expanse of the sea from their vast pedestal, a kind of throne in the remote interior of American lands"].[38] Lines like these are common enough in texts from the nineteenth century, despite Sarmiento's fears about the potential in nature for barbarity. But eventually, the narrator communicates that sense of nostalgia for pastoral landscapes so characteristic of the region's literature of nature produced in the second half of the nineteenth century. Here, López uses language much more familiar to the current rhetoric of conservationism than to that of the national narratives of foundational Spanish American texts:

> El mercantilismo prosaico que hoy reedifica nuestras ciudades a su modo, nos priva, poco a poco, de las obras de la naturaleza por las fábricas de la industria, modulando nuestra vida social a otros modos de ser próspera y feliz. Día a día desaparecen nuestras huertas; y con ellas, los naranjos y la madreselva, que desde el fondo de nuestras casas perfumaban nuestras calles. (López, *Novia de hereje*, 345)
>
> [The dictates of common mercantilism that seem to reconstruct our cities in their own likeness, little by little, deny us contact with works of nature and replace them with factories, altering our social existence to fit other models of prosperity and happiness. Day by day, the city gardens disappear and with them, the aroma of orange trees and honeysuckle that emanated from our houses to perfume our streets.]

Phrases in the passage above with references to city residents' lack of access to nature because of the disappearance of *huertas* (in this context, city gardens) are notable for expressing a common lament among environmentalists today: the despair that non-commercial models for society have all but disappeared. But such sentiments are even more remarkable coming as they do in the context of an Argentine literary and political moment where Sarmiento's ideas about the imperatives of development and the primary role of the city for the identity and future of Argentina had been so dominant. López's nostalgia in the above passage demonstrates that Sarmiento's political ecology was not as widely embraced as the canonical stature of *Facundo* may indicate; rather, several of his contemporaries, even fellow intellectuals of the same ideological stripe, were publishing texts with significant deviations from Sarmiento's ideals.

As suggested by the subtitle to Argentine Alberto Blest Gana's *Martín Rivas: Novela de costumbres político-sociales* (Martín Rivas: A novel about political and social customs) (1862), it is not surprising that there is almost no mention of the world outside of the plazas, salons, and other interior spaces frequented by the wealthy and powerful. However, even when his novels make almost no mention of nature, there is at least something in the way of the positive effect of extra-urban space as in the following description of the "provincial" from Blest Gana's *La aritmética en el amor* (The arithmetic of love) (1860):

> Lector, ¿has vivido en la provincia? Allí cesan la agitación y los cuidados con que en las grandes poblaciones la dura necesidad nos espolea; más plácido es allí el aire que se respira, más sencillos los amores de los jóvenes y más francamente viejas las señoras que pasan de los cincuenta.[39]
>
> [Reader, have you ever lived in the country? There, the assault of a constant hustle-bustle in the city and concerns about what needs to get done there simply cease; the air is calmer, easier to breathe, young love seems simpler, and to be frank, women in their fifties seem older somehow.]

Similarly, *El capitán de patricios* (The patriot captain) by Argentine Juan María Gutiérrez, composed in 1843 but which did not appear until 1874 in the *Revista del Río de la Plata*, does not exhibit much in the way of ecological ideology, but it does represent a counterpoint to Sarmiento's representation of the evils of nature. While in *Facundo*, the immense pampa was portrayed as a poor conductor of the values of civilization, the landscapes that witness the love affair between the Captain and María are realized with literary flourishes approaching the style of the *modernistas* that would become so characteristic of Latin American literary expression starting at the close of the nineteenth century. Nature is described as "maestra de los grandes pintores" ["teacher of the great painters"], its flowers "formando figuras regulares y festones, ocupaban los espacios dejados en las paredes por las puertas y las ventanas y perfumaban el aire con esa fragancia tan exquisita como pasajera que todavía no sabe imitar el arte del destilador ni del perfumista" ["formed regular and festive figures by occupying the spaces left in walls by doors and windows and perfumed the air with a fragrance as exquisite as it was fleeting and which cannot be imitated, even by all the arts of distillers and perfumers"].[40] The idealization of nature which is forever an element of environmentalist activism takes the form of a certain persistent philosophy of nature in this text whose tenets are occasionally fleshed out in the story. Ideas that would not seem out of place in an essay on Green Studies are articulated by the characters and narrator throughout the novella.

The narrator affirms that "el paisaje, según lo ha dicho alguien que lo entendía, no está fuera sino dentro de nosotros" ["the landscape, as has said one who understood it, is not located outside of us but within"]. The Captain expresses the idea that "el ideal de la vida consiste en cultivar un campo propio, sin abandonar el cultivo de la inteligencia. La nobleza del porvenir ha de tomar por escudo de armas un arado y un libro entrelazados con un gajo de palma" ["the ideal life includes the cultivation of one's own land without forgetting the cultivation of one's intellect. The nobility of the future must adopt as their coat of arms a plough and a book, crested by a palm bough"].[41] As much as Sarmiento famously deplores the corrupting influence of nature against the ideals of civilization, here images of beautiful nature inhabit and elevate humankind and are to be idealized in the symbols of nationhood. Thus, the aims of *El capitán de patricios* are similar to Sarmiento's *Facundo* concerning the foundation of national ideals, but their articulations of nature are diametrically opposed.

Later Literary Political Ecologies from the Southern Cone

In the next chapter, I consider the nineteenth century literature and poetry of other Spanish American nations; but several of the later and lesser-known examples of literature from the Southern Cone were most invariably affected by, produced immediately subsequent to, and the most direct inheritors of the cultural categories of the federalist/*unitario* political milieu of the early nineteenth century. The second wave of Argentine, Chilean, and Uruguayan authors produced literature in a period subsequent to the oppositional influence of members of the *Asociación de mayo*. Although some anthologies and introductions refer to the group of writers as a generation of 1880,[42] there is little agreement about who exactly should be included owing to differences in style, setting, considerations of sub-genre, and other such defining characteristics that often confound literary historians (and readers of their histories).[43] For the purposes of this portion of the analysis, I have chosen several of these later authors from the region, ordered chronologically according to the publication date of the relevant texts, and whose representations of nature, country, and city are most amenable to an ecocritical approach: Lucio Mansilla, Bartolomé Mitre, Eugenio Cambaceres, Juan Zorillo de San Martín, Eduardo Acevedo Díaz, Carlos Reyles, Joaquín V. González, and Eduardo Wilde. In each author's work, there remains a certain articulation of nature, inherited from the civilization/barbarism dialectic that so dominated Argentine thought during the nineteenth Century: some embracing Sarmiento's ideals, others rejecting them. Perhaps the best example of the divergence of ideas concerning nature, civilization, and barbarism comes from Argentine Lucio Mansilla's *Una visita a los indios ranqueles* (*A Visit to the Ranquel Indians*) (1870).

If Sarmiento's idea of natural space as a poor conductor of civilization was at the heart of *Facundo*, then Mansilla's journal from his visit to the indigenous Ranqueline homeland of *cacique* Mariano Rosas puts that theory to the test. Mansilla, a colonel in the armed forces of Argentina, was nephew to hated dictator Juan Manuel Rosas who Sarmiento had so relentlessly attacked in *Facundo*. Now, with Sarmiento as president, he sends Mansilla to negotiate a treaty with the *cacique* Rosas in central-western Argentina. Inevitably, many of the prejudices and categories from *Facundo* survive in the narrative of Mansilla's letters, published serially in the *Tribuna* of Buenos Aires. However, the reality of what Mansilla actually experiences among the indigenous/*mestizo*/*cautivo*[44] communities in the lands they inhabit challenges many of Sarmiento's presumptions in *Facundo*.

First, there are several moments in Mansilla's narrative where the land, rather than a conductor of barbarity, is viewed as a refreshing destination and antidote to the overly developed civilized places from which Mansilla has been dispatched. As Mansilla moves into what is called "*tierra adentro*" (the outback), he comments that the paintings he has seen of the pampa are grossly inaccurate and bemoans the fact that most city dwellers live "en la ignorancia hasta de la fisonomía de nuestra patria" ["in total ignorance of what much of our country even looks like"].[45] He finds an escape in undeveloped land where:

> es fácil conciliar el sueño cuando la civilización no nos incomoda, no nos irrita con sus inacabables inconvenientes, cuando no tiene uno más que echarse, cuando no hay ni el temor de desvelarse, quitándose la ropa, o pensando en lo que la justicia y la generosidad humana acaban de hacernos o se proponen hacernos. (Mansilla, *Excursión*, 66-67)
>
> [it is easy to get to sleep when we are not bothered by civilization, nor by its unending inconveniences, when there is nothing more than lying down, without fear of being kept awake, or having to take off clothes, or having to think about what human justice or injustice has been done to do us or plans to do to us.]

There are several moments where Mansilla praises the virtues of the country, of nature, and of wilderness, in particular when he is on the journey home, back to the tedium of city life: "una hora de libertad por los campos es un placer salvaje que yo trocaría mañana mismo por un día entero de esta existencia vertiginosa" ["one hour of freedom in the country is a kind of untamed pleasure that I would gladly exchange for a full day of life in the madding city"].[46] Rather than a conductor of ruin and villainy, the land is juxtaposed in Mansilla's narrative as an antidote to civilization, where immorality and excess are far more evident. Cristina Iglesia observes that "Mansilla se construye en un sujeto en estado de naturaleza. . . . Estamos en el extremo opuesto de Sarmiento, para quien la naturaleza es siempre una amenaza" ["Mansilla constructs himself as a figure of nature. . . . We are at the extreme opposite of the spectrum from Sarmiento, for whom nature was always a threat"].[47] The categories of civilization and barbarity seem to ever hover over the narrative, but ambivalently, now embraced as a comforting and unambiguous dialectic, now challenged as not quite true to the facts.

One moment, Mansilla imagines a railroad project, a manifestation of civilization that will connect Buenos Aires to the Andean border with Chile and argues that it would be "mejor traerlo aquí"["better to bring it here"]; that is, build the railroad through the lands he crosses to get to the indigenous settlements. He faces the Ranqueline challenge to Sarmiento's insistence on buying "la *posesión* de la tierra" ["the *possession* of the land"] by supporting the idea that "la tierra no era de los indios, sino de los que la hacían productiva trabajando" ["the land did not belong to the Indians, but to those that had worked to make it productive"]. Mansilla's ideas are in opposition to the indigenous position that the land should belong to those who are born in it: "¿Cómo no ha de ser nuestra cuando hemos nacido en ella?" ["How can it not be ours when we were born here?"].[48] The debate, where Sarmiento might set the terms as a contest between "civilization" and "barbarity," really reduces the question of political ecology to its essence: who has authority to access the resources of the land? Who defines the parameters of said accession? Who controls the processes of its development? History has demonstrated that, in the end, the ideology of civilization, that is, the Sarmentine model of development, has won out. (There really are no more Ranqueline Indians.[49]) But there are cracks in the model: Mansilla's thoughts may have been pervaded by Sarmiento's political ecology, but after he has made the trip and seen the sights, the conclusions are not always self-evident.

Mansilla appends an epilogue to his collection of journal entries that describes the lands he has visited as deserted space that cries out for development; they have "un porvenir grandioso, y con la solemne majestad de su silencio, piden brazos y trabajo" ["a grand future and the solemn majesty of their silence seems to beg for hands to work them"].[50] This is the party line; but a second tack emerges from the text that both challenges such an idea and complicates the simplistic characterization in Sarmiento's *Facundo* of the land as nothing more than an empty host to barbarism. His time among the Ranquel validates the indigenous community's perspective of land stewardship by at least offering its representation in the text. But more than that, Mansilla's idealization of life in the country anticipates a central tenet of current environmentalist thought: the idea of getting back to the land, to the country, and to the wild as antidotes to the excesses of urbanized civilization. There are problems with the idealization of wilderness,[51] but in the context of nineteenth century Argentine literature, any deviation from Sarmiento's strict categorization of nature as barbarity counts as evidence that Sarmentine notions about nature were not supreme. Indeed, Mansilla's praise for the Argentine outback represents a powerful challenge to the imagery in *Facundo*. Despite Mansilla's overt claim in his epilogue to the highly dichotomized vision of Argentine nature as represented in Sarmiento's text and eventually in Sarmiento's presidency, in *Una visita a los indios ranqueles*, the categories of civilization and barbarity break down when conceptions of land authority and political ecology are left as ambivalent as they are in the wake of the colonel's visit to the places claimed by the Ranqueline Indians as their own.

Argentine Eugenio Cambaceres employed a strikingly different style than the journalistic accounts in Mansilla's journal; Cambaceres's *Sin rumbo* (Aimless) (1885) has very little in the way of nature writing. And while its tales of moral corruption in the city have some critics calling it an early example of Spanish American decandantism,[52] it does feature a brief but weighty moment when the protagonist Andrés (who has abandoned the daughter of a woman he has raped and flees to the city where he starts an affair with an actress) realizes that he hates the city and longs to return to the pampa and his hacienda there: "Una brusca nostalgia de la Pampa lo invadía, su estancia, su libertad, su vida soberana, fuera del ambiente corrompido de la ciudad, del contacto inefectivo de los otros, lejos del putrílago social" ["A strong nostalgia for the pampa came over him, for his country house, his freedom and liberty, away from the corrupting influence of the city and its stultifying crowding, far from that putrid social scene"].[53] Critics may be divided with regard to the classification of his literary production, but this moment from Cambaceres's novel characterizes the growing discontent with the corruption of city life that I mention above as a sentiment expressed more and more commonly in later nineteenth century texts.

While the previously mentioned *Caramurú* is somewhat conspicuously absent from Doris Sommer's comments on foundational Uruguayan fictions, her discussion of novels by Eduardo Acevedo Díaz and *Tabaré* (1888) by Juan Zorilla de San Martín argues for the latter as the eventual "national" novel for "the degree to which Uruguayan readers embraced *Tabaré* as a collective prehistory."[54] I tend to agree with her as *Tabaré* continues to generate scholarly interest, while there is quite a bit less out there on Acevedo Díaz's novels (or on *Caramurú*).[55] *Tabaré* is filled with sumptuous images of nature and fits well in this chapter, but it also contains an insistent association of nature with the indigenous Charrua so I leave Zorilla de San Martín's text for chapter 6 where I address the ecology of indigenous literature. Furthermore, I do not need to raise the debate here about which of the two authors' work contributed most persuasively to the national identity of Uruguay, but the fact that Acevedo Díaz's work was collected in the 1950s and 1960s series of Classic Uruguayan Authors under the auspices of Uruguay's Ministry of Education at least speaks to the previously canonical stature of his novels and suggests that they were somewhat influential in their time. If that is the case, then ideas about nature and political ecology that can be gleaned from his novels are relevant here.

The four Acevedo Díaz novels discussed briefly below, *Ismaél* (1888), *Nativa* (Native) (1890), *Grito de gloria* (Shout of glory) (1893), and *Soledad* (Solitude) (1894) deal with highly specific regional themes including independence, the conflict to throw off Portuguese imperial influence, and the famed invasion of thirty-three patriots to definitively liberate Uruguay from Brazilian control. *Soledad* differs somewhat from the strict historical fiction of the other three in that it recreates an episode in the life of a gaucho who sets fire to parched lands belonging to the father of his beloved to take revenge on him for refusing to bless their union. The historical and gauchesque novel sub-genres were developed previously and better by other authors (see the discussion of *Enriquillo* in chapter 6 and *Martín Fierro* above), but Acevedo Díaz, more than most other authors of the nineteenth century, took pains to describe the details of nature. This likely had more to do with the literary exigencies of his realist style than with any genuine ecological concern. Nevertheless, these narratives provide a far more complete view of nature than most other works with phrases and even whole passages that break through the realist veneer with representations of nature as a force for good and not corruption.

In *Ismaél* and later in *Soledad* there is the typical condemnation of gaucho characters so central to Sarmiento's critique, but often tempered by praise or at least admiration for the freedom of that lifestyle. The natural setting—the pampa—where the gaucho roamed, ever upon his horse, "en medio de las tinieblas de la ignorancia y del error" ["in the middle of the darkness of ignorance and error"] was also that place where "la libertad indómita tenía por vehículo al potro, por refugio el seno de los bosques, y por tipo genérico al primitivo gaucho de la leyenda heroica" ["the use of the horse meant unconquerable liberty, the forests a sure refuge, and the central character, a primitive gaucho of heroic legend"].[56] The ambivalence between the description of the same place as both "ignorant darkness" and "refuge of freedom" illustrates the challenge to a binary description of barbarous nature and urban civilization in Sarmiento. An unsettling nostalgia and longing for nature pervades several other passages in *Nativa* and *Grito de gloria*: unsettling, because in many ways, the ideology of Acevedo Díaz's historical fiction aligns him quite closely with Sarmiento's negative ideology of nature. Even so, in *Nativa*, there are here and there references to the refreshing and healing power of nature: "siempre se respiraba en los campos el aire puro" ["in the country, there was always fresh air to breath"]; praise for "las impresiones gratas [de] la naturaleza" ["inspiring impressions of nature"]; and, contrary to the idea that nature was a poor conductor of civilization: "[la] naturaleza pródiga . . . daba valor a los hombres en la misma medida que la audacia" ["abundant nature . . . gave to men as much valor as audacity"].[57] If unrestrained praise for nature is scant in Acevedo Díaz, his novels remain remarkable for their notably anti-developmentalist political ecology just a few decades after Sarmiento's full-throated advocacy for unrestrained development.

Political ecology in Acevedo Díaz often takes the form of the condemnation of economic injustice and fears about over-exploitation which may be explained by what Susana Crellis Secco documents as a concern about waves of immigration to Uruguay in the 1870s and 1880s: "las posibilidades reales de sustentación del Uruguay y . . . la situación no siempre cómoda de las grandes masas de inmigración que llegaron al Uruguay desde 1870" ["the very real question of the sustenance of Uruguay in general and . . . the not always comfortable situation of the masses of people arriving in Uruguay from 1870 onward"].[58] Thus, there is concern in *Nativa* about over-exploitation of the fisheries in the northern Uruguayan rivers: "La riqueza pecuaria del país empezaba a ser objeto de latrocinios por parte de los dominadores en la vasta zona del norte; y si bien no había llegado allí el sistema de la confiscación y del despojo, no debía demorar en hacerse sentir de la manera más irritante e inicua" ["A domineering class of landowners began to set their sights on the riches of the country's fisheries in the north; even if they had not yet formalized their attempts at robbery and forced evictions, they did begin to unleash a bothersome and downright evil influence on the region"].[59] As it has played out, such landgrabs indeed resulted in the eventual disappearance of the fishing industry in these northern rivers due to pollution and over-fishing.[60] Later, there is also an expression of distinct concern about the feudal *caudillo* system where the narrator suggests that the notoriously inefficient and capricious program of land (mis)management will ruin it forever: "e implantando en las campañas por medios inicuos un sistema de tiranía, propio a devastar zona por zona, como si hubiese sido el intento evitar que en esa tierra desolada volviese a crecer más la hierba" ["the implementation by foul means of a kind of tyranny in the country would only just destroy it, region by region, as if the idea all along was simply to have that desolate land never green up again"].[61] Finally, in *Grito de gloria*, the consequences of the Brazilian invasion are described as much in ecological as in political terms: "Las campañas antes tan hermosas, rebosantes de vida, estaban ahora mustias, llenas de desolación profunda. Creerías que un ciclón inmenso las hubiese devastado de norte a sur y del este al occidente, sepultando hasta el última rebaño bajo las ruinas del desastre" ["The countryside, so beautiful before and full of life, was now withered and profoundly desolate. You would think an enormous cyclone had hit and covered the land with devastation from north to south and east to west, burying everything, up to the last flock of sheep, under the ruins of the disaster"].[62] Such a fearful vision of the land marks a stark contrast to an explicit articulation of the text's political ecology later in the novel. When one of the soldiers contemplates why they are at war with Brazil and whether the Uruguayan patriots should seek annexation to Argentina if the Brazilians are defeated, he argues in terms of the great potential of the land and its connection to political and military achievements:

> Y veía en sus campos feraces llenos de luz y de verdor eterno, treinta millones de toros; en sus canales escuadras enteras con todas las banderas del mundo; un mar de espigas y de viñas en sus vegas; emporio de comercio en sus playas admirables; solidaridad nacional, leyes justas, historia gloriosa, culto por los mártires y los héroes. (Acevedo Díaz, *Grito*, 191)
>
> [And he saw in those fertile, light-filled, and verdant lands, thirty million bulls; in its canals, entire squadrons of ships flying all the flags of the world; a sea of wheat and vineyards, each on a richly productive plot; veritable markets right on the banks of rivers; national solidarity, just laws, glorious history, and respect for martyrs and heroes.]

While the vision can have flaws as seen in the over-developed landscapes of industrialized societies, it is remarkable to observe in Acevedo Díaz the close ties between the land and the ideology of national identity. Certainly the nationalist sentiments dominate the ever-present insistence on political ecology here as they do in many other nineteenth century texts. But the concern with over-development, resource depletion, deforestation, desertification, conservation, and biodiversity in the pages of Acevedo Díaz's novels represents something unique as these are ecological categories which continue to inform environmentalist activism of the present. In Uruguay itself, the livestock industries of beef and wool production have always been the primary economic engine, but more recently the country has promoted itself as a site for eco-tourism:

> Ecology is of central importance in the Uruguay Natural campaign for attracting foreign investors and tourist dollars. The Uruguayan countryside, once considered hostile and threatening or culturally backward and primitive, has been re-presented as an arcadia of beautiful landscapes and quaint traditions, a natural Eden for the ecologically sensible. The revalorization of the natural has coincided with rising forms of "green" consciousness, reflecting broader socio-political processes.[63]

The ecological elements in Acevedo Díaz's novels did not in their time lead to anything approaching a crusade for conservationism against deforestation and soil depletion; however, the expression of such ideas in the foundational literature of a nation makes for a less conflicted cultural fit as local and national economies begin to incorporate "green" industries such as eco-tourism.

Another Uruguayan working to fix the national identity of the small and comparatively newer nation was Carlos Reyles; his 1894 work *Beba* represents an attempt to reconcile a nostalgia for early foundational ideals expressed as they were in Díaz Acevedo with the modernizing impulses of development as they played out in late nineteenth century Uruguay. In *Beba*, an orphaned girl living in the country outside Montevideo marries Rafael, a man from the city, but eventually leaves him for her uncle who had raised her on his ranch. When their union results in a pregnancy and still-born child, Beba takes her own life, wrapping her body in a huge chain and throwing herself in the Río de la Plata. While scholars have analyzed the incestuous undertones in the novel,[64] I include the text here because, unlike so many of the other tragic love stories that populate the literature of nineteenth century Spanish America, in this case, the central conflict is an opposing view of nature: Rafael loves the city and Beba loves the country. As Beba greets her uncle Tito during a visit with her husband and his family, she says as much: "aquí se respire, se vive . . . a mí me gusta mucho el campo. Créeme, Tito, si de mí dependiera no volvería a la ciudad. Pero a *él* no le gusta el campo, cree que es cosa de salvajes: en cambio a mí me revienta la ciudad con su vida frívola e insípida" ["here, you can breathe and live. . . . I love the country. Believe me, Tito, if it were up to me, I wouldn't go back to the city. But *he* doesn't like the country, he thinks it's for savages: but me, I hate the boring and superficial life of the city"].[65] The conflict between Beba and Rafael and the decision to eventually return to her uncle is set up in the text as a conflict between nature and city, and upsets the Sarmentine dialectic (yet again) as the description of everything associated with nature comes up for praise while the urban environment associated with Rafael is wholly negative. As Sabine Schlickers has observed of cultural values from that time in Uruguay: "La vida sobreagitada en la ciudad moderna produjo la añoranza de lo rural y la evocación nostálgica del pasado que se manifestó pronto en la reivindicación de un 'patrimonio criollo'" ["The highly agitated life in the modern city produced a desire for the country and the evocation of nostalgia for the past that manifested itself in the revindication of a '*criollo* heritage'"].[66] As I suggest in comments on several other nineteenth century texts, this is an intrinsically conservationist phenomenon: as rural environments are transformed into urban ones, the concern is that the former will disappear forever.

But *Beba* also contains at least two other elements that are fruitful for an ecocritical analysis. First, Tito is described as a progressive rancher who uses science to improve nature; his livestock is carefully isolated on his land, the pastures fenced in so that breeding can be tracked and evaluated. The idea is to identify genetic advantages in each generation of animals with the objective to produce cattle most profitable for export. Previously, cattle in Argentina and Uruguay were left to their own devices and most died of old age. Others were captured and butchered for a single choice cut and the rest went unused.[67] In the context of the modernization of livestock agriculture in *Beba*, the uncle attempts to introduce efficiency into the process and makes plain his preference for a political ecology of state support for his carefully managed breeding program:

> Lo que hace falta es que los gobiernos lo entiendan así y se den prisa en secundar los esfuerzos de los criadores progresistas. . . . Entonces entraríamos en una época de prosperidad y engrandecimiento reales, porque lo que no venga de ahí, de nuestra industria natural y espontánea, será progreso falso y por consiguiente efímero.[68]
>
> [The governments must understand the importance of what the progressive breeders are doing and help them. . . . Then we would begin an era of prosperity and true advancement; because what does not come from here, from our own, domestic industrial production, must be false progress, and as such, ephemeral.]

The appeal to government hearkens to the dominance of a statist political ecology in Sarmiento, but, at the same time, it represents a radical departure from that model as well. Here, instead of a plan for transport of raw materials via the natural waterways of Argentina with Buenos Aires as the point of departure for the delivery of these resources to the international market, *Beba* specifically advocates a system of managed development for domestic industries developed in the interior by local residents as a path to true and long-lasting progress. Reyles also explores the idea later in *El terruño* (The small plot) (1916), a novel written in the style of *modernismo*, but with a political ecology that repeats ideas similar to those in *Beba*:

> El progreso de nuestro amado país pende del progreso de la campaña; . . . La campaña, aunque no lo digan los doctores: todos nos nutrimos de ella, desde el

> presidente de la República hasta el último gaucho. Y bien: mientras en las ciudades discursean y tragan viento o papan moscas, ocupémonos nosotros en doblarle el vellón a las ovejas y el peso a las vacas. . . . Los rodeos y las majadas son las únicas cosas serias del país.[69]
>
> [Progress in our beloved nation depends upon progress in our more rural lands. . . . The country, although perhaps our scholars may not see it that way: we all live off of it, from the president of the republic to the least of the gauchos. And so: while there is a lot of empty talk, shooting the breeze, and wasting time in the cities, we are busy trying to increase wool yields and livestock production. . . . And the cattle-drives and round-ups are the only truly important things in our nation.]

The ironic description of life in the cities and utter distaste for the cultures spawned there is a classic component of *modernista* literary expression as shall be seen in chapter 3. But it seems that in Reyles, whether writing in any of a number of literary modes, the idea that agricultural production is central to Uruguayan identity and culture remains prominent in his novels.

The second noteworthy moment in *Beba* involves the love for nature by the protagonist herself as a precursor to ideas about ecofeminism that have been developed more recently. In Reyles's novel, the description of female perspectives on living in a city is a central element:

> No sé qué va a ser de mí en Montevideo; me espanta sólo el pensar en las horas de aburrimiento que allá me esperan. ¡Si yo pudiera dedicarme a una tarea que me ocupara todo el día! Pero no, las mujeres estamos imposibilitadas para toda suerte de empresas; nuestra única tarea consiste en procurar a casarnos, y después tener hijos.[70]
>
> [I don't know what will become of me in Montevideo; I am horrified just thinking of the hours of boredom that await me there. If only I could find something to do that would keep me busy all day! But no, we women are thought to be incapable of any industrious activity; our only job is to get married, and then have kids.]

City life is here described by Beba, where she has nowhere to go and nothing to do, as restrictive of women's freedom in a way that living in the country where there is much to occupy the time does not. In the country, it is natural for women to work all day at tasks that contribute to the success of the farm or ranch. Thus, Beba articulates her preference for life in the country in the form of the expression of a right relationship to the natural world, one comprised of useful work rather than the boredom of a limiting matrimonial and maternal life. To echo Karen Warren's ecofeminist philosophy, I find that *Beba* gives "voice to a felt sensitivity, an emotional disposition, or . . . a sensitivity to conceiving of oneself as fundamentally being in relationships with others including the nonhuman world."[71] While the expectation that women take charge of domestic duties was not being challenged by the text, there was nonetheless the expression of preference for life in the country as a place that does not hinder women as happens in large cities where right relationships with the nonhuman world are disrupted. While *Beba* is a late entry in the corpus of nineteenth century texts, its importance resides in its early expression of ideas about the value of nature, its political ecology that preferences the role of government in the exploitation of nature, and most especially, for its expression of a kind of proto-feminism that explicitly connects ideas about well-being to emotional dispositions and sensitivities felt uniquely in the country rather than the city.

With an unambiguous narrative of sublime nature, Argentine Joaquín V. González's *Mis montañas* (My mountains) (1893) represents still another in the counter-tradition to Sarmiento's concrete and inflexible categories of civilization and barbarity. Stylistically similar to the folkloric *tradiciones* genre (made famous in Perú by Ricardo Palma),[72] González recounts episodes from his childhood with inspired narratives of nature. These include entire chapters devoted to the condor, to hunting the *guanaco* (a South American ostrich), to time spent in the country schoolhouse, to the climatic characteristics of winter in the global South, to the strange rituals of the Chaya (a festival that originated with an indigenous celebration of thanksgiving to land deities for agricultural fertility), and to several other phenomena unique to the Argentine province of the Rioja. Literally from the first page, González calls up these child-like experiences of nature: "buscando reposo, después de rudas fatigas, de esas que rinden el cuerpo y envenenan el alma, quise visitar las montañas de mi tierra natal" ["seeking rest, after much hard labor of the kind that exhausts the body and poisons the mind, I wanted to visit the mountains in the land where I was born"]. His native land is something akin to "los grandes templos, donde el esfuerzo material no basta, si de lo íntimo del ser no brota al mismo tiempo el sentimiento religioso, el arrebato mística" ["grand temples, where physical effort doesn't count for much if it is not accompanied by religious or even mystical sentiments that come from the very depths of the soul"];[73] the narrator's reverence for sacred nature is expressed through a sublime language of religious, spiritual, even mystical belief. As I propose in the introductory comparison between Bello's poetry and Cané's novella, environmental ideas in nineteenth century Spanish American literature revolve mainly around two axes: Sarmiento's political ecology of development accompanied by his harsh condemnation of nature as a force of barbarity and the contrary redemptive description of nature as a place of escape, liberty, emotional sensitivity, and sublimity in several other subsequent fictions. González's text falls squarely within the latter category with textual moments that describe mountains as "la más alta y etérea poesía de la naturaleza" ["the highest and most ethereal poetry of nature"], nature as a place with "un alma sensible . . . dócil a las caricias de sus hijos" ["a sensitive soul . . . responding gently to the caresses of its children"], and humans presented with a natural world as "[una] sorpresa sublime, una súbita revelación del arte inconsciente pero inimitable de la inteligencia ignota, creadora de la belleza originaria" ["a sublime surprise, a subtle, unconscious revelation of inimitable art that proceeds from an unknowable intelligence, creator of original beauty"].[74] While it is something to describe the element of the sublime as a component of nineteenth century Spanish American literature as in these quotations, it is quite a bit more surprising to have such expressions in Argentine nineteenth century literature where Sarmiento's barbarous nature had cast so long and ubiquitous a shadow. An ecocritical examination that casts a wider gaze upon the literary context of that century finds exceptions to the hegemony of Sarmiento's categories in Argentina and the rest of the Southern Cone in well-known works like *Martín Fierro* as well as in ones discussed in the last several pages that have garnered far less attention: *La novia del hereje*, *El capitán de Patricios*, novels by Acevedo Díaz and Carlos Reyles, and in *Mis montañas* by Joaquín V. González.

Finally, Eduardo Wilde's *Aguas abajo* (Waters under the bridge), written in 1914, comes at a time when the bulk of the texts that make up the corpus of the *fin de siècle* and even those of early twentieth century *modernismo* had already been written. So it seems anachronistic to include Wilde's novel in a chapter about nineteenth century literature when *Aguas abajo* was written fifteen years into the twentieth. But as these dates and periods are often arbitrary assignments invented by literary scholars, I take the liberty to include Wilde's work in this chapter owing to the author's association with many Argentine authors of the previous century as well as to similarities with several other texts in the *tradiciones* sub-genre of the nineteenth century. The novel contains short reflections from the perspective of a fictional Boris functioning as Wilde's autobiographical voice. Also, perhaps for its late publication date, we find more of a sense of loss, similar to that of present environmentalist representations of nature. There is not much more nature in *Aguas abajo* than what I cite in the following lines, but the sentiment seems so characteristic of present thinking that it is appropriate to include Wilde here as an epilogue to ecological Southern Cone literature of the nineteenth century:

> Amaba mucho los bosques, las praderas, las montañas i Colinas. . . . Cincuenta años después todavía veía en su mente los árboles, los paisajes, los arroyos, las peñas, i evocaba la sensación que el arrullo de las palomas ó el grito de las aves producía en sus oídos, i se deleitaba con la música melancólica, suave, sin ritmo ni tonalidad precisa, de los rumores engendrados en la naturaleza por las cosas que se mueven rozándose unas con otras á favor del violento ó del agua corriente. ¡Quién diera entonces al revivir estos recuerdos la dicha de volver á

Palala, con la aptitud de sus sentidos infantiles, para gozar con todos ellos, de los dones de una escena virgen, inmodelada, primitiva, aún no contrahecha por la civilización, que quita á todas las cosas de este mundo su encanto poético, empezando por privar al espectador de sus gustos sencillos i de sus aptitudes orgánicas para saborear las frutas, sentir intensamente los olores, respirar el aire puro, bañarse en la luz de los cielos i beber con ansia el agua clara de las vertientes, tras de una fatiga sana, hija joven de la marcha por prados, montes i llanuras; por los bosques donde la impresión del ambiente es de paz i de quietud. Los árboles estáticos dan la idea de la concordia; . . . I todo eso dura hasta que la extrema vejez llega ó la dureza del hombre cruel, agresivo, hunde su hacha en el tronco indefenso.[75]

[He loved the forests, the meadows, the mountains and hills. . . . Fifty years later, he could still envision the trees, landscapes, streams, crags and evoke the sensation that the cooing of doves or the loud cawing of birds produced in his ears, a soft, melancholy music without precise rhythm or tone, a rumor of sound made in nature by things rubbing together in the wilderness or by the running water. What he wouldn't give to relive those moments and return to his beloved estate, Palala, like a child, to take pleasure in the youthful joy of frolicking in untouched, primitive, virgin nature, unsoiled by corrupting civilization that takes the poetic joy out of everything by denying the spectator the simple happiness of organic connections to the taste of fruit, the intense experience of smell, the breath of pure air, the bath of light from the heavens, and the drinking of crystalline water from springs, after an invigorating walk through mountains, plains, and woods; through forests where the experience of the environment is peace and quietness. The ecstatic trees give off a feeling of harmony; . . . and all this lasts until old age or the cruelty of brutish and aggressive men bury their axes in defenseless trunks.]

The mythical Palala assumes the contours of a place infused with strong ecological meaning for the narrator; its trees, rivers, birds, gentle breezes, rolling hills, perfumed scents, and delicious fruits evoke every physical sense. The fact that for the narrator, his childhood home recalls these strong sensory memories represents a sense of place that challenges the primacy of industrialization and development for relegating such places to memory and eliminating them from the material space of the Argentine nation. Recall the contrapositive corollary of Alexandra Kogl's definition of place from the introduction: "space + meaning = place"[76] as equivalent to "place – meaning = space." When the narrative of pastoral bliss cited above makes reference to the superiority of undeveloped land, "aún no contrahecha por la civilización," and the durability of such a place until "la dureza del hombre cruel, agresivo, hunde su hacha en el tronco indefenso," the novel makes an explicit connection between development, deforestation, and the way in which such things erase the contours of place. Before, the bliss of a beloved country house offers sensual inputs that assign meaning to the land—birdsong as lullaby, trees that represent peace and tranquility, the entire scene as provider of purity, innocence, and organic aptitudes. Now, these are accessible only through memory; civilization—through the erasure of such places—deprives humankind of similar pleasures. The erasure takes the form of development and urbanization: Shawn William Miller finds that from 1600 to the present Latin Americans went from 40 percent living in cities to 80 or 90 percent (92 percent in Uruguay).[77] By the time *Aguas abajo* was published in 1914, most readers only had access to nature as a rhetorical experience of the ecosublime in the strictest sense given to it by Lee Rozelle: that is, an experience mediated by literature. The textual experience of sublime nature, so common throughout nineteenth century Argentine and Southern Cone literature is accompanied by a fairly mature expression of ecological principles that point to the more explicitly environmentalist fictions that would appear in the decades to come: a love for trees, hills, mountains, bodies of water, the desire for a return to a virginal state of nature, an unflinching condemnation of civilization for its abuse of nature, a desire for unpolluted wilderness, the condemnation of deforestation. These ideas, and others from the several examples of Southern Cone nineteenth century literature examined above, are also expressed in novels, stories, poetry, and essays from the slightly different political, cultural, and social contexts in other regions of Spanish America during the nineteenth century. It is to a consideration of those texts that I now turn.

NOTES

1. Andrés Bello, *Poesías de Andrés Bello* (Barcelona: Casa editorial Maucci, 1909), 67–68.
2. Bello's work appears in the major undergraduate introductory anthologies of Spanish American literature including Raquel Chang-Rodríguez and Malva E. Miller, eds., *Voces de Hispanoamérica: Antología literaria* (Boston: Heinle & Heinle, 1996) and John F. Garganigo et al, eds., *Huellas de las literaturas hispanoamericanas* (Upper Saddle River, NJ: Prentice Hall, 2002)—as well as in several other stand-alone collections of Spanish American poetry, including at least five anthologies dedicated just to Bello's work.
3. After my own fruitless search, the reference librarian at my college was able to come up with a poorly reproduced microfilm of a reprint from 1930: Miguel Cané, *La familia de Sconner* (Buenos Aires: Imprenta de la universidad, 1930). Needless to say, Cané's novella does not appear in any anthology of any kind.
4. Cané, *La familia de Sconner*, 416.
5. Doris Sommer, *Foundational Fictions: The National Romances of Latin America* (Berkeley: University of California Press, 1991), 6.
6. Ibid., 7, 42.
7. Jennifer L. French, *Nature, Neo-Colonialism, and the Spanish American Regional Writers* (Hanover, NH: Dartmouth University Press, 2005), 13.
8. Jonathan Bate, "From 'Red' to 'Green,'" in *Green Studies Reader: From Romanticism to Ecocriticism*, ed. Laurence Coupe (New York: Routledge, 2000), 169.
9. William Blake, "Nature as Imagination," in *Green Studies Reader*, ed. Coupe, 16.
10. John Ruskin, "Landscape, Mimesis and Morality," in *Green Studies Reader*, ed. Coupe, 27.

11. Stephen C. Behrndt, "Remapping the Landscape: The Romantic Literary Community," in *Comparative Romanticisms: Power, Gender, Subjectivity*, ed. Larry H. Peer and Diane Long Hoeveler (Columbia, SC: Camden House, 1998), 24.

12. See Introduction, note 43.

13. Michael Lowy and Robert Sayre, *Romanticism Against the Tide of Modernity* (Durham, NC: Duke University Press, 2001), 23.

14. Sommer, *Foundational Fictions*, 4, 24.

15. *Foundational Fictions* was first published in 1990 when ecocriticism was in its infancy, not to mention that Sommer's project was not ecocritical but an historical literary study.

16. Shawn William Miller, *An Environmental History of Latin America* (New York: Cambridge University Press, 2007), 59.

17. Luis Vitale, *Hacia una historia del ambiente en América Latina* (Mexico City: Editorial Nueva Imagen, 1983), 81, 84.

18. Domingo Faustino Sarmiento, *Facundo* (Madrid: Ediciones escolares, 2001), 37.

19. The political conflict between *federalistas* and *unitarios* in Argentina was a dispute among various factions concerning the political structure of the new Argentine government. The *federalistas* favored a weaker central government and strong protections for the autonomy of the provinces while the *unitarios* were associated with a political philosophy of strong central government, based in Buenos Aires. The former were often associated with charismatic *caudillos* (regional strong-men) and supported by gauchos and other indigenous elements; the latter with non-Spanish European models of government and supported by formally educated elites and a developing urban middle class.

20. Lelia Area, "Geografías imaginarias: El *Facundo* y la *Campaña en el ejército grande* de Domingo Faustino Sarmiento," *Revista Iberoamericana* 67, nos. 194–95 (January–June 2001): 96.

21. Ibid., 97.

22. Raymond Williams, *Marxism and Literature* (New York: Oxford University Press, 1977), 18.

23. Sarmiento, *Facundo*, 27, 47, 254.

24. Henry David Thoreau, *Walden* (Boston: Houghton, Mifflin and Company, 1897), 60.

25. Sarmiento, *Facundo*, 123, 265, 281.

26. Esteban Echeverría, *La cautiva. La Guitarra = Elvira* (Buenos Aires: La Cultura Argentina, 1916), 65–66.

27. Fernando Operé, "*La cautiva* de Echeverría: El trágico señuelo de la frontera," *Bulletin of Spanish Studies* 53, no. 5 (2003): 553.

28. Alejandro Magariños Cervantes, *Caramurú* (Buenos Aires: Imprenta del Orden, 1865), 2, 112.

29. José Hernández, *El gaucho Martín Fierro* (Buenos Aires: Librería Martín Fierro, 1894), 13.

30. Sommer, *Foundational Fictions*, 111.

31. Hernández, *El gaucho Martín Fierro*, 2.

32. Susana Rotker, *Captive Women: Oblivion and Memory in Argentina*, trans. Jennifer French (Minneapolis: University of Minnesota Press, 2003), 98.

33. Ibid., 109.

34. Rosa Guerra, *Lucía Miranda* (Buenos Aires: Imprenta Americana, 1860), 22–23.

35. Alberto del Solar, *Huincahual* (Paris: Garnier hermanos, 1911), 56.

36. A group of progressive, anti-federalist Argentine writers that met in a bookstore and discussed European, mainly French writers and intellectuals, these latter characterized by a tendency toward liberal politics and an interest in education. Doris Sommer, in *Foundational Fictions*, cites the following list as inspirational to the group: "Cousin, Guizot, Lerminier, Quinet, Villemain, Saint-Simon, Leroux, Lamennais, . . . Mazzini, Tocqueville, among many others," 88.

37. Nina Gerassi-Navarro, *Pirate Novels: Fictions of Nation Building in Spanish America* (Durham, NC: Duke University Press, 1999), 83.

38. Vicente F. López, *La novia del hereje* (Buenos Aires: La Cultura Argentina, 1917), 146.

39. Alberto Blest Gana, *La aritmética en el amor* (Santiago de Chile: Zig-Zag, 1950), 24.

40. Juan María Gutiérrez, *El capitán de patricios* (Buenos Aires: Imprenta de la Universidad, 1928), 4, 9.

41. Ibid., 9–10.

42. E. Anderson Imbert, *Historia de la literatura hispanoamericana* (Mexico City: Fondo de cultura económica, 1997), 1:325; Kessel Schwartz, *A New History of Spanish American Fiction* (Coral Gables, FL: University of Miami Press, 1971), 1:93; Eugenio Alonso Martín et al., *Siglo XIX*, vol 2. of *Manual de literatura hispanoamericana*, ed. Felipe B. Pedraza Jiménez (Berriozar, Navarra: Cenlit Ediciones, 1991), 283–85.

43. Generally, I avoid the use of defining "generations" and classificatory "isms" for nineteenth century texts: the terms are slippery and inconsistently used among scholars. Eventually, it will be unavoidable and in those cases, I will give some literary context to the "ism" or generation to which I refer. In this case, the generational date has been set at 1880 for the historical fact of final Argentine unification following the failure of one last attempt by Buenos Aires and its then-governor Carlos Tejedor to secede from the republic.

44. *Mestizo* refers to individuals of mixed Spanish/indigenous heritage; *cautiva* to non-indigenous individuals kidnapped or taken captive by native tribes.

45. Lucio Mansilla, *Una visita a los indios ranqueles* (Buenos Aires: Espasa-Calpe, 1943), 58.

46. Ibid., 242.

47. Cristina Iglesia, "Mejor se duerme en la pampa: Deseo y naturaleza en *Una excursión a los indios ranqueles* de Lucio V. Mansilla," *Revista iberoamericana* 63, nos. 178–79 (January–June 1997): 187–88.

48. Mansilla, *Una visita*, 57, 176, 231.

49. "Pueblo indígena," under "Población," Instituto Nacional de Estadística y Censos, http://www.indec.gov.ar/ (accessed November 16, 2011). Total indigenous population is less than 2 percent according to the Argentine National Census Institute's latest statistics from 2005.

50. Mansilla, *Una visita*, 293.

51. William Cronon, ed., *Uncommon Ground: Rethinking the Human Place in Nature* (New York: Norton, 1995). The essays by Carolyn Merchant, "Reinventing Eden: Western Culture as a Recovery Narrative," 132–59; Candace Slater, "Amazonia as Edenic Narrative," 114–31; and Cronon, "The Trouble with Wilderness," 69–90, critically examine notions of wilderness, Edenic discourse, the nature of nature, etc.

52. Zoila Clark, "Rasgos modernistas y naturalistas en *Sin Rumbo* (1885) de Eugenio Cambaceres," *Hispanófila* 154 (2008): 47–58.

53. Eugenio Cambaceres, *Sin rumbo* (Buenos Aires: Ediciones Jackson, 1944), 110.

54. Sommer, *Foundational Fictions*, 240.

55. The MLA database returns far fewer hits on Magariños Cervantes's *Caramurú* and all of Acevedo Díaz's work combined than it does on *Tabaré*. My personal experience corroborates this disparity in interest: the editions of *Ismaél, La nativa, Grito de gloria*, and *Soledad* that I borrowed from the University of Notre Dame library (printed as part of the Collection of Uruguayan Classics series published in the 1960s) had uncut pages which I separated at home. No one had read these particular books since the library had acquired them.

56. Eduardo Acevedo Díaz, *Ismaél* (Montevideo: Biblioteca Artigas, 1953), 107, 108.

57. Eduardo Acevedo Díaz, *Nativa* (Montevideo: Biblioteca Artigas, 1964), 32, 44, 249.

58. Susana Crellis Secco, "La nación uruguaya en la literatura: Acevedo Díaz y la creación de una nación," *Alba de América: Revista Literaria* 20, nos. 37–38 (2001): 420.

59. Acevedo Díaz, *Nativa*, 104–5.

60. Evoy Zaniboni Filho and Uwe H. Schulz, "Migratory Fishes of the Uruguay River," in *Migratory Fishes of South America: Biology, Fisheries, and Conservation Status*, ed. Joachim Carolsfield, Brian Harvey, Carmen Ross, and Anton Baer (Ottawa: World Bank and International Development Research Center, 2003), 157–94.

61. Acevedo Díaz, *Nativa*, 242

62. Eduardo Acevedo Díaz, *Grito de gloria* (Montevideo: Biblioteca Artigas, 1964), 3.

63. Daniel Renfrew, "In the margins of contamination: lead poisoning and the production of neoliberal nature in Uruguay," *Journal of Political Ecology* 16 (2009): 89.

64. Sabine Schlickers, "La novelística naturalista-criollista uruguaya: *Beba* de Carlos Reyles y *Gaucha, Campo* y *Guri* de Javier de Viana," *Arrabal* 4 (2002): 179; Mariano López, "Stress, psicoanálisis y sentimiento de culpa en *Beba* de Carlos Reyles," *Texto Crítico* 12 (1986): 180–82.

65. Carlos Reyles, *Beba* (Montevideo: Biblioteca Artigas, 1965), 20.

66. Schlickers, "La novelística," 177.

67. Miller, *Environmental History*, 93.

68. Reyles, *Beba*, 117.

69. Carlos Reyles, *El terruño* (Montevideo: Agencia General de Librería y Publicaciones, 1927), 64.

70. Reyles, *Beba*, 139–40.

71. Karen J. Warren, *Ecofeminist Philosophy: A Western Perspective on What It Is and Why It Matters* (Lanham, MD: Rowman & Littlefield Publishers Inc., 2000), 103.

72. A sub-genre that was part folktale, part recording of popular cultural practices, part biographical sketching of notable personalities of the time.

73. Joaquín V. González, *Mis montañas* (Buenos Aires: William Jackson, n. d.), 1, 238.

74. Ibid., 152, 91, 155.

75. Eduardo Wilde, *Aguas abajo* (Buenos Aires: Fondo nacional de las artes, 1999), 71–72.

76. Alexandra Kogl, *Strange Places: The Political Potentials and Perils of Everyday Spaces* (Lanham, MD: Lexington Books, 2008), 15.

77. Miller, *Environmental History*, 168.

Chapter Two

Foundations from Topography: Literary Political Ecologies of Nineteenth Century Andean, Amazonian, Caribbean, and Central American Literature

The progression of ecological thought is easiest to trace in Argentine and other Southern Cone literatures: the political divide between federalist and *unitario*[1] ideologies makes for convenient literary categories in that the political leanings of writers from Argentina, Chile, and Uruguay in the first part of the nineteenth century were also characteristic of the texts that they produced. The tendency among the writers of the *Associación de mayo*[2] to incorporate Domingo Faustino Sarmiento's political ecology in their work makes for a readily apparent contrast when subsequent writers depart from that model. Whereas Sarmiento's harsh dialectic puts nature, the gauchos, and indigenous groups on the side of barbarity, a good amount of fiction and poetry by other authors challenge the ideological hegemony of the political ecology in *Facundo* (1845), even from within the temporal and literary bounds of nineteenth century Southern Cone literature as I lay out in chapter 1. But the tendency is not confined to Argentina, Uruguay and Chile: in the following pages, I identify nineteenth century poets and authors from other regions in Spanish America that feature political ecologies which not only challenge the validity of the civilization/barbarity dialectic, but in several cases are simply not bounded by such categories.

The environmental facts on the ground are similar in other regions to what was seen in the Southern Cone nations with regard to the way in which ecological issues played out economically and politically. In Argentina and Uruguay, the development of livestock-based industries—the rendering of cattle into beef, jerky, and leather; the production of wool, fleece, cashmere, and other textiles from sheep—established that region as an export-based economy after independence. In the newly independent nations, the export of various natural resources that began to become the object of industrial demand in the nineteenth and early twentieth centuries made Latin America particularly attractive to European and North American economic interests as a source for raw materials. Tin, copper, and silver; plant-based crops such as potatoes, corn, cacao, and coffee; and various other relatively "new" resources—recently discovered tree species for timber and rubber, Peruvian guano and nitrates for fertilizers, and other now-in-demand agricultural, mineral and forest products—seemed freely available for any economic power willing to invest and develop. As Shawn William Miller put it, "before there were banana republics, there were copper republics, cattle republics, and coffee republics."[3] The newly independent nations inevitably fell victim to a kind of neo-colonial dependence that was often more economically limiting than the previous political status as colony of empire; wars of independence left the region with large debts, mostly owed to England and France. As these and other European powers were being transformed by the Industrial Revolution, there was great demand for what the quarries, mines, jungles, and coastal islands of Central and South America could supply. The outstanding debts were leveraged to gain access to these resources at favorable rates.

Contemporaneous to the efforts in tropical latitudes to harvest resources required for industrialization, a developing belief in the principle of environmental determinism justified the patently inequitable terms of contracts negotiated by European and North American entities. This was the opinion, according to Miller, that "nature, more than race, decide[d] human barbarism and decadence."[4] For Domingo Faustino Sarmiento, exposure to nature in general and the pampa in particular were reliable indicators of an individual's propensity for barbarism. In the region bounded by the Tropics of Cancer and Capricorn which includes parts of every Spanish American nation except for those of the Southern Cone, the notion of environmental determinism invoked the stultifying effects of the "torrid zone." However, despite the prevailing belief in the debilitating powers of tropical climates among Europeans and the resultant prejudice against native, slave, and Creole populations, many authors articulate representations of that reality as an ideal and Edenic place. In the literature of Spanish America, the battle between competing notions of the region as either degenerate or ennobling seems to come down on the side of the characteristically romantic conception of nature as sublime. And where it does not, these novels and poems articulate political ecologies that advocate local control and condemn globalized resource markets, slavery, and the ecological disasters of large-scale metropolitan development.

POETRY

Like various moments from the poetry of Venezuelan Andrés Bello and the rhymed verse in José Hernández's *Martín Fierro* (1872, 1879), I consider similarly ecological lyrical expression by other Spanish American poets beginning with the Colombian José Eusebio Caro, followed by work from the Bolivian Manuel José Cortés, the Guatemalan Juan Diéguez Olaverri, and the Cubans Juan María Herredia and José Jacinto Milanés. According to Colombian literary scholar Carmen Galvis Yepes, the poetry of José Eusebio Caro has not received the critical attention that it deserves and has mainly been studied for its formal qualities or for Caro's status as the first romantic poetic in Colombia.[5] However, I find that the content of his poems reiterate the sense of the sublime with elements such as odes to a tree or to a volcano—el Chimborazo—(now part of Ecuador). In the latter, the greatness of Colombia is compared to the awesome heights of the volcano:

> ¡Oh monte-rei!—Pues bien: a su alta cumbre
> ¡Colombia! ¡entonces tu estandarte eleva!
> Fíjalo allí, i al ancha muchedumbre
> Que el bajo mundo lleva
> Muéstralo, i grita: "Aquí mi gloria alumbre!"[6]
>
> [Oh, King Mountain! Upon its height,
> Colombia, raise your flag!
> Raise it there as if in the light.
> While those below look to that high crag,
> Display it there and declare the glorious rite.]

Not only does Caro's poetry affirm the value of nature, it goes so far as to associate the glory of a country with the impressiveness of its topography, an affirmation diametrically opposed to the fears about nature expressed in *Facundo.* Additionally, the devout Roman Catholic tendencies in Caro's poetry also find their expression in poetry about the jungle:

> Medita esa selva, su fin i su oríjen medita;
> ¿En ese desierto quién pudo quererla sembrar?
> ¿Quién, dime, quién pudo tener la paciencia infinita
> De uno por uno sus árboles todos plantar?
> No! nadie, sin duda! La selva por sí se ha formado;
> De un árbol primero los otros vinieron en pos,
> Que a aquel primer árbol el jérmen fecundo fué dado
> De miles de robles, de miles de selvas por Dios!
> (Caro, *Poesías*, 166-67)

[Think on this jungle, on its beginning and end;
Who, in this desert, had the heart to sow?
Who could make that vast place to human will bend?
One by one, such trees to grow?
None under heaven! This jungle was formed by no man;
All the others followed from just one tree:
From one small seed, this forested land,
Thousands of oaks: a forest by God's decree.]

Although the sacred sublime has been an element of nature writing and a common element of ecocriticism, here we find the invocation of God as creator of the densely forested jungle. When he wrote, the threat of deforestation was still many years away and to imagine that Caro had something like that in mind when he wrote his poetry would be anachronistic. However, the foundational nature of these nineteenth century texts at least allow the observation that something along the lines of the inherent value of nature (as product of the creator God) was part of the national imaginary of Latin American countries. Therefore, as efforts to combat deforestation in Latin America were instituted in the twentieth century,[7] ideological justifications could be founded upon the raw data of forest loss but could also find echo in images like these from Caro's poetry.

Yet even in examples of nineteenth century poetry that have all but disappeared from Spanish American literary studies, the representation of nature comprises a critical dialogue with Sarmiento's characterizations. The poem "A la naturaleza del oriente de Bolivia" (Ode to east Bolivian nature), by poet and historian Manuel José Cortés, could hardly be less familiar, even to serious students of nineteenth century Spanish American literature. It appeared, and can still only be gotten, in the August 13, 1859 edition of the Chilean literary magazine *La semana*;[8] yet, its imagery of sacred nature from the speaker's first experience of the jungle at the eastern Bolivian Amazon is as representative an example of the ecosublime as any other Spanish American text. With references to things like "pintadas y bellas mariposas" ["beautifully painted butterflies"] in a "magnífico, encantado panorama" ["magnificent, enchanted panorama"],[9] the speaker confesses to a kind of religious ecstasy:

Al rápido fulgor del rayo ardiente
cual inmensos fanales que colgara
de Dios la mano en el celeste dombo,
Mas nada iguala el cuadro que contemplo
en éstasis divino embebecido.
(Cortés, "A la naturaleza," 203)

[The fleeting fire of burning lightning
whose immense beacons hang

from the hand of God under cover of sky,
incomparable in my mind's eye
and divinely drunk from the ecstasy.]

The emphasis on the spiritual experience of the contemplation of nature fits under the categories of the ecosublime from Jonathan Bate and Lee Rozelle: a transcendent experience of nature mediated by literature. The poem expresses many of the categories of romantic nature writing that have been co-opted by contemporary ecological rhetoric: the supremacy of nature, its possibilities for spiritual renewal, its ennobling qualities:

¡El hombre. . .! ¿Qué es el hombre aquí delante
de este grandioso cuadro?
En el espacio un punto imperceptible,
I en el tiempo un instante
Mas la razón, presente de Jehová
Engrandece al mortal. ¡Naturaleza!
Ella admira tu pompa i tu belleza:
Admira, mas no adora; porque solo
Delante de tu autor se postra muda
I en santo acatamiento le saluda.
(Cortés, "A la naturaleza," 204)

[Humankind. . .! How can humans compare
to this great painting?
They are small specks in the infinite universe
And in time, an instant for earth to bear.
But reason, gift of Jehovah
Ennobles humankind. Nature!
Human reason admires your pomp and stature:
Admires, but does not worship;
For reason, before Nature's author, is mute
It can only kneel, humbly, and salute.]

Such imagery is not particularly surprising in the context of nineteenth century world literature, but the emphasis in the above passage—the awe of sublime nature as not just a religious experience, but one that can be appreciated through reason—is an instructive counter-weight to *Facundo* concerning the potential for moral depravity that was at the root of the representation of nature in Sarmiento's text.

Other poets, such as Guatemalan Juan Diéguez Olaverri, also echo the chorus of praise to Latin American nature dispelling the coldly pragmatic Sarmentine dialectic. In "A los Cuchumatanes" (1857), Olaverri describes sublime ecstasy in the peaks of the sierra in his native Guatemala:

¡Oh cielo de mi Patria!
¡Oh caros horizontes!
¡Oh azules, altos montes

oídme desde allí!
La alma mía os saluda,
cumbres de la alta Sierra,
murallas de esa tierra
donde la luz yo vi!
. . .
En tanto que la sombra
no embargue el firmamento,
hasta el postrer momento
en vos me extasiaré;[10]

[Oh, skies of my country—
On the horizon,
I see blue mountains rising—
Hear me now!
My soul greets,
Mountain heights,
Walls of sight,
Where light is seen!
. . .
Much as the shadow
Cannot stop the firmament;
This, my final testament,
In endless brilliance I delight.]

Other poets that display the representation of these sublime tendencies in their work include most prominently Cuban José María Heredia with his poem "A Niágara" (1824) and its several references to the lofty "grandeur" as well as the "terrors," "fearful beauty," and "wrath of elements"[11] in the falls of northern New York state. Also, in poetry by compatriot José Jacinto Milanés, the most common motif is an expression of love for nature and disbelief at those who do not share in this adoration:

¿Puede haber cosa más bella
que de la arrugada casa
saltar, y en la fresca grama
del campo estampar la huella?
. . .
No hay que gozarla en ciudad:
en todo horizonte urbano
se estaciona de antemano
triste vaporosidad.
Luego ved tanto edificio
alto, serio. . . Angustia dan:
El alba, el sol allí están
como sacados de quicio.[12]

[Can there be anything more fine
than in the outdoors to find

space without lines
grasses with footprints left behind?
. . .
This can't be got in the city
Urban places
are limited spaces
sad, smoky, and gritty.
A great edifice
Tall, without bliss . . . Locales and sadness
Sunrise, sun, and madness
A world without an Atlas.]

Here we have two elements of the ethics and ideology in classic nature writing: praise for undeveloped land in the context of condemnation of urban living. Other examples from Milanés's poetry criticize the marginalization of indigenous communities in the wake of the conquest through condemnation of a class-based economic system that favors individuals of Spanish descent over *mestizos* (individuals of mixed Spanish/indigenous heritage) and slaves. Like Milanés, all of the other poets surveyed just above have significant moments of sublime interaction with the mountains, jungles, trees, and other elements of the Latin American countryside; and each also hints at future ecological concerns suggested by these poetic representations of the landscape: deforestation, urban sprawl, air pollution, and environmental injustice.

ECUADOR

In *La emancipada* (Emancipated woman) (1863) by the Ecuadoran Miguel Riofrío , as in many other of the nineteenth century national romances (to use Doris Sommer's terminology), the story concerns a woman, Rosaura, to be married against her will to a man she does not like. Riofrío's novel does not depart from this well-established motif, but as Fernando Balseca has pointed out, part of the project in Riofrío's novel is to represent the ways in which "la naturaleza inspira las acciones humanas más nobles que, por contraste, en la ciudad están condicionadas por el flujo, la rapidez y la inestabilidad. La proposición es que la patria también pasa por el campo, por el paisaje rural" ["nature inspires the noblest human actions, but in the city, on the other hand, human behavior is conditioned by a constant state of flux, hurry, and instability. The idea is that a nation's identity must also be characterized by the country and by its rural lands"].[13] An example comes in the form of Rosaura's comments concerning how her mother had taught her to love God by contemplating the natural beauty of Ecuador:

> Mi madre me enseñó a conocer a Dios, llevándome a las colinas de nuestro pueblo y diciéndome con acento cariñoso: "Mira la hermosura de estos campos, escucha el cantar de los pajarillos, observa ese cóndor perdiéndose entre

> las nubes, . . . él que hizo esto es un Espíritu que no se puede ver." . . . Yo aprendía a adorar a Dios porque era Padre, porque era bueno y porque había hecho cosas tan grandes y tan hermosas.[14]
>
> [My mother would take me to the hills around our town and, in a loving tone, she taught me that to know God, I should consider the beauty of the country, listen to the birdsong, observe the condor flying high above the clouds. . . . "[H]e who made all of this," she would say, "is a Spirit that no one can see." . . . I learned to worship God as Father, but also because He was good and because He had made such beautifully wondrous things.]

The association between the value of nature and the right attitude toward God is located in Ecuadoran literature, right from its very first example.[15] And very few works from the nineteenth century corpus feature a representation of human/nature interaction that so aptly assume what has come to be described as ecofeminist characteristics. Here, the image of a mother that takes her daughter to a mountaintop to spiritually connect with fields, hills and birds fits the contours of ecological feminism as specified in the introduction, particularly inasmuch as the passage from *La emancipada* is an instance of a sensitivity to "conceiving of oneself as fundamentally being in relationships with others including the nonhuman world" and the expression of "caring attitudes in relation to the nonhuman natural environment."[16]

The most well-known of the nineteenth century Ecuadoran novels must be Juan León Mera's *Cumandá* (1879). The story is similar to *Caramurú* (1850) by Uruguayan Alejandro Magariños Cervantes in that the love between the indigenous Cumandá and Carlos, the son of an Ecuadoran priest, crosses cultural and ethnic boundaries. Furthermore, Mera's novel is similar for its positive representation of the jungle, with references to "[la] música de los reptiles y bichos," and "el seno tenebroso de la selva" ["the music of reptiles and other creatures" and "to the dark heart of the jungle"].[17] Along with Colombian Jorge Isaacs's *María* (1867, discussed more below and in chapter 4), these are among the first literary representations of the Amazonian region in Spanish American literature. I devote the whole of chapter 4 to the *novela de la selva*[18] and will comment more extensively on *Cumandá* there.

Finally, twenty-five years after the publication of Mera's novel, Ecuadoran Luis Martínez published his only novel *A la costa* (To the coast) (1904), four years into the new century. The novel's two parts narrate a contrast between the sierra environment of the national capital Quito and the coastal region near Guayaquil in the context of the tragic story of the family life of Salvador Ramírez, a sometime lawyer who leaves the city penniless in search of a new life down the coast. His education as a lawyer affords him favor with his future father-in-law and also with the owner of the coco-growing estate and he is assigned to run the commissary. His love for the virginal

Consuela is a foregone conclusion of the romantic literary legacy to which *A la costa* is indebted as is Salvador's tragically early death. But what is remarkable about Martínez's novel is the keen awareness of landscapes, particularly the difference between the sierra of Quito and the interior coastal lowlands to the south and west of Guayaquil. In general, the latter are held up for praise and compare favorably to the former.

In these lowlands, the fruits of the earth are abundant and described as "los dones que generosa brinda al hombre la naturaleza en esas regiones" ["gifts generously given to men by nature in those regions"]. Also, the decision to escape the city for the country is described for its healthful effects on the body: "el ejercicio violento del cuerpo al aire puro del campo, la contemplación del agreste paisaje, dabanle cierta duplicada energía moral, de que antes nunca había gustado" ["vigorous exercise in the fresh air of the country and in wild landscapes provide a certain degree of energy that seemed absent before"].[19] Then, there are implicit conservationist tendencies in passages such as the following:

> Los matapalos, las balsas, los ceibos, las palmeras, *orgullo de la selva en la cual dominaron durante siglos* caían al suelo aplastando con su mole a los arbolillos y resalvas. El campo presentaba el aspecto informe de *una ciudad destruida por un terremoto*. Troncos inmensos, *ramas destrozadas*, amontonadas al acaso, algunos árboles todavía en pie, pero *mutilados*, con las antiguas ramas tronchadas. . . . Las copas de los grandes árboles parecen *rotas columnas de un templo que ha perdido la cúpula*. (Martínez, *A la costa*, 222)

> [*The pride of the jungle where they had dominated for centuries*, the matapalo, ceibo, balsa, and palm trees fell to the ground and their weight crushed the smaller trees and saplings. The land seemed like *a city destroyed by an earthquake*. Huge trunks with their *limbs destroyed*, piled any which way, some trees still standing but *mutilated* with ancient branches lopped off. . . . The treetops seemed *like broken columns in a church that has lost its dome*.]

I have italicized several moments from the above passage for the negative implications of images like cities destroyed by earthquakes, church ruins, and mutilated trees which communicate a sense of loss, accentuated by the language of architectural beauty such that destroying trees is like destroying ancient works of art or sacred temples. The suggestion—that clear-cutting for agriculture should not be done—remains implicit, but adopts similar contours to current explicit condemnation of the kinds of deforestation represented in *A la costa*: regret at the loss of forests, emotionally negative language to describe the transformation of trees into lumber, the assertion of the inherent value of nature through the use of sacred imagery.

The political ecology of the novel overtly favors economic development in Ecuador's coastal region, a position emphasized in most scholarship on the novel. For example, Emanuelle Sinardet argues that, "[la] inserción de largas descripciones de paisajes andinos y costeños que, lejos de ser meros ejercicios estilísticos, asumen valores altamente ideológicos" ["the insertion of long descriptions of Andean and coastal landscapes, far from merely being exercises in literary style, actually assume strongly ideological values"].[20] The ideology referenced by Sinardet is the liberalization of economic policies that progressive Latin American thinkers and politicians had pursued throughout the nineteenth century as key to development of the region. Although Sinardet does not use the terminology of political ecology, the novel's ideology of development advocates large-scale production and exportation of cash crops: coco, mango, papaya, citrus fruits, banana. These industries are not problematized in the narrative, except in those moments of nostalgia for the lost grandeur of ancient trees and seemingly inadvertent negative descriptions of deforestation. For the time, even in the first decade of the twentieth century, these glimpses of potentially ecocritical narrative are all there, but they remain brief, implicit, and connotative. And, as in the poetry by José Eusebio Caro, when environmentalist efforts against deforestation eventually do develop in Latin America later in the twentieth century, representations of the destruction of matapalo, ceibo, and balsa trees in Martínez's novel will have been among the first to implicitly suggest that such practices should be discontinued.

VENEZUELA

Similar to the Ecuadoran novels of the previous section, nineteenth century Venezuelan texts also include several representations of sublime nature, both brief and more sustained. The earliest of these is contained in the poetry of Andrés Bello which I discussed at the beginning of chapter 1. Later, the short story "La sibila de los Andes" (Prophetess of the Andes) (1840) by Fermín Toro opens with an ode to "las cumbres majestuosas y enriscadas de los Andes" ["the majestic and rocky peaks of the Andes"][21] and in the sociological essay, *El llanero* (Plainsman) (1846), Daniel Mendoza writes that nature, "como todas las hermosas, es esquiva y zahareña" ["is elusive and haughty like all beautiful things"] and describes Venezuela's fertile plains as "maravillosa[s,] . . . son verdaderas Arcadias de riqueza" ["marvelous, with all the virtues of a veritable Arcadia"].[22] These moments reinforce the idea of nature as a good place to be. But *Anaida* (1860), by compatriot José Ramón Yepes, prefigures something more like the literary expression of dread in the face of unforgiving nature as in *La vorágine* (*The Vortex*) (1924) by Colombian José Eustasio Rivera and in the short stories of Uruguayan Horacio Quiroga (both discussed in chapter 4).

Yepes's novella features the jealous rivalry between Turupen and Aruao for the love of the princess Anaida among the indigenous Zapara of Venezuela. While the focus of the story emphasizes the intrigues of courtship and love rivalry among the Zaparos, there are the inevitable elegiacs of nature—"el aspecto imponente, selvático y agreste de la naturaleza americana" ["the imposing, rustic and jungle-like aspects of [South] American nature"], "bellos y nuevos paisajes. ¡Es tan hermosa la naturaleza americana!" ["beautiful and new landscapes. How beautiful is [South] American nature"]—; but which are counterbalanced by a sense of isolation and even terror in the face of something so powerful: "en las selvas americanas se muere como se vive: solo, ignorado de los hombres: del desamparo del desierto se pasa a la soledad de la tumba" ["in [South] American jungles you die like you live: alone, no one knows where you are: the helplessness of the wilderness becomes the solitude of death"]. The narrator, at one point, wonders, "¿Quién podrá pintar los terrores del hijo de las selvas, cuando esa naturaleza se desencadena sañuda con la tempestad, palpita con el rumor desconocido de los sacudimientos subterráneos o se cubre de sombras en mitad del día con los fenómenos del cielo?" ["Who can even imagine all the terrors of the jungle, when nature unleashes a fearsome storm or trembles with unfamiliar strains of subterranean tremors or suddenly goes dark in the middle of the day because of the tricks of light in the sky?"].[23] Only the Zapara seem equipped to flourish in a place characterized as much by death as by life; with such a stark contrast and as Yepe's novel is one of only a few nineteenth century Venezuelan texts to feature a representation of indigenous communities, it is worthwhile to pause and consider their appearance in *Anaida*.

Bladimir Ruiz finds that "el indígena venezolano aparece equiparado al mundo natural, a una naturaleza que el ojo civilizador desea dominar, proceso este en el cual el registro y la descripción juegan un papel fundacional" ["Venezuela's indigenous [Zapara] are portrayed as well-suited for life in the natural world, a world that civilization would like to dominate; the portrayal represents a process in which the literary act of representation and description of nature play a fundamental role"].[24] So these fearsome images need not necessarily read like the condemnation of nature from Sarmiento; the passages quoted above do not construe nature as a path to barbarity, yet they do have nature as so frightening it is a wonder that groups like the Zaparo—for whom the jungle is home—can survive at all. The ecological aspect of such passages mirror positive representations: whether the jungle, the coast, the plains, or the mountains are thought of as terrible, or beautiful, or both, the implication is that such places should be left alone. In the case of positive representations, no one should destroy something that is so beautiful. For representations of nature as a source of terror, the conclusion that such places should not be disturbed is congruent, a kind of implicit nature protectionism: while

Yepes's jungles are not thought of as needing conservation, they should certainly be avoided for their potential to cause the human visitor harm.

COLOMBIA

Ecological aspects of Colombian nineteenth century literature are most apparent in *María* by Jorge Isaacs, a text whose representation of the jungle I leave for further comment in chapter 4. However, given the text's foundational nature, I should mention here that Isaacs's novel, a love story between cousins Efraín and María, also represents an alternative to Sarmiento's utilitarian, pragmatic approach to the land with the use of what may be called "emotional landscapes." From the beginning of the novel, Efraín associates the beauty of nature with his love for María and since the text is fictionally configured as his memoirs after her death, the sense of nostalgia is strongly evoked throughout the text. In those parts of the book where the love between María and Efraín blossoms, the narrative portrays nature as a comforting mother in sorrow and a warm, smiling presence in joy; the forests, plains, and rivers function to better describe María's beauty. When things are not going so well, the *selva* becomes much more ominous. Both of these cases follow the tendency to anthropomorphize nature inherited from European romanticism, an inheritance that was to have lasting literary influence into the twentieth century.

One of the more unusual of the nineteenth century texts under consideration here is *El moro* (Blackie) (1897) by Colombian José Manuel Marroquín (another author, like Sarmiento and Bartolomé Mitre, who also eventually ascended to the office of a national presidency). Like many other of the novels cited above, Marroquín's text makes reference to the sublime beauty of nature: the plains of Colombia are a place where "la naturaleza parece haberse querido aparejar en esa planicie sin fin un escenario digno de sus magnificencias" ["nature seems to have wanted to prepare in that endless space a magnificent scene"].[25] But with a story that unfolds in the first person voice of a horse, the novel finds inspiration in something quite distinct from anything that had come before in Spanish America: concern about animal rights from the 1877 publication of Briton Anna Sewell's *Black Beauty*. With a similar title and a similar device for narrative focalization, Marroquín's novel also shares many of the same ideological features as *Black Beauty* including condemnation of any of several cruel practices as related to the human use of horses. These include the overuse of bearing reins, painful bits, blinders, spurs, and whips; physical alterations to the horse including tail cropping and poor shoeing practices; and generally abusive treatment including underfeeding, overwork, over-exposure to the elements, and other injurious practices associated with livery coaching, draft work, hard riding, and the use of horses for war.

In addition to the obvious advocacy for animal rights, critics like Moira Ferguson perceive anti-slavery elements in Sewell's novel: "The contours of this narrative echo popular contemporary narratives that use familiar tropes of slavery. Formulaically, they describe kidnapping, grieving families, middle passage atrocities, being sold and forced to work for free in a foreign land."[26] Marroquín's *El moro*, like *Black Beauty*, similarly features at least two distinct ethical emphases in the narrative. As a nineteenth century novel, it was as concerned with Colombian political, cultural, and national identity as any of the other texts from that century. But also, as the earliest Spanish American novel to feature a narrative from the point of view of an animal, the novel implies that these identities are wedded to ideas about ethical animal treatment and the protection of nature in general. For example, human cruelty toward nature is expressed through the desire of the horse to escape to where there are no humans: "¿Por qué, me dije, no he de huir de los hombres y de los lugares que ellos habitan, para ir a vivir a mis anchas en alguna comarca que no se halle infestada con su odiosa presencia?" ["Why, I [the horse] said to myself, shouldn't I flee from humans and the places they inhabit to live in some place not infested with their odious presence?"]. The horse complains about "los malos tratamientos que me habían hecho sufrir los que me sacaron del pantano" ["the bad treatment endured at the hands of those who removed me [the horse] from the marsh"] and languishes under a sadistic, abusive, one-eyed owner whose idea of fun is to light animals on fire, a man whose greatest passion was "su afición a martirizar a los animales" ["to make animals suffer"].[27] These examples from the novel suggest that animal rights as an ideological discourse is germane to an ecological analysis of Spanish American fiction.

While the theoretical bases, political advocacies, and interdisciplinary paradigms of what has come to be called "Critical Animal Studies" is beyond the scope of the analysis here, it is useful to briefly reference certain aspects of that approach as they are related to textual analysis. According to the Institute for Critical Animals Studies (ICAS) mission statement, one aspect of their work:

> Deconstructs and reconstructs the socially constructed binary oppositions between human and nonhuman animals . . . but also looks to illuminate related dichotomies between culture and nature, civilization and wilderness and other dominator hierarchies to emphasize the historical limits placed upon humanity, nonhuman animals, cultural/political norms, and the liberation of nature as part of a transformative project that seeks to transcend these limits towards greater freedom, peace, and ecological harmony.[28]

The literary creation of an animal character from whose perspective the suffering of horses can be articulated satisfies many of the objectives cited above. As the analysis proceeds, particularly in chapters 4 and 8, I consider several examples of the narrativization of animal perspectives to elucidate political ecologies, to express the urgency of conservation, to expand the notions of place and space that include the way in which animals fit into and determine these categories, and eventually, to articulate explicitly environmentalist ideologies. I include three novels by Chilean Luis Sepúlveda, the most well-known and widely published of the recent Spanish American environmental novelists where such perspectives are extensively explored; but the former Colombian president and novelist José Manuel Marroquín's horse-story anticipates these developments by nearly one hundred years.

PERU

In Peru, *El padre Horán* (Father Horán) (1848) by Narciso Aréstegui, like a good many of the other fictions of the century, treats nationalist themes in the context of romantic relationships among the moneyed elite. In its pages, nature makes a fleeting appearance, and in such a way as to almost become commonplace in the way it is represented: as a sublime, idealized refuge. The climate of Peru is described as a characteristic that attracts people in general "por la salubridad de nuestro clima" ["for the healthfulness of our climate"]; furthermore, "la naturaleza majestuosa y sublime, se presentaba . . . con un aspecto muy sombrío" ["sublime, majestic nature appeared . . . somewhat somber"].[29] The sublime quality of nature is, as has been seen, a standard element of Spanish American nineteenth century literature. However, there are at least two other tendencies in Aréstegui's novel which, while not central themes, at least represent secondary elements that anticipate what would later become a fundamental environmentalist belief: political ecology must be informed by environmental justice.

With regard to environmental injustice, the novel makes that plain when Padre Horán visits the poverty-stricken Bríjida and can barely stand the filth: "¡Qué fetidez! . . . Pero Bríjida . . . ¿cómo vives en una casa tan hedionda?" ["What a stench! . . . Bríjida . . . how can you live in such a nasty house?"].[30] Only the poor are subjugated to such conditions; the wealthy classes are free to pursue their whims at parties, in public debuts, and in several other vapid manifestations of *limeña* society from the mid-nineteenth century recounted in the novel. However, several pages on, the narrator takes a strong stand against the plight of poverty among the Peruvian indigenous class: "Nuestra misión al tomar la pluma, es la de manifestar las miserias de esa porción de individuos de nuestra especie. . . . La de los legisladores es sin duda echar una ojeada paternal sobre la suerte de los pobres indios" ["Upon taking up the pen, our mission must be to make plain the miseries suffered by this portion of individuals of our very own species. . . . The mission of our legislators must be to truly care about the condition of our poor Indians"].[31] The narrative places unequivocal responsibility for the suffering of indigenous individuals upon the state and the environmental policy espoused by the text is made plain: due to the absolute ecological depravity of living conditions among the poor, it behooves the state to relieve their suffering and clean up the urban environment in which they live. In the final section of this chapter, I conclude with an exploration of the Mexican novel *Santa* (1903) which explores the problem of urban contamination at greater length, but here Aréstegui has Padre Horán expressing dismay at Lima's filth. While it is a minor element in *El padre Horán*, the theme was to become a major element of the environmental awareness both in *Santa* and in the prose and poetry of the *modernistas* surveyed in chapter 3.

In *Los amigos de Elena* (Elena's friends) (1874) by Fernando Casós, the isolated emphasis on ecological issues is similar. The novel has been largely forgotten in current literary scholarship and probably rightfully so: it contains detailed accounts of society parties, a fictionalization of epistolary correspondence among the nineteenth century *limeña* upper class, and tedious reproductions of what certain government ministers might have said in meetings about politics, laws, and economics. The recreation of social events, cultural peculiarities, and other customs in the novel allows it to at least ride the coattails of far better work by Ricardo Palma in what became the *tradiciones* genre of Spanish American literary history.[32] While there may be Fernando Casós aficionados out there who will dispute me on that judgement, I prefer to instead engage his novel on one particular point that makes it relevant to these pages: the economically calamitous guano industry that has its literary debut in his novel.

The references to guano are particularly irresistible to an ecocritic because of the ecological disaster ushered in by the industry in Peru soon after Alexander Humboldt's visit in 1802 and its subsequent acceptance by Europeans as a particularly effective fertilizer. The emphasis in Casós is almost entirely upon the political and economic pitfalls of the guano industry, a state of affairs that sees him complain in the prologue that he is convinced that "el maldito guano ha hecho nuestra sociedad tan insensible, que para nada ha de server" ["the accursed guano has made our society so callous as to be useless"].[33] The critique is common enough among environmental activists who attack the exploitation of nature as motivated purely by economic gain without concern for corollary effects. And in Peru, the connection between ecology and economy was to have a direct and devastating effect. In his novel, Casós has an English banker and advisor to the Peruvian president suggest that "con el producto del guano, hay para todo" ["with guano, there will be enough for anything we want"], but eventually, the English banking houses are granted the sole concession for selling the product on the open market. As Brett Clark and John Bellamy Foster observe, "in the 1840s, Peru was still in debt to Britain for monies borrowed during the fight for independence from Spain. Guano offered an avenue for Peru to meet its debt payments and gain foreign exchange through the sale of guano contracts."[34] While the arrangement worked for several decades, the narrator expresses dismay at the arrangement: "aquella noche memorable como infausta, el gobierno peruano . . . quedaba vendido para siempre" ["that memorably infamous night, the Peruvian government . . . had sold itself out forever"].[35] In the novel, the critique is more political and cultural as the industry is blamed for governmental corruption and defects in Peruvian society, but as Miller writes: "Not until the 1870s did anyone show concern that the resource might fail. . . . When the first rumors of exhaustion were finally verified, the boom evaporated as quickly as it appeared."[36] The exploitation of guano eventually became a full-scale ecological disaster as documented by Clark, Foster, and Miller in the form of the historical and economic analyses cited above. But by the time *Los amigos de Elena* was published in 1874, the ecological consequences of over-exploitation had already come into focus for Latin America; in Casós's novel, it took the form of a literary condemnation of economic neo-colonialism and the promotion of a political ecology that prefers domestic Peruvian management over the granting of concessions to foreign banking.

The guano-extraction industry in Peru was eventually brought back under national control and subsequently, the advent of synthetic fertilizers significantly reduced the demand for guano of any kind. With the industry currently on the rebound because of the growing consumer preference for natural, organic and not synthetic agricultural inputs, Casós's warnings about what is now called a "resource curse" or "paradox of plenty" are still apt. And as shall be seen in subsequent chapters, the literature of the region had also early featured the representation of and, in several cases, proposed solutions to the environmental crises associated with various other industries of raw material export. The effects of toxic gases and slag disposal in and around mines, chemical use on fruit plantations, the deforestation of jungles, ground- and water-contamination from oilfields, and several other instances of environmental degradation are all central concerns for a significant portion of Spanish American literary production since the late nineteenth century.

THE ISLANDS OF THE CARRIBEAN

Spanish-Cuban Gertrudis Gómez de Avellaneda's *Sab* (1841) represents one of a few Spanish American anti-slavery texts from the nineteenth century. Her novel contains several descriptions of the beauty and value of nature as in several of the other novels and poems considered here, but the text also contains ideas about sustainability that would not become an enshrined environmentalist principle until more than one hundred years later. The idealization of nature—"los campos fertilísimos de aquel país privilegiado" ["the highly fertile fields of that privileged country"], "todos aquellos bienes reales, tan próximos al idealismo, que la naturaleza joven, y superabundante de vida, prodiga al hombre" ["all of these benefits provided by a young and super-abundant nature, despite seeming like merely idealizations, are very real"][37] —comes in the context of an explicit preference for small-scale consumption. In the novel, Carlota, future beneficiary of her father's Cuban lands, is to marry Enrique, son of an English entrepreneur whose wealth is in decline. But she is also loved by Sab, a slave of Carlota's family with a noble mind and pure heart. Much of the novel concerns a psychological exploration of slavery from the perspective of Sab in the context of his frustrated love for Carlota, but there is also an inherent condemnation of the commercialization brought to the island by English entrepreneurs. The moment is brief but direct, expressed by Carlota in a conversation with her fiancé: "¡Cuán fácilmente pueden ser dichosos dos amantes tiernos y apasionados! En esta pobre aldea, en esta miserable casa, con una hamaca por lecho y un plantío de yucas por riqueza, yo sería dichosa contigo, y nada vería digno de mi ambición en lo restante del universo." ["How easy for two young and passionate lovers to find happiness! I really would have no other ambition and I could be forever happy with you in this tiny town, in a shack, with a hammock for a bed, and a plot of yucca plants as our only asset"].[38] Ever the English businessman, Enrique does not respond to an idea so contrary to his ambitions, but later he recognizes his love for Carlota by muttering to himself: "¡No hay remedio! Esta mujer será capaz de volverme loco y hacerme creer que no son necesarias las riquezas para ser feliz" ["There's nothing for it! This woman could make me so crazy that I would believe that wealth is not the key to happiness"].[39] The sentiment of small-scale sustainable living is emphasized earlier in the same conversation and is connected to the novel's anti-slavery position when Carlota recalls the fate of Cuba's original inhabitants:

> Aquí vivían felices e inocentes aquellos hijos de la naturaleza: este suelo virgen no necesitaba ser regado con el sudor de los esclavos para producirles: ofrecíales por todas partes sombras y frutos, aguas y flores, y sus entrañas no habían sido despedazadas para arrancarle con mano avara sus escondidos tesoros. (Gómez de Avellaneda, *Sab*, 203)
>
> [Here those innocent children of nature lived happily: this virgin soil never needed the sweat of slaves to make it produce: rather, it was filled with shade and fruit, water and flowers, and its innards had never been torn apart by greedy hands seeking hidden treasures.]

While the elimination of slavery fits within the context of human ecology and would find a place here for that reason alone, the abolitionist ideology in *Sab* takes the form of advocacy for agricultural sustainability and small-scale production as represented by Cuba's original inhabitants and by Carlota's claim that happiness should not have to be connected to mining, wealth, and ambition but to hunting/gathering, love, and passion. And what is more, gendered readings of *Sab* have found in Avellaneda's work a seminal example; as Doris Sommer indicates, "feminist readers of Spanish American literature are giving her the kind of attention that amounts to a gender-specific claim on her work."[40] As *the* canonical feminist title of nineteenth century Spanish American literature, re-analysis of *Sab* from an ecofeminist perspective seems appropriate inasmuch as the novel "gives expression to a variety of ethical attitudes and behaviors overlooked or underplayed in mainstream Western ethics" and "has argumentative significance by suggesting *what counts* as an appropriate conclusion to an ethical situation."[41] When Carlota defines a hammock and yucca-garden as an alternative possible definition for wealth, Enrique laments that she might actually convince him that such a lifestyle could qualify as such. In other words, Carlota's perception of wealth departs from the mainstream and suggests alternative economic behaviors practiced primarily by women or indigenous communities. Also, when Carlota laments the poor treatment of slaves as compared to the seemingly effortless hunter/gatherer model of Cuba's original inhabitants, she categorizes both slaves and Indians as moral agents "that count" when they suffer discrimination, disease, and death as by-products of the Spanish/European desire to sack the land of its hidden treasures. The ecofeminism of such sentiments resides precisely in the fact that *Sab* recounts a woman's experience alongside the articulation of a political ecology that favors non-mainstream alternatives to development and includes the interests of all members in Cuban society as deserving of consideration.

In the Puerto Rican context, *La peregrinación de Bayoán* (Bayoán's pilgrimage) (first part 1863, second part 1873) by Eugenio María de Hostos and *La charca* (*The Pond*) (1894) by compatriot Manuel Zeno Gandía similarly emerge from an historical moment previous to independence from Spain. Nonetheless, there is a strong sense of *patria* (homeland, fatherland) that infuses both novels and that does not connote or promote strong connections to Spain. Instead, in each of these cases, the sense of *patria* is ascribed to the colony and the authors are very much thought of as Puerto Rican and not Spanish.[42] Both works have their share of the representation of nature as a sublime refuge, nurturing place for recovery, and object of extreme nostalgia for those who once lived there but now have left. In Hostos's novel, nature is "prodigiosa, y admira, y embelesa, y al ánimo contemplativo le da una vida de sencillos placeres, de inocentes alegrías, de inagotables bendiciones" ["marvelous, worthy of praise, it enthralls the contemplative soul with sensuous pleasures of innocent happiness and endless blessings"]; also, "la naturaleza es una madre cariñosa, siempre dispuesta a calmar nuestra agitación" ["nature is like a loving mother always ready to calm our worries"]; and "la contemplación de la naturaleza es una como transfusión de vida" ["contemplating nature is like receiving a transfusion of new life"].[43] In Zeno Gandía, the vivid colors of the countryside in bloom are described as "proyectados al mundo de los sueños por la mano de la primavera" ["cast upon the world from the dreams of Spring]"; the dream imagery continues in another passage: "la Naturaleza, entonando himnos de eternal poesía; el suelo, en la copiosa dehiscencia de inagotable riqueza; los seres, gozando el privilegio de tanta dicha. Todo desde la distancia lo veía embellecido por el ensueño" ["Nature, singing hymns of eternal poetry; the earth, unleashing unending riches; humans, enjoying the privilege of so much pleasure. From a distance, everything seemed like a beautiful dream"].[44] Other than the unique content of the images—nature as blessing, mother, curative transfusion, and the stuff of dreams—none of the selected passages cited above include anything unique to the tendency to emphasize the sublime qualities of nature that has been so strong in the tradition. They simply provide more evidence that nineteenth century Spanish American literature repeatedly expresses the value of nature; either in its potential for development in the political ecology of texts by writers like Sarmiento or for its non-instrumental value as a thing of beauty or as the inherently precious creation of a Supreme Being. The latter idea is expressed several times in these two novels as seen above and in countless other novels and texts from the nineteenth century discussed in this chapter. However, *La peregrinación de Bayoán* and *La charca* also both contribute something unique to the content and conventions of the literature from that century: the language of contamination in Hostos's novel and ideas about sustainability and conservationism in Zeno Gandía's.

The narrative of *La peregrinación de Bayoán* is tirelessly registering complaints about the inferiority, pollution, and destructive qualities of the city. As in many others of these nineteenth century novels, the story concerns the tragic love between a man and the inevitably doomed object of his affection. In this case, Bayoán keeps a diary about his love for a certain Marién in the context of travels among several Caribbean islands, including his home on Puerto Rico, and the European mainland. In one entry, Hostos has him write: "adiós, ciudad, como las ciudades de Europa, repugnante" ["Farewell, city, you are like the cities of Europe: repugnant"]. Later, he explains his disgust: "Y yo, en las ciudades de esta patria querida, no respiro; la indignación me sofoca; me digo que está profanada por el hombre esta naturaleza augusta" ["And I cannot breathe in any of the cities of my beloved homeland; the indignation suffocates me, the indignation at the way in which such beautiful nature has been profaned by humankind"]. Finally, the most resounding criticism of the ill effects of city-construction is accomplished in

Bayoán's diary in the form of commentary about a possibly uninhabited island the ship is passing during one of his many travels at sea: "Si está habitada, yo me tapo los ojos para no saberlo: no quiero ver lo que he visto en todas partes: . . . egoísmo, es decir, sociedad; codicia, es decir, comercio, agricultura, industria; felonía, es decir, leyes sociales" ["If it's inhabited, I cover my eyes so I don't have to see it: I don't want to see what I have seen everywhere else: . . . selfishness, that is, society; greed, that is, commerce, agriculture, industry; crime, that is, laws"].[45] The condemnation of development—"commerce, agriculture, industry"—in the context of a text that is so critical of cities and so elegiac when it comes to nature and the country exposes an implicit ecological undercurrent in the novel that approaches the equally ideological expression of a desire for independence that marked almost all of Hostos's work. Enrique Ríos describes him as a messenger of "la libertad soñada para las Antillas y en busca de apoyos para la independencia" ["liberty for the West Indies and always in pursuit of support for the idea of independence"].[46] But something else permeates the novel: the belief that cities are centers of decay and the country far preferable, a position that implies many of the principles of present environmentalism: nature protectionism (good places should be conserved), environmental justice (urban environments should be reformed and improved for their negative impact on so many people), deep ecology (the identification with nature by Bayoán is insistent in his diary).

While the criticism of urban space is counterbalanced in Hostos's novel by the repeated representations of the superiority of many natural places, some of which are cited above, Zeno Gandía's *La chacra* does not contain scenes of degraded city-scapes. Instead, the novel deals with the many social implications of the coffee-growing industry on Puerto Rico and there are subtle invocations of the value of sustainable development. Agriculture is not condemned whole-scale with commerce and industry like in *La peregrinación de Bayoán*; rather, a vision of the use of the land with an eye toward the future of the island informs the husbandry of the forward-thinking *hacendado* (large landholder) Juan del Salto. While the novel emphasizes the tragic and more dramatic stories of the abuse of *campesinos* (peasants) by avaricious and unjust foremen (probably because these elements make for more entertaining storytelling), the briefer attention paid to del Salto's progressive ideas about agriculture evokes ideas about sustainability which should not be ignored. In general, del Salto is described as a man interested in "remedios para los daños, medios para preparar lo porvenir, recursos para conservar lo bueno, hoces para cercenar lo malo; y en toda aquella labor de su cerebro había un fondo de infinito amor filial, de cariño profundo por la bendita tierra nativa" ["cures for the ills and plans for the future of his country, a man who tried to conserve what was good, and cut away the bad; and in all that thinking, there remained a love for his fellow person and a profound caring for his blessed native land"].[47] The connection between a philosophy of forward-thinking development and love for the land takes a practical and very specifically conservationist form in del Salto's coffee-growing cultivation. The owner is intimately involved with the production of coffee, develops the land for long-term use rather than quick profits, workers are treated well. A sustainable economic model, although only a small part of the story, is treated as an ideal backdrop upon which the interpersonal drama of the rest of the novel plays out. In both of these early Puerto Rican texts, the political ecology is clear: the corrupting influence of the city must be rectified; the development of the country must be achieved through responsible means.

MEXICO

The perception that there is a shortage of ecological writing in Spanish American nineteenth century texts is one that I challenge in these pages; however, its absence is certainly characteristic of nineteenth century literature from Mexico. While the dialogue among the Argentine founding fathers was rich with references to the role of the land in the development and future cultural identity of that nation, there are several significant examples of canonical Mexican texts with almost no representations of nature. Considered the first Spanish American novel, *El periquillo sarniento* (*The Mangy Parrot*) (1816) by José Joaquín Fernández de Lizardi takes place in a Mexico almost entirely devoid of landscapes. The ambivalence of the narrative toward non-urban space is reflected in Periquillo's claim that, "asi estoy contento en el campo como en el campo, y divertido en la ciudad como en la ciudad" ["I'm happy in the country when I'm in the country, and in the city when I'm in the city"].[48] Published only six years after Miguel Hidalgo's famous proclamation of independence but set during the colonial period, the novel owes more to the Spanish picaresque tradition than anything native to the American continent. Similarly, *El cerro de las campanas* (The bells toll in the hills) by Mexican Juan A. Mateos and *Calvario y tabor* (Calvary and drums) by Vicente Riva Palacio both published in 1868 and *Clemencia* (Clemency) (1869) by Ignacio M. Altamirano are sweeping works of historical fiction whose detailed recounting of the French intervention make barely any reference to the landscapes in which these military and political battles occurred. In Mateos's novel, there is reference to the imposing slopes of Xinantécatl over Toluca, "destacándose con la majestad de un monumento" ["majestic and imposing, like a monument"];[49] in Riva Palacio's text, the narrator observes that "pocos paisajes habrá más pintorescos sobre la tierra" ["there are few more picturesque places on earth"][50] than the canal economies around Lake Chalpa; and Altamirano's *Clemencia* contrasts the beauty of nature to the invading French troops: "La naturaleza toda parecía elevar un himno a Dios, solemne y dulce. Y en medio de esta alegría del cielo y de la tierra, debajo de este manto infinito de zafiro y de luz, atravesaba aquel cortejo militar silencioso y terrible" ["All of nature seemed to sing solemn, sweet hymns to God. And in the middle of this heavenly and earthly joy, amidst bright flashes of sapphire, the cruel and terrible military processions"].[51] The last quotation represents an example of a technique that Altamirano would also use in *El zarco*, published posthumously in 1901. The quiet natural beauty of the small town of Yautepec where "el río y los árboles frutales son su tesoro" ["the river and the fruit trees are its treasure"][52] is shattered by the bands of bandits that operated throughout Mexico in the years after the French intervention. But these moments represent exceptions rather than central elements from the quoted novels.

Mexican José López Portillo y Rojas's *La parcela* (*The Parcel of Land*) (1898) is similar; despite the fact that the story is literally about feuding families and a conflict over a piece of land between their properties, the land itself barely makes an appearance. There are the standard (as has been seen above) references to "el encanto que producía aquella naturaleza fresca y exuberante" ["the magic of a young and exuberant kind of nature"]; there is even a brief moment of explicit conservationism: "Don Pedro había prohibido por muchos años cortar leña en aquel sitio, y aun ahora que comenzaba a explotarlo, hacíalo de modo que no se destruyese el bosque—con el propósito de conservarlo siempre hermoso y tupido" ["Don Pedro had for many years strictly forbidden anyone to fell trees for firewood, and even now that he had begun to cut those woods for timber, he did it so as to not harm the forest: the idea was to preserve it for posterity"].[53] Yet, as with many of the other Mexican novels of the nineteenth century, on the whole there is very little in the way of nature writing or the ecosublime. These positive descriptions of nature express a certain preference for mountains, volcanoes, flowers, gardens, forests, rivers, and undeveloped land, but the sheer scarcity of such descriptions in the novels by Mateos, Riva Palacio, Altamirano, and Portillo y Rojas express a certain ambivalence toward nature: its absence in these (and other) nineteenth century Mexican works affirms the overarching emphasis on politics, interrupted only occasionally by the appearance of nature in those pages.

But *Santa* by Federico Gamboa represents a different case altogether; first, for its late publication date (three years after the turn of the twentieth century), it cannot be counted among those fictions like *Facundo* and *María* or *Amalia* (1851) by Argentine José Marmol, or *Tabaré* (1888) by Uruguayan Juan Zorilla de San Martín that Doris Sommer found to be so crucial to the forging of a national identity in the nineteenth century; it may perhaps be considered a foundational fiction 2.0 (like late work from the Southern Cone—Joaquín V. González's *Mis montañas* (My mountains) [1893], Eduardo Wilde's *Aguas abajo* (Waters under the bridge) [1914], or Carlos Reyles's *El terruño* (The small plot) [1916]). That is, Gamboa's work, like that of the Argentine authors of the late 1800s, certainly suggests a more heterogeneous character for the new nations' residents, cultures, and societies; yet the implicit ideology in *Santa*, as in much other contemporaneous work, still seems directed toward an audience of citizens from the newly formed nation. Gamboa's novel portrays the corrupting nature of large cities with the protagonists' unhappy lives functioning as object lesson and warning.

The story concerns the expulsion of a young girl, Santa, from her family home, her subsequent employment as prostitute, her inevitable decline, and predictably, her tragic death at the close. The novel contrasts the healthful benefits of country living versus the contamination of cities through a flashback that gives Santa's backstory from before her hiring at the brothel, a familiar one apparently, among the young women who end up in the oldest profession: "la historia vulgar de las muchachas pobres que nacen en el campo y en el campo se críanal aire libre, entre brisas y flores; . . . al cuidado de la tierra, nuestra eternal madre cariñosa" ["that vulgar story of young girls from the country who grew up in the fresh air, free among the flowers . . . cared for by the generosity of mother nature"]. Later, but still before her fatal decision to head for the city, the novel finds Santa "en casta y meditativa comunión con la naturaleza" ["chastely in communion with nature"].[54] These moments are contrasted with the brutal tragedy of prostitution, poverty, corruption, and filth in Mexico City. While Gamboa's novel explicitly functions as a didactic condemnation of the social problems of urban life, on one occasion, the ideology becomes ecological; Gamboa's description of the worst of the houses of prostitution is compared to the polluted water that inundates the city:

> En ellas rodaba Santa, en los sótanos pestilenciales y negros del vicio inferior, a la manera en que las aguas sucias e impuras de los albañales subterráneos galopan enfurecidas por los oscuros intestinos de las calles, con siniestro glu glu, de líquido aprisionado . . . en ángulos y oquedades sospechosos y hediondos. . . . Allá van, a escape, por las cloacas y letrinas, . . . golpeándose contra insensibles paredes tapizadas de barro y limo que las estrangulan, deforman y encauzan, que casi han de contemplarlas con las cicatrices que las inmundicias han grabado. . . . Allá va el agua, incognoscible, sin cristales en su lomo, sin frescor en sus linfas; conduciendo detritus y microbios, lo que apesta y lo que mata; retratando lo negro, lo escondido, lo innombrable que no debe mostrarse; . . . allá va, expulsada de la ciudad y de las gentes, a golpearse contra los hierros de la salida, a morir en el mar, que la amortaja y guarda, que quizá sea el único que recuerde que nació pura. (Gamboa, *Santa*, 307)

> [This was where Santa was wallowing: given over to vice in fetid basements; like putrid and impure sewer water, she faded away with a kind of nasty glug, glug, as water in a trap . . . stinking and with no escape . . . flowing down

> through the very guts of the streets, from drains and latrines . . . splashing against unfeeling mudded walls that strangle the waters like the scars of unclean canals. . . . She bled away like an unknowable pus-like filth that flows out in a parade of stench and microbes, blackly menacing and deadly, hidden, shameful and unnamable; . . . cast out from the city, passing through iron-cast drains and away from human contemplation, to end up eventually in the sea, the only place perhaps where the memory of purer origins can be recalled.]

The reference to these polluted waters would have had a strong resonance for inhabitants of Mexico City living in the first decade of the twentieth century. The capital, unlike other metropolises that were located near large bodies of water, was situated in a natural, low-lying basin: the Valley of Mexico. When residents disposed of waste in Lake Texcoco, "rather than dissipating and disappearing, the filth accumulated, concentrated, and during the rainy season, came back to haunt them."[55] As the garbage became more voluminous for the sheer quantity of urban migration to Mexico, a migration that was characteristic of human demographics throughout Latin America, the solution, effected by dictator Porfirio Diaz, was the construction of the "Great Canal." The project ultimately provided a waterway to drain all of Mexico City's sewage and other water-borne refuse. But it was built to the open air and as the water-level in Lake Texcoco slowly descended, it exposed the accumulated waste from the lake to the eyes and nostrils of the city's inhabitants. This was the historical context for Gamboa's references to sewers, drains, and the sinister "glug, glug" that characterized the basement where Santa lived.

Such literary imagery on its own perhaps does nothing more than indicate that Gamboa's representation of Mexico City at the turn of the twentieth century was characteristic of residents' experience from that time. But the powerfully negative imagery here also represents one of the earliest expressions of toxic discourse in Spanish American fiction, a narrative of pollution, contamination, and corruption that comprises what Lawrence Buell would call the language of environmental toxicity. According to Buell, toxic discourse contains four different types of defining characteristics: the promotion of awareness about toxicity, the representation of a world without refuge from toxic penetration, the gothic representation of the consequences of toxicity, and the expression of outrage against the hegemonic centers of power perceived as responsible for the toxicity.[56] The passage cited above contains the first three of these four elements with its filth of unseen sewers, its description of the ubiquitous stench of unclean water, and its darkly ominous tone evoked by phrases like "sótanos pestilenciales," "los intestinos de las calles," and "las cicatrices que las inmundicias han grabado." As shall be seen in chapter 9 where I deal with explicitly environmentalist texts, the tendency has not changed much in Mexican literature. There, I consider two novels by noted eco-activist Homero Aridjis that describe a future Mexico

City so badly polluted that the narrative innovates a dark imagery of environmental apocalypse.

The ecological aspects of *Santa* represent more of a condemnation of city living than the praise of the beauties of nature. Consequently, these criticisms, for the very toxic discourse they employ, serve as an implicit proposal for alternatives; whatever is broken must be fixed, whatever is fouled must be cleaned. In the actuality of environmental policy, it only took the Grand Canal project in Mexico a decade or so to demonstrate the engineering shortcomings of a project to eliminate seasonal flooding from a lagoonal ecosystem: as Lake Texcoco drained, Mexico City's soil began to compact and the urban elevation declined by 7.5 meters.[57] The ecological curse of inadequate sewage disposal was replaced by something worse: as the earth on which the city was built began to condense and sag, the heavy precipitation of the rainy season caused flooding that was just as bad as before, only now accompanied by widespread architectural instability. Foundations that had previously been set five meters below the ground were now flush with the surface and the buildings constructed upon them liable to go over. The concerns about water contamination in *Santa* serve as a fitting bookend to the ecology of nineteenth century Mexican fiction and to nineteenth century Spanish American literature as a whole; while political ecology is still strongly present, the echo that results from the authors' advocacy begins to adopt the more recently familiar language of environmental degradation.

As praise for sublime nature is tempered by the unsettling experience of urban living, Spanish American authors begin to more and more openly advocate for a state of affairs that does not have the Sarmentine ideology of unfettered development as the sole inspiration. In the next chapter, I examine the way in which dissatisfaction with the modernization of the region finds its expression in ecological elements of literary *modernismo*. The common scholarly understanding of that period in Spanish American literature is that the *modernistas'* rejection of the industrialization and commercialization of society and culture finds its expression in the production of highly formal and emphatically aesthetic novels, stories, and poetry. However, I propose that such dissatisfaction is not as decadently removed from the actual concerns of the physical world as a cursory reading of *modernista* tales of fairies, gnomes, and hashish may suggest. Instead, these authors demonstrate concern in their highly aesthetic texts for real environmental degradation as it grows more acute in the first decades of the twentieth century. While the *modernistas* do not always succeed in persuading the powers that be to adopt policies that embody their rejection of the excesses of urban industrialization, they certainly set the table for a serious consideration of alternative policies that begin to be adopted in later decades. It was early in the twentieth century, for example, when the government of Peru was finally convinced of the unsustainability of unfettered exploitation of guano and finally closed down the islands from which it came. It was also in the first half of the 1900s that Latin American nations first began to establish national parks and wilderness refuges, something that cannot have happened without the kind of political pressure that mounts after awareness of environmental degradation becomes widespread, at least among political actors. My argument in these pages is that part of the awareness comes as a result of the effect that the literature of the region had upon readers. As I argue in chapters 1 and 2, there was a limited but obvious ecological current in these novels, stories, and poetry; these elements must have contributed to a general awareness of environmental degradation given the uniquely political influence of literature in Spanish America. And I shall further the same argument in chapter 3: the *modernista* tendency to strongly condemn the dynamics of developmental economics from the closing decades of the nineteenth century also prepared the theoretical ground for the development of environmentally-aware literature that began to appear more extensively as the twentieth century wore on.

NOTES

1. See chap. 1, note 19.
2. See chap. 1, note 36.
3. Shawn William Miller, *An Environmental History of Latin America* (New York: Cambridge University Press, 2007), 106.
4. Ibid., 107.
5. Carmen Galvis Yepes, "Jose Eusebio Caro y su poética de la muerte del padre," *Estudios de Literatura Colombiana* 7 (July–December, 2000): 60.
6. José Eusebio Caro, *Poesías de Caro i Vargas Tejada* (Bogotá: Imprenta de Ortiz, 1857), 63.
7. See Miller, *Environmental History*, 191–227 for a discussion of the development of such efforts in Latin America.
8. Manuel José Cortés, "A la naturaleza del oriente de Bolivia," *La semana*, August 13, 1859, 203–4. I quote from a digitized version from the Google Books Project in a hard-bound collection of the journal's first 49 issues apparently scanned from a hard-back anthology in the University of Texas library.
9. Ibid., 204.
10. Juan Diéguez Olaverri, "A Los Cuchumatanes," *El Jardín* 1, no. 21 (June 14, 1908): 168.
11. José María Heredia, "Niagara," in *Poetry of Niagara*, ed. Myron Thomas Pritchard (Boston: Lothrop, 1901), 34–38.
12. José Jacinto Milanés, *Obras de Don José Milanés* (New York: Juan F. Trow, 1865), 3–4.
13. Fernando Balseca, "En busca de nuevas regiones: la nación y la narrativa ecuatoriana," *Procesos: Revista ecuatoriana de historia* 8 (1995–1996): 149.
14. Miguel Riofrío, *La emancipada* (Guayaquil: Editorial de la Universidad de Guayaquil, 1988), 73.
15. Antonio Sacoto, Introduction to *La emancipada*, by Miguel Riofrío (Guayaquil: Editorial de la Universidad de Guayaquil, 1988), 7. Sacoto indicates that while Juan Mera's *Cumandá* was long considered the first Ecuadoran novel, Riofrío's novel pre-dates Mera's by sixteen years.
16. Karen J. Warren, *Ecofeminist Philosophy: A Western Perspective on What It Is and Why It Matters* (Lanham, MD: Rowman & Littlefield Publishers Inc., 2000), 103–4.
17. Juan León Mera, *Cumandá* (Boston: DC Heath, 1932), 115.

18. See Introduction, note 43.
19. Luis Martínez. *A la costa* (Madrid: Ediciones de Cultura Hispánica, 1992), 204, 142.
20. Emanuelle Sinardet, "*A la costa* de Luis A. Martínez: ¿La defensa de un proyecto liberal para Ecuador?" *Bulletin de Institut francais des etudes andines* 27, no. 2 (1998): 288.
21. Fermín Toro, "La sibila de los Andes," in *Biblioteca de escritores venezolanas*, ed. José María Rojas (Caracas: Rojas Hermanos Editores, 1875), 455.
22. Daniel Mendoza, *El llanero* (Madrid: Editorial América, 1918), 21, 31.
23. José Ramón Yepes, *Novelas y estudios literarios de José Ramón Yepes* (Maracaibo: Imprenta Americana, 1882), 31, 37, 18, 53.
24. Bladimir Ruiz, "Incorporaciones y exclusiones al proyecto de nación: El negro y el indio en la narrativa decimonónica venezolana," *Revista Iberoamericana* 49, no. 205 (2003): 892.
25. José Manuel Marroquín, *El moro* (Bogotá: Editorial Oveja Negra, 1985), 116.
26. Moira Ferguson, "Breaking in Englishness: *Black Beauty* and the Politics of Gender, Race, and Class," *Women: A Cultural Review* 5, no. 1 (1994): 38.
27. Marroquín, *El moro*, 29, 10, 63.
28. Stephen Best et al., "Introducing Critical Animal Studies," *Journal for Critical Animal Studies* 5, no. 1 (2007): 5.
29. Narciso Aréstegui, *El padre Horán* (Cuzco: Universidad Nacional del Cuzco, 1954), 1:70, 1:160.
30. Ibid., 2:4.
31. Ibid., 2:14.
32. See chap. 1, note 72.
33. Fernando Casós, *Los amigos de Elena* (Paris: La Librería Española de E. Denné Schmitz, 1874), 1:viii.
34. Brett Clark and John Bellamy Foster, "Guano: The Global Metabolic Rift and the Fertilizer Trade," in *Ecology and Power: Struggles over Land and Material Resources in the Past, Present, and Future*, ed. Alf Hornborg, Brett Clark, and Kenneth Hermele (New York: Routledge, 2012), 74.
35. Casós, *Los amigos de Elena*, 2:201, 2:323.
36. Miller, *Environmental History*, 151.
37. Gertrudis Gómez de Avellaneda, *Sab* (Habana: Instituto Cubano del Libro, 1973), 132, 152.
38. Ibid., 205.
39. Ibid.
40. Doris Sommer, *Foundational Fictions: The National Romances of Latin America* (Berkeley: University of California Press, 1991), 114.
41. Warren, *Ecofeminist Philosophy*, 103–4.
42. Eugenio María de Hostos, *Obra literaria selecta* (Caracas: Biblioteca Ayacucho, 1988); Manuel Zeno Gandía, *La charca* (Río Piedras, Puerto Rico: Ediciones Huracán, 1999). For my research, I read reprints of both novels from Spanish American publishing houses. *La charca* was republished by Ediciones Huracán whose mission, according to the back cover material of the edition I was reading, was to "reactualizar los autores y los textos puertorriqueños consagrados en la literatura puertorriqueña" ["take new account of long-canonized Puerto Rican authors and texts"]. Biblioteca Ayacucho, an entity of the Venezuelan federal government, published the volume of work by Hostos which I consulted to read *La peregrinación de Bayoán*. Ayacucho's mission is similar to Huracán's except that the former's focus is more broadly on the classics of the Spanish American continent (and its islands). See the mission statement on Ayacucho's website at http://www.bibliotecayacucho.gob.ve.
43. Hostos, *Obra literaria selecta*, 27, 59, 176.
44. Zeno Gandía, *La charca*, 38, 138.
45. Hostos, *Obra literaria selecta*, 35, 95, 76.
46. Enrique Ríos, "El grito de rebeldía de Hostos's *Bayoán*: Perfiles del 'peregrino de la libertad' ante el 98," *Historia y Comunicación Social* 3 (1998): 169.
47. Zeno Gandía, *La charca*, 57.
48. José Joaquín Fernández de Lizardi, *El periquillo Sarniento* (Mexico City: Editores Mexicanos Unidos, 2002), 57.

49. Juan A. Mateos, *El cerro de las campanas* (Mexico City: Editorial Porrúa, 1971), 25.

50. Vicente Riva Palacio, *Calvario y tabor* (Mexico City: Editorial Porrúa, 1985), 234.

51. Ignacio M. Altamirano, *Clemencia y La navidad en las montañas* (Mexico City: Porrua, 1966), 211–12.

52. Ignacio M. Altamirano, *El zarco* (Habana: Casa de las Américas, 1976), 5.

53. José López Portillo y Rojas, *La Parcela* (Mexico City: Imprenta de El Tiempo, 1904), 175, 174.

54. Federico Gamboa, *Santa* (Madrid: Cátedra, 2002), 95, 106.

55. Miller, *Environmental History*, 142.

56. Lawrence Buell, *Writing for an Endangered World* (Cambridge, MA: Harvard University Press, 2001), 30–54.

57. Miller, *Environmental History*, 146.

Chapter Three

Green Modernism

While European romanticism was still an indirect influence, Spanish American Modernism (*el modernismo*) is considered the first truly Spanish American literature. That is, unlike the mid-nineteenth century foundational novels that so freely appropriated romantic discourses by the likes of Chateaubriand, Byron, and Shelley,[1] the *modernistas* do not so much follow European literary models as subscribe to what Micheal Lowy and Robert Sayre describe as an essentially romantic ethos: "a reaction against the way of life in capitalist societies."[2] This could not be more apt as an expression of the *modernista* ethos; Nicaraguan Rubén Darío expresses the exact sentiment in the prologue to a subsequent edition of *Prosas profanas* (*Profane Prose*) (1896) published in Madrid in 1901 in which he explains the content of his poetry: "Mas he aquí que veréis en mis versos princesas, reyes, cosas imperiales, visiones de países lejanos o imposibles: ¡qué queréis!, yo detesto la vida y el tiempo en que me tocó nacer" ["Please note that in my poetry, you will see princesses, kings, royal things, visions of far-away, impossible places: What? I hate this life and the times to which I have been born"].[3] The reaction to "life in capitalist societies" from Darío (and other like-minded writers of the time) was to develop unique paradigms of representation, imagery, and narrative: a form that came to be called *modernismo*. These authors and their texts have generated enormous scholarly interest, particularly for the innovative unconventional style, but somewhat less for the relationship of the content of poetry and prose to political ideology, socio-cultural values, and, relevant to the analysis here, ecological concerns of late nineteenth and early twentieth century Spanish America. As the nineteenth century wanes and the twentieth dawns, ecological catastrophe was manifest as a combination of internal migration toward the cities and a national foreign policy of resource exploitation that rapidly approached their maximum capacities:

> La fragilidad de esas áreas áridas, intervenidas ya el siglo pasado [el siglo XIX] por sobrepastoreo, se vio agravada por sequías extremas. . . . Ya a fines del siglo pasado todo el polígono de las secas presentaba grandes extensiones con notorios procesos de erosión y además con una vegetación deteriorada.[4]
>
> The fragility of these arid areas, already altered in the previous century [the nineteenth], was exacerbated by extreme droughts. . . . By the end of last century, the whole of the drought-afflicted areas had developed extreme levels of soil erosion and a serious decline in viable vegetation.

Soil erosion, deforestation, water contamination, species extinction, urban migration, and several other ecological challenges had become the ecological reality for residents of especially urban Latin America at the turn of the twentieth century. These were all intensified by the desperate culmination of early exploitations of the region's natural resources: the fencing of the Argentine plains, the depletion of guano in Peru, the decline of fisheries in northern Uruguay, the near disappearance of certain tree species in the Amazonian jungle, and the general fouling of the air in and around urban metropolises. These are the touchstones of the development of an early ecological ethos in late nineteenth century Spanish American literature, but they also represent the point of departure for the literature of the *modernistas* who associate these phenomena with the prevailing sentiment of the times in which they were writing and which they wholeheartedly rejected.

Unfortunately for the development of an ecological sensibility in Latin America, the *modernista* spirit in certain of its manifestations may have been guilty of a kind of proto-fascist imposition of environmental attitudes from an early stage of its genesis and among some of its authors. Leopoldo Lugones, whose poetry and short stories are considered from an ecocritical perspective below, eventually became an advocate for the kind of conservative and reactionary beliefs found in fascist and even Nazi ideology. For example, Fermín Adrián Rodríguez compiles a summary of authoritarian and overly nationalist positions evident in several of Lugones's short stories:

> Hay que alfabetizar a las masas y enseñarles a hablar español, hay que adiestrarlas en el servicio militar obligatorio para sujetarlas férreamente a un ideal de nación y construir con ellas un cuerpo nacional normalizado. La escuela, la prensa popular, el ejército son espacios de disciplinamiento y supervisión donde se trazan las líneas de fuerza que distribuyen en el mapa de la identidad nacional lo normal y lo anormal, lo humano y lo animal, lo argentino y lo extranjero, lo sano y lo patológico, lo moral y lo inmoral, lo inofensivo y lo peligroso.[5]

> [We must educate the masses and teach them to speak Spanish and prepare them for compulsory military service so as to create a loyalty to the nation and a citizenry as strong as the stiffest iron. Schools, the press, and the army are spaces for discipline such that the lines of national identity are unerringly drawn, separating strength from weakness, normal from abnormal, and rightly distinguishing between human and animal, Argentine and foreigner, healthy and pathological, moral and immoral, and between harmless and dangerous.]

And Mariana Amato claims that Lugones's short fictions and his eventual political theories are related: "These heterogeneous texts—the fantastic fictions compiled in 1906, on the one hand, and the philo-fascist essays published three decades later, on the other—intersect at their respective biopolitical imaginations."[6] Lugones's eventual conservative political ideology illustrates something of a quandary, contextualized as it is here in a history of ecological ideas in Spanish American literature. If I choose to include a writer like Lugones in the corpus of proto-environmental literature, I cannot hide from the ecological sensibilities that took on strains of fascist and even Nazist overtones. And despite the distasteful political baggage, Lugones's representations of the natural world that warn about its destruction and of fearful technologies that foretell environmental degradation are part of the heritage of ecology in Spanish American literary expression.

Furthermore, the inclusion of authors like Lugones and others like Paraguayan Natalicio González, an *indigenista* writer I consider in chapter 7, whose political views may be less than "politically correct," is useful to illustrate the ways in which the development of environmentalist ideas can range over theoretical territory that may sometimes move in more conservative, even reactionary directions. As Shawn William Miller notes, the environmental movement in Latin America "started with the state. And environmental policies descended from both the upper right and the upper left."[7] It should not surprise that diverse political foundations for ecological policies were characteristic of Spanish American *modernism*: the European romanticism from which it was partly inspired was also found by Lowy and Sayre to have informed certain aspects of Nazi ideology. They note that historiographers of the period perceive "Romantic political ideologies specifically as a preparation for Nazism" and that "Nazi ideologues were unquestionably inspired by certain Romantic themes."[8] Therefore, some *modernistas*, via ideological pathways manifest in their work, inspired as it was by inherently conservative romantic tendencies, embraced certain ideas that seem distastefully arrogant, exclusionary, and occasionally outright racist in retrospect. However, it remains important to highlight what may be called a kind of conservative conservationism in these early works because, like other works that I survey in chapters 1 and 2 and 4–7, this also foreshadows some less progressive but nevertheless distinctly concrete aspects within the current environmentalist debate. Some neo-imperialist dimensions to present ecological thinking or the advent of divergent views within the environmentalist community itself are each considered in chapter 10, critically in Nicaraguan Gioconda Belli's *Waslala*, but embraced in the Gaia trilogy by Chilean Fernando Raga. The novels by Belli and Raga are merely the most recent manifestations of a critical dialogue within established literary cultures of nature representation, conservationism, and political ecology inaugurated by *modernistas* like Leopoldo Lugones.

All of these highly political considerations notwithstanding, the enduring reputation for the authors of the literary period has been for the production of art for art's sake: poetic prose, fanciful stories, pure aesthetics. But in essays and poetry by authors such as Cuban José Martí and Uruguayan José Enrique Rodó, or even in the short stories by Darío, politics are never far from the surface, particularly as a continuation of the challenge to prevailing categories of civilization and barbarity from Argentine Domingo Faustino Sarmiento. It was nowhere more apparent than in Martí's essay *Nuestra América* (*Our America*) (1891) where the author issues a resounding repudiation of the nineteenth century's entrenched Sarmentine dialectic: "No hay batalla entre la civilización y la barbarie, sino entre la falsa erudición y la naturaleza" ["There is no conflict between civilization and barbarity, only between false learning and nature"].[9] Martí re-imagines the conflict as a pair of opposites that set bad education against the dictates of nature, and in the rest of the essay deals with the articulation of a philosophy of political identity that encourages the creation of new forms of governance that, like *modernista* literature itself, reflect Latin American realities rather than imitate models from Europe and the United States. However, what I will attempt to establish in the following pages is that the ecological reformulation of the dialectic is also indicative of the spirit of the *modernista* sense of literature, nature, and society.

José Enrique Rodó was not the first *modernista*, nor the most well-known, nor the most influential. However, his ideas concerning the importance of beauty, grace, creativity, and cultured living represent the central principle of Spanish American *modernismo*: the creation and appreciation of art, literature, and poetry was an end unto itself. Nowhere are these principles more eloquently stated than in Rodo's essay *Ariel*, published (conveniently enough for literary historians) at the very turn of the century in 1900. Rubén Darío's influential *Azúl* (*Blue*), a collection of poetry and short stories that are considered classics of *modernismo*, was published in 1888 and José Martí, the aforementioned literary precursor and eventual contributor to the corpus of *modernista* texts, died in 1895, five years before the publication of Rodó's essay. Nevertheless, the chronologically untidy development of literary *modernismo* is somewhat more symmetrically captured in *Ariel* with its expression of the values of the literary philosophy of the *modernistas* from right at the turn of the century. In Rodó's essay, these values, embodied in the ennobling spirit of the character Ariel from Shakespeare's *The Tempest* (as opposed to the beastly and savage Caliban), are characteristic of the *modernista* spirit, but also share many features with current ecological thought.

The analysis in Rodó's essay is structured around two touchstones: the value of the appreciation of beauty and the ignoble effects of utilitarian thinking. The appreciation of beauty allows individuals to more effectively avoid vulgarity, distinguish right from wrong, develop virtue, and promote justice. Utilitarian thinking can lead to a stultifying backwardness that stunts intelligence, promotes unhappiness, and prevents spiritual development. The narrator of *Ariel*, construed as Próspero (another of the characters from *The Tempest*), declares that among all human facilities, "es el sentimiento de lo bello, la visión clara de la hermosura de las cosas, el que más fácilmente marchita la aridez de la vida limitada" ["it is the appreciation of all that is fair and the cultivation of a clear notion of beauty that can most readily undo the dullness of a boring life"]. The appreciation of beauty entails moral qualities that elevate human endeavor and represents "un motivo superior de moralidad [que] autorizaría a proponer la cultura de los sentimientos estéticos como un alto interés de todos" ["a morally superior reason to think of the development of a sense of aesthetics as the most important thing for all individuals"].[10] The essay associates a sense of aesthetics with justice, mercy, dignity, charity, truth, faith, hope, and love. When, in the last two chapters, I cited numerous evocations of the sublime in the narrative of nature in several of the foundational nineteenth century texts, quite a few of those passages render nature in these same terms. Shared characteristics do not necessarily imply that a similar rhetoric of nature will be a central element of Spanish American *modernismo*, but Rodó makes the connection explicit when he turns to the critique of utilitarian thinking in the second part of the essay: "La inculpación de utilitarismo . . . se funda en parte, sobre . . . sus titánicos esfuerzos por la subordinación de las fuerzas de la naturaleza a la voluntad humana y por la extensión del bienestar material" ["What is wrong with utilitarian thinking . . . can be blamed, in part, upon . . . its thorough-going insistence upon the subordination of nature to human will for the purpose of material well-being"].[11] Rodó's explicit criticism of utilitarianism invokes an objection to the conception of the maximization of happiness through an expansion of "material well-being." The condemnation of the subordination of nature to human will in *Ariel* invites an ecocritical reevaluation for all *modernista* prose and poetry precisely because Rodó's essay is so representational of the genre's values, ideology, and beliefs. However, keeping in mind Lugones's fascist tendencies, Rodó's condemnation of utilitarianism also represents a kind of elitist thinking which could also have troubling consequences for the development of ecological ideas in *modernista* ideology. As Diego Alonso observes, in Rodó:

> La democracia es aceptada plenamente a condición de que haya logrado conciliar su pasión igualitaria con el reconocimiento de nuevas y legítimas jerarquías. . . .
>
> . . . Independientemente del carácter universal y desinteresado que Rodó atribuye a la estética, ésta le permite establecer la diferencia entre una minoría culta de elegidos . . . y una masa que desconoce los fundamentos mismos de la belleza.[12]
>
> [Democracy is fully accepted, but on the condition that it can reconcile its egalitarian passions with recognition of certain new, legitimate hierarchies. . . .
>
> . . . Independent of the universal and disinterested character that Rodó attributes to aesthetics, it is just that which allows him to establish the boundaries between a select, cultured minority . . . and the masses that have no idea about the concept of beauty itself.]

The emphasis on an enlightened aesthetics from an elite class of artists was characteristic of the *modernistas* and, as I undertake an ecocritical reading of their work, the conservation of nature as an implicit ideological subtext, emerges with similar prejudice. That is, *modernista* political ecology tends to qualify the negotiation of access to resources as reserved to an enlightened elite and only valid inasmuch as it satisfies aesthetic ends. Therefore, as I develop the ecology of *modernismo* in the rest of the chapter, I make reference to the way in which, as the first truly *Spanish American* literature, these texts anticipate some of the present concerns and debates in political ecology and environmentalism. But when it confines nature protectionism to the will of a small elite, it also anticipates some of the critical flaws that would beset later ecological thinking.

The actual social and ecological realities within which the *modernista* authors were living are characteristic of urban migrations and the accompanying ecological degradation that these demographic upheavals implied. I numbered Mexican Federico Gamboa's *Santa* (1904) as among the last of the novels in the tradition of the nineteenth century foundational fictions but was especially concerned to highlight the way in which that novel documents the sordid physical conditions of city living. The state of affairs which seemed somewhat uncharacteristic for the rest of the nineteenth century texts had for the *modernistas* become the standard account of city-living throughout Latin America, North America, and Europe. As Cathy Jrade notes, "The Spanish American writers of the end of the nineteenth century—most of whom lived in the urban capitals of their countries and/or traveled extensively in Europe—believed that they were confronting, in a noble struggle, the most acute issues of modern life."[13] As was seen for many nineteenth century texts, this often took the form of a kind of nostalgia for rural living as people began to move to the cities. But by the end of that century, the massive migrations led to a kind of human infestation upon slices of land incapable to absorb them. Waste disposal, inadequate shelter, and toxic living conditions became endemic. And outside of the cities, natural resources were slowly but inevitably exhausted; in his environmental history of Latin America, Luis Vitale enumerates deforestation, mono-cultured pastures for livestock, the drastic decline in fisheries, and the near extinction of whale species, sea walruses, the Bolivian chinchilla, and other mammalian species at the hands of the hide, lantern oil, and fur industries. The result, as Vitale writes, is that "la economía de exportación configure las características esenciales del subsistema urbano latinoamericano. Las ciudades se crearon en función de la economía agrominera exportadora" ["the export economy configures the essential characteristics of urban Latin American living. The cities were created in the wake of an agro-mining export economy"].[14] The harsh social, political, economic, and cultural effects of these changes which were begun in the nineteenth century come to a head in the first decades of the twentieth. And it was against the environmental and economic brutality of these decades that the *modernistas* reacted so strongly.

RUBÉN DARÍO, JOSÉ MARTÍ, AND THE ECOLOGICAL POETRY OF *MODERNISMO*

In the introductory comments of the last few paragraphs, I reference the work of Rubén Darío along with José Martí and José Enrique Rodó; Darío's work is quintessential of the *modernista* aesthetic: a profound skepticism toward the culture of the region, nation, and moment; an equally profound sense of the beautiful; and a rejection of the didactic conventions of Spanish American literature with the subtle incorporation of an alternate ideology in highly stylized prose and poetry. Darío wrote in every literary form—essay, poetry, and short story—except for the novel. On the occasion of the Spanish tragedy from their disastrous war with the United States in 1898, Darío's short essay "El triunfo de Calibán" (The triumph of Caliban) pre-dates Rodó's seminal *Ariel* by two years and employs the same rhetorical reference to characters from *The Tempest.* In Darío's text, the criticism is leveled against the culture of the United States and that country's tendency to steamroll, cement, and erect. Darío's complaint often takes an ecological form as he expresses a sense of foreboding at the "abrumadoras ciudades de hierro y piedra" ["overwhelming cities of iron and stone"] that dominate the northern country's landscape and an exhortation to prepare a defense against "tentáculos de ferrocarriles, brazos de hierro, bocas absorbentes" ["railroad tentacles, arms of iron, ever-sucking mouths"]:[15] ecological in the sense that a description of cities engulfed by trains, iron, stone, and steel implies that the cultural influence from the North will involve fundamental changes to the land. Such is Darío's greatest fear, already then being realized in many of Latin America's cities.

His disgust with these developments impels Darío to produce a body of work that attempts an evocation of distant places as an antidote to and condemnation of his social and cultural moment. If it's planes, trains, and automobiles that Darío can't stand, then it's the allure of nature that captivates his poetic attention:

> ¡Pradera, feliz día! Del regio Buenos Aires
> Quedaron allá lejos el fuego y el hervor;
> Hoy en tu verde triunfo tendrán mis sueños vida,
> Respiraré tu aliento, me bañaré en tu sol.
> Muy buenos días, huerto. Saludo la frescura
> Que brota de las ramas de tu durazno en flor;[16]
>
> [The joy of the fields! Far from regal Buenos Aires
> And its fire and smog and not enough room;
> Here in the triumph of this greenery, my dreams come true,
> I breathe in your scent, I bathe in your sun.
> Good morning, garden! I greet the flowering branches
> Of your peach tree in bloom.]

The "fire" and "smog" of Buenos Aires are cheerfully abandoned for the green triumphs of nature. These ideas about the escape from a contaminated city to a purified nature are repeated in several of Darío's short stories from *Azúl*. In "El rubí," (The ruby) the production of artificial gems is described as a process of scientific trickery: "¡Pues el conjuro!, fusión por veinte días de una mezcla de sílice y de aluminato de plomo: coloración con bicromato de potasa, o con óxido de cobalto. Palabras, en verdad, que parecen lengua diabólica" ["The cheap trick! Bond silica gel to aluminate of lead: color with potassium dichromate or cobalt oxide. Words that truly seem like some kind of devil language"].[17] Diabolic terminology is the scientific language of urbanized industrialization, a state of affairs disdainful to the poet from "El Rey Burgués," (The bourgeois king) another story from *Azúl*: "He abandonado la inspiración de la ciudad malsana, la alcoba llena de perfumes, la musa de carne que llena el alma de pequeñez y el rostro de polvos de arroz" ["I've left the inspiration of the unhealthy city, the tiny rooms filled with perfumes, and a fleshly muse that seemed to fill the soul with bile and my face with her makeup"].[18] Unfortunately for the poet, the bourgeois king has no idea what to do with someone like him and he is left outside in the elements to turn the crank of a music box whenever the regent passes by. The scorn (inherited from the romantics) for a newly wealthy class of merchants and their tasteless sense of beauty was common among the *modernistas* and explains the ubiquitous use of nature in their fiction and poetry; no one can argue with the beauty of trees, flowers, and (the *modernistas*' favorite image) a graceful swan floating in placid waters. In a written address to a group of poets of Spain, Darío says as much: "He cantado, en mis diferentes modos, el espectáculo multiforme de la Naturaleza y su inmenso misterio" ["In several different ways, I sing a song of the spectacle of nature's many and immense mysteries"].[19] In his poetry, there are countless examples of this; one is from "Coloquio de los centauros" (Colloquium of the centaurs) (1901):

> Himnos a la sagrada Naturaleza; al vientre
> De la tierra y al germen que entre las rocas y entre
> Las carnes de los árboles, . . .
> . . .
> Cada hoja de cada árbol canta un propio cantar
> Y hay un alma en cada una de las gotas del mar.[20]
>
> [Hymns to sacred nature; from beneath
> The soil, seeds among the rocks, bequeath
> The flesh of trees, . . .
> . . .
> A song for every leaf in every tree,
> A soul for every drop of water in the sea.]

Scholars view the use of sacred, Edenic images as a strategy to infuse the representation of nature with a sense of the divine, an element that has had a long history in the annals of ecological writing.[21] But it is the sentiment expressed in the last of the lines quoted above that contains the most relevant idea for an ecocritical analysis. The idea that each and every part of nature has an incalculable value (even leaves and drops of water) represents a poetic pre-figuration of the idea of the "economy of the absurd" from Fernando Mires's essay on the discourse of nature discussed in the Introduction. In the serious, meticulous, and dry language of sociology and political science, Mires argues that the true cost of the capitalization of nature can never accurately be determined for the lack of value categories that can do the calculus. Darío makes the same point, but in the poetic language of sacred land, melodies, souls, sacraments, flowers and trees. If each leaf has its own song and each drop of water its own soul, the value of such things can never be determined. While the purposes and modes are vastly divergent, the poetry of Darío and the essay by Mires point to exactly the same conclusion about the capitalist "economy of the absurd"; it is unsustainable, unjustifiable, economically infeasible, and a danger to the natural world. The discomfort expressed by Mires for the incorrect calculations of the capitalization of nature is quite similar to what is at the root of the *modernista* alienation from their societies at the turn of the last century. Darío said it best:

> El progreso moderno es enemigo del ensueño y del misterio, en cuanto a que se ha circunscrito a la idea de utilidad. Mas, no habiéndose todavía dado un solo paso en lo que se refiere al origen de la vida y a nuestra

> desaparición en la inevitable muerte, el ensueño y el misterio permanecen con su eterna atracción.[22]
>
> [For as much as it imposes notions of utilitarianism, modern progress is the enemy of dreams and mystery. But, because this utilitarianism cannot understand metaphysical things such as the origin of life or what happens after death, dreams and mystery will always remain attractive to humankind.]

These sentiments, particularly the rejection of modernity and progress, are foundational to *modernista* thinking and represent what I will develop in the rest of this chapter: industrialization, city-life, and the bourgeois values that are associated with them are roundly rejected for reasons that are strikingly similar to the ideas that undergird current environmentalist ideology; dreams and mystery (and as we shall see in the authors considered below, time spent in the country among forests, trees, mountains, plains, and the sea) are always preferable to the progress represented by cities.

Like Rubén Darío, José Martí also employed various forms to express his ideas; I have already discussed the essay "Nuestra América" for which he is perhaps most well-known, but Martí was a novelist and poet as well. Martí's poetry was similar to the work by Darío in that the country is almost uniformly preferred to the city, with the latter usually described in no uncertain and negative terms.[23] In "Amor errante" (Errant love) from *Ismaelillo* (1882), the poet finds relief in the sea air:

> Los aires frescos
> Limpian mis carnes
> De los gusanos
> De las ciudades.[24]
>
> [The fresh air
> Cleans my body
> From the worms
> Of the city.]

In *Versos libres* (*Free Verse*) (1882), there is more negative reaction to the city: "¡Me espanta la ciudad! ¡Todo está llena / De copas por vaciar, ó huecas copas!" ["I fear the city! It is full /of cups to be emptied or empty cups!"].[25] Rejections of urban space can even take rhetorical forms as Martí employs the imagery of poetic injury as the result of writing about the city:

> Cuando va a la ciudad, mi Poesía
> Me vuelve herida toda, el ojo seco
> Y como de enajenado, las mejillas
> Como hundidas, de asombro: los dos labios
> Gruesos, blandos, manchados; una que otra
> Luta de cieno—en ambas manos puras
> Y el corazón, por bajo el pecho roto

Como un cesto de ortigas encendido:
Así de la ciudad me vuelve siempre:
Mas con el aire de los campos cura
Bajo del cielo en la serena noche
Un bálsamo que cierra las heridas.
(Martí, *Ismaelillo*, 164–65)

[When poems about the city I write
It seems the very words get hurt, eyes
Go dry, cheeks turn white
And sunken with fright:
Lips blister, swollen and stained;
Pure hands smeared with mud;
In the chest, a heart pained
As if in a basket of thorns.
That's what the city does to my poetry:
But the fresh air of the country cures
Like a balm that heals all wounds
Under a serene sky and night of stars.]

Poetry gets a beating in the city, but the country is a balm; the rhetorical meta-textuality in the poem elicits an evocation of the city as doubly bad. Not only do these images conjure a sense that the city is not suitable for tranquility, but that it is also not suitable as an object for poetry itself. The ecological aspects here have implications for real action (because of their predilection for violence toward the products of human endeavor, the construction of cities must be avoided; the curative airs of the country must be conserved for their benefits to humanity), but also for the production of literature. If the representation of contaminated cities and polluted urban environments can do rhetorical harm to poetry, then such things must be cleaned up or avoided so they will not damage the production of art.

But the representation of the benefits and the criticism toward the contamination of nature are not always evoked in Martí's poetry in imagery that is so structurally complex. And it helps to know that Martí, exiled to the United States in the 1880s, fostered a relationship with several of the seminal thinkers now associated with the ecological movement:

> El desarrollo del interés en la conservación de los recursos naturales en los Estados Unidos en la segunda mitad del siglo XIX se vio asociado en múltiples sentidos a aquella corriente democrática de la cultura norteamericana . . . de intelectuales y reformadores sociales como Ralph Waldo Emerson, Henry David Thoreau, Walt Whitman y Henry George, con los que Martí demostraría desde muy temprano una intensa afinidad.[26]
>
> [The development of interest in the second half of the 19th century for conservation of natural resources in the United States was seen as having to do with democratic currents in U.S. culture, . . . especially among intellectuals and

social reformers such as Ralph Waldo Emerson, Henry David Thoreau, Walt Whitman, and Henry George with whose ideas Martí would quickly demonstrate an intense affinity.]

An affinity toward nature, which is one of the principal ideas associated with the authors, poets, and intellectuals cited above, is an ideal also quite regularly expressed in Martí's work. In "Canto religioso" (Religious song) from *Versos libres* the value of nature is evoked in unequivocally spiritual terms:

Y sigo a mi labor, como creyente
A quien unge en la sien el sacerdote
De rostro liso y vestiduras blancas—
Practico: en el divino altar comulgo
De la Naturaleza.
(Martí, *Ismaelillo*, 158)

[I continue my work, like a believer
With temples anointed by a priest
Of plain face and wearing white garments—
I believe: and take communion
Upon the divine altar of nature.]

In Poem 3 from *Versos sencillos* (*Simple Verses*) (1891), the religious aspects of nature are further developed:

Busca el obispo de España
Pilares para su altar;
¡En mi templo, en la montaña,
El álamo es el pilar!
. . .
¡Díganle al obispo ciego,
Al viejo obispo de España
Que venga, que venga luego,
A mi templo, a la montaña!
(Martí, *Ismaelillo*, 155)

[The bishop of Spain seeks
Pillars for his altar;
In my temple, in the mountains,
The only pillars are trees: a stand of poplar!
. . .
Tell that blind bishop,
That blind old bishop from Spain,
To come, to come right away,
To my temple in the mountain!]

Both of these examples make the spiritual aspects of nature plain; it does not take complex ecocritical analysis to infer an obligation of conservation from something as valuable to human existence as the Eucharist to Christians or a temple to the devout. Kay Milton observes that "wholeness, beauty and mystery will form the basis of sacredness wherever these are the most valued, the most emotionally powerful qualities" and that "the question of why some people value nature and natural things more highly than do others often leads, in environmental discourse, to observations about sacredness."[27] According to Milton's analysis, sacred images of nature like those in Martí have become one of the key elements of ecological texts; and they aptly describe the poetry of Spanish American *modernismo* as a whole: innovative imagery and ingenious use of language to create poetic allegories of love, spirituality, and nature.

In poetry by Uruguayan Julio Herrera y Reissig, for example, there are references to the virginal qualities of the land, a similarity to mythical Arcadia, its barely audible heartbeat. These evocations of the sublime are similar in content to any number of the descriptions of nature cited in chapters 1 and 2, but the *modernista* aesthetic, as seen in the following lines from Herrera y Reissig, renders them in terms far more elegant and refined than anything from the nineteenth century: "un suspiro de Arcadia peina los matorrales" ["an Arcadian sigh adorns the brush"]; "Como una virgen, mística de tramonto, se arroba . . . / Aves, mar, bosques: todo ruge, solloza y trina" ["Like an enchanted virgin that seems to disappear in the mountains . . . / Birds, sea, forests: everything roars, sobs, and chirps"];[28] these are representations of nature that could only come from the pen of a Spanish American *modernista*. However, the specter of intrusive modernization seems to always hover over the evocation of nature in the poetry of Herrera y Reissig. As Cathy Jrade observes:

> While most *modernistas* believed that poets could translate the book of nature flawlessly by becoming attuned to the rhythms of the universe and by opening their souls completely and sincerely to the perfection of existence, Herrera y Reissig . . . hints at the unavoidable presence of social and artistic obstacles that hinder the pursuit of that goal. The imagined medieval purity of the provinces . . . is made to confront the intrusive realities of life and the inevitable distortions of the ideal.[29]

It comes as no surprise that certain representations are haunted by the threatening presence of a contaminating urban environment:

> ¡Pálida virgen, ebúrnea, cándida, mística, santa,
> La tierra es un incensario de intacta, inhollada nieve
> En donde, trémula y casta, sutil, impoluta y leve,
> La niebla, incienso con alas, vuela, gira y se levanta![30]
>
> [Pallid virgin, ashen, wan, mystical, saintly,
> the land is like a censer of pure, white snow
> from which a subtly, clear mist does blow,

like incense on wings that spins and whorls so faintly.]

Here the land is not just virginal, mystical, and sacred, but also idealized as a place free from pollution, the only smoke it emits articulated as winged mists floating from pure white snow. But the presence of the polluted, smoky, perverted, broken, and tread-upon lingers just on the other side of meaning from terms like "vírgen," "intacta," "inhollada," and "impoluta" as if they were about to become violated, spoiled, trampled, and contaminated. The longing for unsullied and undisturbed natural places is made the most explicit in Herrera y Reissig's poetry in "El guardabosque" (The forest ranger) from *Los éxtasis de la montaña* (The ecstasies of the mountain) (1904–1907). The title literally refers to a profession dedicated to the protection and cultivation of forests, the forest ranger here a high priest of the woods:

Él se mira en su selva como un padre en su hija.
Y aunque cruja la nieve y aunque el cielo se enfosque
la primera instantánea del oriente lo fija
como a un genio hierático, Sacerdote del bosque.[31]

[He sees something of himself in the jungle as father to daughter.
Though the snow howl and the sky lower its hood,
something like a photograph in a pool of water
captures his stern genius: the high priest of the wood.]

The poetry has gone over from escape-to-the-country praise for the sublime qualities of nature to an expression of love for nature: the woodsman is construed as the father of the forest, and one that identifies himself with the characteristics of the woods, his "daughter." In the introduction, I reference Kay Milton's idea that nature protectionists are inclined toward the land in terms of connection, adoration, and love: "Anyone who identifies with natural things, who sees them as part of themselves, is therefore likely to feel inclined to protect them."[32] And, as Karen Warren characterizes it, ecofeminism "gives expression to a variety of ethical attitudes and behaviors overlooked or underplayed in mainstream Western ethics."[33] Nowhere are such sentiments more explicitly expressed in the poetry of Spanish American *modernismo* than in a comparison of the affection for the forest by the "*guardabosque*" as similar to that of the genetic, familial, unconditional love of a father for his daughter.

While the poets considered thus far occasionally express explicitly ecological and even proto-environmentalist ideas, some counted among the *modernistas* have produced a rich oeuvre generally, but little in the way of a representation of nature that invites an ecocritical response. Mexican Amado Nervo has volumes of poetry, but only occasional references to nature of the type that features ideological connotation. In his work, the poet declares "cada vez hallo a la Naturaleza / más sobrenatural, más pura y santa" ["each time more than the last, I find nature / more mystical, more pure, more holy"][34] and rejects "la Horda / científica [que] destruye cuanto la especie humana / supo crear" ["the scientific/horde [that] destroys whatever humankind / was able to create"].[35] These lines express the *modernista* admiration of nature and rejection of the modernized world, but they are not characteristic of Nervo's poetry as a whole. Mexican compatriot Manuel José Othón's poem "Himno de los bosques" (Hymn to forests) (1891) marks the typical *modernista* dissatisfaction with city life and preference for nature:

> En este sosegado apartamiento,
> lejos de cortesanas ambiciones,
> libre curso dejando al pensamiento
> quiero escuchar suspiros y canciones.
> ¡El himno de los bosques![36]
>
> [In this gentle place
> far from the city and its ambition,
> I let my mind wander apace
> songs and sighs its main intuition.
> The hymn of the forest!]

And fellow Mexican Luis G. Urbina observes the suffering of a single tree, scarred by axes and wildfires, in "El poema del lago" (The poem of the lake) (1906):

> ¿Qué dice tu nervioso gesto de *selva* oscura
> árbol vetusto y seco sin una verde rama?
> Con cicatriz de hachazos y quemazón de llama,
> como un espectro tiendes tu sombra en la llanura.[37]
>
> [What is this somber, dark jungle gesture:
> ancient tree, withered, a memory of green?
> Where your burned out bark and hatchet marks seem
> to ghostly, cast shadows on the plain and fester.]

The counterpoint in these quotations between sublime nature and trees withered by fire and axes represents something akin to the imagery of nature in explicitly environmentalist works; but these lines make up an ideological representation of nature that is not the norm in the broader corpus of the poetry by Othón and Urbina.

Similarly, poems by Uruguayan Delmira Agustina as well as in others of the most well-known *modernistas* including Mexican Salvador Díaz Mirón, Cuban Julián del Casal and Bolivian Ricardo Jaimes Freyre do not contain an explicitly ecological ideology. Obviously, each of these poets has verses that reference nature, but the tendency is often to frame human endeavor, emotion, and relationships. The following line from Freyre's "Vesperal" in the collection *Anadiomena* (Goddess of love) (1917) is typical:

> Una tarde a la luz del sol poniente

Marco de tu belleza parecía
Y el resplandor que entre la fronda huía
Brilló en tus ojos y ciñó tu frente.[38]

[Twilight: sun's rays among the fronds leak
And your beauty seem to frame.
Their gleam not the same
As the brilliance of your eyes and cheek.]

The invocation of sun reflected on the leaves of ferns serves merely as a frame for a memory that comes to the poet as a woman's beauty is recalled in the lines of the poem. There is a redeeming quality to the nature represented here (and above), but one that is secondary to the human presence. I do not mean to claim therefore that the *modernista* poets and authors were proto-ecological *as such*; the sparse appearance of nature in Nervo, Othón, Urbina, Agustina, Díaz Mirón, del Casal and Freyre demonstrate that this was not the case. However, there were several (such as in Darío, Martí and Herrera y Reissig above) where urban decay is condemned in ecological terms and where nature finds pervasive representation and is invoked for its own sake.

Perhaps the one who most clearly expressed these ideas in his poetry was Peruvian Manuel González Prada who produced a body of work which was itself indirectly informed by the War of the Pacific, an armed conflict among Peru, Bolivia, and Chile for control of the nitrite, saltpeter, and guano deposits on the islands and in the Atacama desert along the three countries' borders in the 1880s. González Prada was promoted as high as lieutenant colonel in the Peruvian army that was soundly defeated by the Chileans. The humiliation of the Treaty of Ancón that awarded the victors the formerly Peruvian province of Tarapacá inspired González Prada to famously declare: "¡Los viejos a la tumba, los jóvenes a la obra!" ["The old to the grave, the young to work!"] in a speech that was censored by the government but widely reproduced in Peruvian newspapers of the time.[39] As Eugenio Chang-Rodriguez observes, "Ahora había llegado el momento del examen de conciencia, de la siembra de nuevas semillas para la reconstrucción moral y física del país, para la redención de la patria" ["The time had come to take account, to imagine new landscapes for the moral and physical reconstruction of the country"].[40] And Juan Guillermo Gómez García finds that González Prada, "desilusionado de las tradicionales formas de agrupación política, desemboca en el movimiento obrero que se enfila en las luchas sindicales bajo las banderas del anarquismo" ["disillusioned by traditional political arrangements, finds himself aligned with the labor movement and union struggles under the flags of anarchism"].[41] As Chang-Rodríguez and Gómez García demonstrate, González Prada's texts were profoundly influenced by political concerns and attempted to break with the past and embrace the future. In other words, he embraces the ethos of *modernismo* characterized by Darío's declaration—

"yo detesto la vida y el tiempo en que me tocó nacer": the Spanish American *modernistas* sought new forms of literary expression as statements of disdain for the bourgeois values of their contemporary modern world. González Prada's work contributes to that ideology through the expression of a vision for a new and future Peru, shorn of the influence by feudal lords over an antiquated economic and political model, demonstrably bankrupt after the War of the Pacific. However, from an ecocritical perspective, González Prada's highly aesthetic work also articulates many elements characteristic of current environmentalist texts: the ecosublime, toxic discourse, concern about deforestation, and a defense of the inherent versus the instrumental value of the natural world.

In "Le tour du propietaire" (The tour of the property owner) (1911), the poet accuses the wealthy landowner for his apathy toward nature: "Nada escucha el burgués. . . . Nada huele el burgués. . . . Nada mira el burgués" ["The bourgeois man hears nothing, . . . smells nothing, . . . sees nothing"].[42] Instead, what the poet knows to be the sublime sounds, odors, and sights of nature are wasted on that particular type of person:

> Naturaleza, en vano
> Almibaras tu almíbar, hermoseas tu hermosura,
> Y como regio don ofreces a los hombres
> Tu desnudez olímpica y gloriosa.
> El burgués, el binario de Harpagón y de Tartufo
> No te comprende a tí, la madre sin hipócritas remilgos,
> A tí, la pródiga divina;
> A tí, la gran pagana.
> (González Prada, *Exóticas*, 107)
>
> [Nature vainly,
> Preserves its fruits, shows off its beauty,
> And like a regal gift for men,
> Flashes Olympic glory.
> The bourgeois, a double for Harpagon and Tartuffe
> Could never understand you, mother without false pretenses.
> Never, oh divine prodigal;
> Never, oh great pagan.]

The references to Moliere's Harpagón (the Miser) and Tartufo (the Imposter) represent the preference for money over nature, a preference that González Prada perceives as central to Peruvian society. And these criticisms in many cases take the form of an imagery of polluted cities, poisoned air, and deforested land, dismay at "the mud and misery of streets and plazas" and steam-powered development accompanied by "belching smoke," "screams of iron," and "infected air."[43] Negative images of the city implicitly suggest a preference for the country, and indeed, explicit and repeated elegies to nature are found throughout González Prada's poetry. The following lines from "Laude" in *Presbiterianas* (1909) are characteristic:

Monte, selva, mar y llano
Alcen himno tan pagano
Que hasta el pecho del anciano
Se estremezca de alegría.[44]

[Forest, jungle, plain and sea,
Raise a hymn of felicity.
Pagan, yes, but even the elderly,
Tremble as they hear it, with joy.]

Love for nature contrasts with the poet's exasperation at the excesses of the industrialized city and González Prada repeatedly warns that to forget about nature is to commit grievous fault. These are warnings that the engineers of modernization would do well to heed in the wake of planetary warming, oil spills, and tsunami-caused meltdowns.

ECOLOGICAL ELEMENTS IN *MODERNISTA* NOVELS

The *modernista* perspective that was the focus of the last section derives as much from an aesthetic disposition in the poems as from specific experiences in an ever more modernized and politicized world. The irony of the term *modernismo*, coined by Rubén Darío to "establish a mode of discourse commensurate to the new era that Spanish America had entered,"[45] resides precisely in the fact that the *modernista* oeuvre represents a rejection of the values of modernized, bourgeois society. For essayists like Rodó and Martí, the critique was more political and sociological than literary, but several other writers, such as José Asunción Silva, Manuel Díaz Rodríguez, Clímaco Soto Bordo, Leopoldo Lugones, and Clemente Palma, adopt a critical posture towards the politics and societal values of their day as material for novels and short stories that were as aesthetically realized as the poetry. While Darío must be the most obvious candidate for the title of most quintessentially *modernista* among the *modernistas*, and Martí figures as the first among them, Colombian José Asunción Silva's only novel *De sobremesa* (*After-Dinner Conversation*) (1896, published 1925) represents an example of Spanish American *modernista* prose that offers a sketch of the ideological contours of the movement in the form of a novel.

Silva's novel, the reading of the diary of one of four dinner guests during the *sobremesa* (the traditional Spanish-American custom of an after-meals discussion time), features numerous moments that both define the tenets or central characteristics of *modernismo*, but also suggests an implicit connection between the ideals of the movement and ecological principles. As critic Rolando Pérez has noted about *modernismo* in general, most scholars "have failed to see that even behind some of the so-called 'ornamental' works of literature lays a social critique of modernity, from a very unique Latin American perspective"[46]; in the analysis of *De sobremesa* and of the other novels in this section, I identify the ecological aspects of the *modernista* critique among the ornamental imagery and aesthetic emphases in the texts. In Silva's novel, the protagonist José Fernández dedicates his time to the search for Helena, a woman he has met only once, but who threw him a bouquet of flowers from the balcony of her room in a hotel where both were staying. From this small gesture, Fernández expends a lifetime of obsession searching for her and the description of his search provides a reliable characterization of *modernismo*: first, the rejection of bourgeois values: "¿La vida real? . . . Pero ¿qué es la vida real, díme, la vida burguesa sin emoción y sin curiosidades? . . . [No.] Mi pasión [es] más honda: el deseo de sentir la vida, de saber la vida, de poseerla" ["Real life? . . . But what's real life? Tell me. An emotionless bourgeois life without curiosity? . . . [No,] my deepest passion is to live life to the full, really feel it, really know it"]; also, the importance of pleasure: "La vida no es para saber, es para gozar. Goza, gozar es mejor que pensar" ["Life's not for thinking, it's for feeling. Take pleasure, pleasure is better than thinking"]; and the attraction of the abnormal: "Lo anormal me fascina como una prueba de rebeldía del hombre contra el instinto" ["I am fascinated by abnormality because it's proof of human rebellion against instinct"]; finally, the principle of love as the most ennobling of all pursuits: "Es un amor sobrenatural que sube hacia ti como una llama donde se han fundido todas las impurezas de mi vida. Todas las fuerzas de mi espíritu, todas las potencias de mi alma se vuelven hacia ti como la aguja magnética hacia el invisible imán que la rige" ["My love for you is like a supernatural fire fueled by a melted conglomeration of all the impurities of my life. All the strength of my spirit, all the potency of my soul yearns for you like an iron file that cannot resist the invisible magnet that draws it in"].[47] I cite these elements of the novel because they seem to most accurately summarize the emphases of *modernismo* (as well as *decadentismo* which I consider below) after the most obvious definition of the literary period's aims: the production of art for its own sake rather than (or as well as) the political, didactic, and nationalist emphases of the foundational nineteenth century texts. *De sobremesa* elucidates these ideas in the obsession of the protagonist José Fernández for the idealized figure of Helena, but the ideology that emerges from the text also supports an ecological position as the inevitable product of *modernista* ideals which are so characteristic of the novel.

One ecological aspect in *De sobremesa* is contained in what is explicitly rejected and abandoned. In his diary, the protagonist Fernández imagines a process of personal enrichment that culminates in a vision of dictatorial control over his home nation of Colombia: a doubling or tripling of his wealth; research into the fabulous economics of North American capitalism and the mystifying tribal culture of native Colombian indigenous inhabitants; the use of propagandistic persuasion through the press to arouse a reactionary political assertion of power with himself at the head, first as president then as conservative dictator; finally, after all these projects are happily complete, a comfortable retirement to a country estate, the reading of philosophy, and the establishment of a legacy similar to Bolívar or Napoleon. While it is a chilling vision of the raw ambitions of a moneyed Latin American elite, the ecological consequences of such endeavors are unequivocally laid out in Fernández's diary:

> Monstruosos fábricas donde aquellos infelices encuentren trabajo y pan nublarán en ese entonces con el humo denso de sus chimeneas el azul profundo de los cielos que cobijan nuestros paisajes tropicales; vibrarán en los llanos el grito metálico de las locomotoras que cruzan los rieles comunicando las ciudades y los pueblecillos nacidos donde . . . a la hora en que escribo, entre lo enmarañado de la selva virgen extienden sus ramas las colosales ceibas, . . . como una red aérea los hilos del telégrafo y del teléfono agitados por la idea se entenderán por el aire. (Silva, *De sobremesa*, 174)
>
> [Monstrous factories, where unhappy workers may find labor and bread, will soon create an impenetrable smog from the smoke of chimneys that will block out the blue skies that cover our tropical landscapes; the metallic vibrations of locomotives screaming over rails that connect the cities to new towns built on sites that, . . . as I write, are still covered by virgin jungle and colossal ceiba trees, . . . up above, as if trembling from the transmission of new ideas, telephone and telegraph wires will crisscross the skies like a net.]

From the perspective of ecological conservation, the description is disastrous with the belching smoke of polluting factories and deforestation as the consequence of the construction of railway lines and communications infrastructure. In the novel, therefore, the surrender of these dreams for the sake of passion, pleasure, love, and all the other qualities associated with *modernismo* requires a more complete articulation of the movement's aims. Pérez emphasizes Fernández's abandonment of his grand scheme characterizing it as the author Silva's recognition that "capitalism—modernity's handmaiden—turned everything from Eastern religions to precious jewelry, into equally exchangeable fetishes of consumption."[48] The novel's representation of the bankruptcy of such pursuits is characteristic of the *modernista* critique of over-development, industrialization, and the accompanying "fetishization of consumption" to use Perez's turn of phrase. My argument here is that along with the critical perspective toward capitalism comes a corresponding advocacy for more sustainable economic and social models: models founded on love, beauty, and passion rather than raw economic gain. My conclusion can perhaps be construed as a hastily constructed contra-positive from the premises of the abandonment of Fernández's grand plan in favor of the pursuit of an idealized Helena. But there are several other explicit moments in the text that corroborate an ecological analysis of *De sobremesa.*

First, there is the description of urban living as entrapment in the "ciudad monstruo" ["monstrous city"] as well as a criticism of the over-development of Niagara falls, "profanadas por los ferrocarriles y por la canallería humana que va a divertirse en los hoteles que las rodean, son un lugar grotesco cerca de la majestad de templo del agreste sitio" ["profaned by train lines and human villainy, by tourists that infest grotesque hotels surrounding the majesty of the temple-like rural site"].[49] Finally, the text even approaches tones of advocacy for environmental justice when it condemns the abuse of workers by possessors of capital: "¿Tú no sabías nada de eso, obrero que con las manos encallecidas por el trabajo haces todavía la señal de la cruz y te arrodillas para pedir por los dueños de la fábrica donde te envenenan los vapores de las mezclas explosivas?" ["You knew nothing of this, workers, who with work-calloused hands, still make the sign of the cross to bless the owners of factories which spew chemical vapors that poison the very air you breathe as you pray for them?"][50] While the abandonment of Fernández's development dream does not in itself constitute a proto-ecological current in the novel, the representation of polluting factories, toxic cities, and over-development suggests that while environmentalist ideas were certainly not explicit elements of the novel, such moments approach advocacy for conservationism and environmental justice.

Like many other *modernista* writers, Venezuelan Manuel Díaz Rodríguez was strongly influenced by his travels to Europe as a young man, but his novels and short stories are among the most illustrative of that kind of experience as a source for their content. The literary conceit allows Díaz Rodríguez to write what he knew, but it also allows for a context within which to express a critique of Latin American societies that the *modernistas* believed were possessed of a brutish and unrefined culture at the turn of the last century. In *Ídolos rotos* (Broken idols) (1901), for example, the disillusioned sculptor Alberto Soria considers the moment when he returns to Venezuela from Paris: "Su imagen de la patria no era la misma que guardaba en el corazón cuando arribó a sus costas, . . . cuando a través de las ventanillas del tren vio surgir la belleza del paisaje nativo, original y soberbio, desconocido u olvidado" ["The idea of his country had changed from the one he had in his heart when he first arrived on its borders, . . . when through the windows of the train he perceived the impressive beauty of forgotten or unknown native landscapes"].[51] The *modernista* ideal of aesthetic beauty is expressed here in the description of overpoweringly beautiful landscapes and later, in a rhetorical move typical of Spanish American *modernismo*, that beauty is articulated as the product of something like a classic Greek sculpture or landscape: "la belleza de la tierruca, al través de su propia serenidad, encantaba sus ojos como la belleza de una estatua blanca y serena, de contornos limpios" ["in his mind's eye, he was enchanted by the beauty of that little bit of land which seemed to contain all the beauty of a clean-lined, serene white statue"].[52] Nature writing of the kind that praises the beauty of Spanish American nature is not unique to Díaz Rodríguez among the *modernistas*, but the criticism of the corruption of cities with which such descriptions are juxtaposed in the novel takes the form of political ecology, something not commonly present in these writers' texts. In *Ídolos rotos*, the city is a place devoid of undeveloped green space, and governmental corruption is explicitly offered as the reason why:

> El césped en algunas partes no existe: apenas quedan rastros de haber existido en el borde de anchas peladuras que son como una calvicie de la tierra. En otras partes, la calvicie comienza y no se la estorba, o por incuria del Municipio, o por la escasez de sus rentas. . . . Tal vez por iguales motivos, algunos árboles, y de los más hermosos, languidecen y mueren. (Díaz Rodríguez, *Ídolos rotos*, 82)
>
> [In some areas, the grass had simply ceased to exist: there were a few blades growing around a vast bare patch like a huge bald spot on the earth. In other places, these barren patches continue to expand unopposed, who knows whether because of municipal negligence or a lack of tax revenue to support such projects. . . . Perhaps this also has something to do with why some of the most beautiful trees languish and die a slow death.]

Ironic criticism of hapless governance and kleptocracy would sound similar in any contemporary environmentalist critique of the inability of bureaucratic entities to effectively protect the natural world. In the novel, Díaz Rodríguez has the protagonist Alberto thoroughly disgusted with that state of affairs:

> En sus ojos había toda la tristeza del mundo. Y cuando muy tarde, esa noche, volvía a su casa, hallóse viendo y considerando, si no con verdadero odio, con algo muy parecido al odio verdadero, los hombres, las cosas, todo lo de aquella ciudad estrecha y mezquina, de conciencia, como sus calles, angosta y sucia. (Díaz Rodríguez, *Ídolos rotos*, 92)
>
> [He was faced with all the sadness of the world. And later that night, returning home, he would find himself thinking and hating, if not with true hatred in his heart then something much like it, men, objects, and everything else in that crowded, miserable city, a city that, likes its streets, was narrow-minded and dirty.]

The frustration with his urban milieu takes the form of a kind of hatred toward the things of the city and for its atmosphere of moral bankruptcy, a bankruptcy that is explicitly compared to the filthiness of its streets. The only thing that comes in for more strident criticism than the city in the novel is the societal consequences of war; and in *Ídolos rotos* this too takes the form of a condemnation based upon, among other things, the ecological catastrophes of armed conflict.

In the year before the publication of *Ídolos rotos*, a civil war broke out in Venezuela—the Liberal Revolution of Restoration. It is beyond the scope of the present analysis to go into the historical and political details of the hostilities, but it forms the backdrop of the latter portions of Díaz Rodríguez's novel. It is not overly difficult to imagine that the *modernista* writers were highly disillusioned by the politics of warfare which was the case with the narrative message of *Ídolos rotos*; there can be few things more contrary to the beauty aesthetic of Spanish American *modernismo* than the facts on the ground of military combat. The disillusionment in Díaz Rodríguez's novel takes the form of an ecological as much as a social, cultural, or political critique:

> ¿Qué les importaba que la guerra segase en flor innúmeras vidas inútiles, devastase los campos, talase los bosques, destruyese el humilde conuco del montañés labrador y el hato del llanero, cuando las vidas de ellos estaban en salvo, cuando la hacienda de ellos estaba en seguro y su capital político . . . [e]n vez de padecer y disminuir con la guerra, más bien se acrecentaba[?] (Díaz Rodríguez, *Ídolos rotos*, 139)
>
> [What did it matter to politicians that countless lives were cut down in their prime, or that the land would be left desolate, or the forests cut down, or the small plot of the humble mountain laborer or plainsman destroyed when they themselves were safe, production was up, political capital assured, . . . and instead of pain and loss, they only got richer?]

The rest of the twentieth century in South America, Europe, and the Pacific was, because of various wars, as ecologically destructive a time as any that had been seen before what with the use of mustard gas, leaded ammunition, nuclear bombs, Agent Orange defoliant, weaponized uranium; the vast deforestation and human and animal extinction that resulted from military campaigns; and the permanently indisposable waste of toxic war materiel. Yet here, as early as the first year of the war-plagued new century, Díaz Rodríguez foresees the cynical corruption, human cost, and ecological destruction of war.

I find Manuel Diaz Rodríguez's decidedly *modernista* novel a highly interesting example as a work of fiction that prepares the way for later ecological work in the Venezuelan tradition. Certainly, the representation of war in *Ídolos rotos* as an ecological catastrophe has the immediate pacifist effect of implicitly advocating for an end to armed conflict in general. However, the ecological aspects of the advocacy as well as the novel's overt appeals to state and municipal authorities to correct ecological oversights prepare the way for the explicitly ecological petro-novels in the *indigenista* and regional traditions that appear in later decades. As the environmental devastation of the ecosystems around Lake Maracaibo becomes acute with the petroleum industry's exploitation of oil reserves located underneath the lake-bed, later authors begin to explore the social, economic, and ecological issues that were (and are) at stake: *Mene* (1936) and *Sobre la misma tierra* (The very same land) (1944) by Venezuelans Ramón Díaz Sánchez and Rómulo Gallegos as well as *Mancha de aceite* (Oil stain) (1935) by Colombian César Uribe Piedrahita.

In *Diana cazadora* (Diana the huntress) (1915) by Colombian Clímaco Soto Borda, the second part of a *modernista* duality—love of nature/repulsion for the city—is the focus of the narrative as it was in *Ídolos rotos*. The rapid development of Bogotá with its newly installed electric lights, noisy streets, and imposing edifices functions as the backdrop for a story of descent into alcoholism and eventual death suffered by the principal protagonist Fernando. The story is tragically similar to the biblical parable of the prodigal son with Fernando's brother Alejandro as the sober first-born and Diana the incarnation of nameless "friends" on whom the prodigal squanders his wealth. Many critics see the demise of Fernando as the by-product of a newly modernized Bogotá,[53] but I am interested in the way that urban descriptions strike an ecological tone. The city is described as a place without oxygen due to "las emisiones como cucharadas" ["emissions like cockroaches"], impoverished and ugly, and in an image that could only come from the pen of a *modernista,* a place "mientras más pobre y triste fuera, como se quiere a la madre aunque sea una vieja sin dientes, llena de canas y sin una peseta" ["although poor and sad-looking on the outside, was warm on the inside like a toothless, poor, gray-haired, but beloved old mother"].[54] The image of Bogotá as sad, toothless, penniless and old speaks to the urban decay that might describe the decrepitude of any city not properly maintained, but in *Diana cazadora,* the urban imagery is also precisely negative in terms of the distressed natural phenomena that just barely survive the onslaught of concrete and congestion:

> En los espacios toda una vegetación raquítica y desgraciada, de tonalidades lechosas, hija de los gérmenes que desecha el viento, y que sirve de botiquín a los gatos bohemios y a las aves errantes con sus flores escuálidas, sus raíces al aire, sus hojas miserables y desteñidas. (Soto Borda, *Diana cazadora*, 79–80)

> [In the open spaces, there was this decrepit and disgraceful milky vegetation, grown up from wind-blown seeds with squalid flowers, roots blowing in the wind and barely a faded and miserable petal: a home for alley cats and lost-looking birds.]

The picture of urban decay takes the form of a diseased and squalid nature and provides more evidence for the ecological consciousness of the aesthetic critiques in *modernista* representations of Spanish American society.

Among the last of the *modernista* novelists, Chilean Manuel Latorre published *Zurzulita* (1920) contemporaneously with *Diana cazadora*, but incorporated a unique approach to the dissatisfaction with the corruption and contamination of the city so characteristic of *modernismo*. The novel features an orphaned young man, Mateo, scion of a moneyed family and inheritor of a vast estate in the country. Although he knows almost nothing about agriculture, he decides to escape the city for the country:

> El campo era la salvación: el aire puro y vigorizante, las ásperas labores harían nacer en su cuerpo una voluntad poderosa de vivir, ya que la inacción, la vida sin ideales, la satisfacción de todos sus apetitos habían concluido por convertirlo en un mecanismo mohoso que se derrumbaba penosamente. Su imaginación viva representábase los beneficios de la vida campesina, sentía ya deseos de conocer aquel rincón de montaña y de ver producir la tierra con el esfuerzo de su mano.[55]

> [The country was his salvation: some hard work up in its pure, invigorating air would excite in him a strong desire to live as compared to how now, all his vices had resulted in feeling more like a moldy old soul, easily tired out. His vivid imagination pictured all the benefits of country living and he felt a strong urge to visit that little corner of the world and, by the labor of his own hands, see the earth produce crops.]

The idea of an escape to the country is nothing new for the *modernistas*, but *Zurzulita* is unusual for the way in which it has the adaptation of Mateo to country living as a process of the creation of place. The city is a locale with which Mateo has no discernible affection but for the satiation of destructive appetites; yet the country is referred to in terms of salvation and healing and in terms of place: a "*rincón*" or "little corner of the world" in the mountains from which the work of his hands can produce a real benefit. The novel emphasizes that physical residence is key to accessing healing benefits and as Mateo spends more and more time in the country, the language of place becomes more pronounced: "Advertía que cada rinconcito de bosque, cada pájaro, cada murmullo de viento tomaban para él ahora un sentido familiar e íntimo que nunca había experimentado" ["Something happened to him there that he had never experienced: he came to realize that for him, each little corner of the forest, each bird, each murmur of the wind was now a familiar, intimate sensation"].[56] The image of a little corner of the world recalls Mateo's first desire to abandon the city for the country, but now the elements of nature that inhabit the forest are familiar and intimate. As Jonathan Tittler points out in his analysis of Latorre's ecological fiction, the representation of the Chilean countryside is commensurate with the author's profound "compromiso con el paisaje chileno" ["dedication to Chile's landscapes"].[57] There is less in *Zurzulita* of the sublime idealizations that had become the standard literary convention of nature in *modernista* texts and more of the simple concept of the natural world as "place." The country becomes his own signified corner to which the citified Mateo can escape as it becomes infused with real meaning and value rather than an idealized abstraction or mythic construct. Both of these are valid as ecological rhetoric, but Latorre's novel represents a less common example of the articulation of place and complements sublime evocations of nature in other *modernista* authors.

ECOLOGY, *MODERNISMO*, AND THE SHORT STORY

Finally, I arrive at the short story, perhaps the classic expression of Spanish American *modernismo*: among the highest profile authors—José Martí, Rubén Darío, Leopoldo Lugones—the latter two were as well known for their short fiction as for their poetry. According to Gabriela Mora, the *modernista* short story can be characterized by several features unique to the genre and to its short-form fictional prose: "atracción y rechazo hacia la ciencia" ["attraction by and rejection of science"]; "el deseo general de retener el misterio" ["the general tendency to maintain a sense of mystery"]; and "el cuidado por la forma del lenguaje" ["care in the use of language"].[58] Alongside these elements, much *modernista* short prose features a kind of descent into decadence where society in general is represented at a nadir; the interest in science that established the fictional premises for many of these stories was also often represented more in terms of scientific capacity for destruction, horror, and evil than for progress. This had an inevitable and negative effect on nature such that "el paisaje natural no existe, o se ve distorsionado u horrible por descomposición" ["natural landscapes did not exist or were distorted or were seen in various states of grotesque decay"].[59] A profound fascination with and fear of scientific development led the *modernistas* to explore the dark side of urban society, the capacity for destruction lurking behind the seeming progress of technological advances, and the likelihood of a nightmarish apocalypse. In the process, these pessimistic visions commonly evoke the way in which the natural world is made to seem a victim of science gone wrong as humans fool with an "omega force," unleash the psychic possibilities of highly polished carbon, attempt to contact the dead, or experiment with mind-altering drugs. These are several of the themes central to the *modernista* short story and in many of the stories that deal with such things, an unmistakably ecological consciousness emerges as inherent to the *modernista* fascination with science, extrasensory experience, and the occult.

Mexican Manuel Gutiérrez Najera's work contains something similar to the narrative of *Diana cazadora* in that many of his short stories include themes concerning the ecological bankruptcy of modernity, particularly customs like gambling, the frantic pursuit of wealth, the incessant noise of urban existence, and the resulting desire for the simplicity of a pre-modern life. In "La novela del tranvía," (Tramway novel) (1883), the city is described as a turtle, "que extiende hacia los cuatro puntos cardinales sus patas dislocadas. Esas patas son sucias y velludas. Los ayuntamientos, con paternal solicitud, cuidan de pintarlas con lodo, mensualmente" ["that extends its dislocated limbs in all four directions, limbs that are hairy and filthy, but that the municipal government patronizingly agrees to make up with mud on a monthly basis"].[60] As early as the late 1800s, the issue of urban sprawl is criticized for the ugliness of its development: the image from "La novela de tranvía" conjures muddy, dislocated turtle's feet. In other short stories by Najera, the preference for country living represents an antidote to such repulsiveness; in "Juan, el organista," (John, the organist) (1898), it is reported that in the country, "las campanas suenan de otro modo que en la ciudad: tocan a gloria" ["the bells sound different than in the city: out there, they toll like splendor"];[61] in the story, Juan leaves his home for the country with his daughter after being cuckolded by his wife: "Huía de México, como se huye de las ciudades apestadas. . . . 'La vanidad es una lepra contagiosa', decía para sí. . . . '[Q]uiero que mi hija crezca en la atmósfera pura de los campos: las aves la enseñarán a ser buena madre'" ["He fled Mexico City like one who escapes a stinking metropolis. . . . 'Vanity is like a contagious form of leprosy,' he said to himself. . . . '[I] want my daughter to grow up in the fresh air of the country: the birds will teach her to be a good mother'"].[62] The sentiment is as characteristic of the *modernista* aesthetic as of current ecological thought: the city is pestilent, the country is pure.

While the Argentine Leopoldo Lugones was as well-known for his poetry as for his prose, his fantastical short stories are of the most interest here (and the most unlike other *modernista* texts considered above). His poetry of nature is ambivalent; in earlier poems, there is what Cathy Jrade calls an "expression of delight in presenting the unexpected and unorthodox and in engaging in . . . iconoclastic playfulness that unsettled readers."[63] That is readily apparent in the following images from *Las montañas de oro* (Mountains of gold) (1897): from "A histeria" (To hysteria): "la selva dolorosa cuyos gajos echaban sangre al / golpe de las hachas, como los miembros de un molusco extraño" ["branches cut from the jungle, pained, bleeding, and /

left looking like a strange mollusk after the work of the hatchet was done"];[64] and from "La rima de los ayes" (A rhyme of sighs):

> Que allí ruge una
> mar de ondas acerbas que enturbian los asfaltos y
> las naftas, y que en ella las almas desembocan los
> tristes sedimentos de sus llagas.
> (Lugones, *Montañas*, 24)
>
> [There roars a sea
> of fetid waves polluted by asphalt and
> gasoline, and where sad souls unleash
> the detritus of their wounds.]

In these poems the ugliness and destruction of bleeding axe marks on trees and seas of asphalt and gasoline into which human souls leave their sediment are examples of particularly unsettling imagery that strongly evokes ecological reactions in readers. This is how, in the analysis of Laurence Buell, acts of environmental engagement "may affect one's caring for the physical world" by making it feel more "endangered or disposable":[65] bleeding axe marks and seas of asphalt and gasoline tend to make whatever is exposed to such things seem more "endangered or disposable."

In a second collection of poetry, *El libro de los paisajes* (A book of landscapes) (1917), Lugones continues the theme. In "Mapamundi" (World map) for example, the poetry features the representation of a kind of cataclysmic nature that is as violent and raw as the moments from *Montañas de oro* cited above:

> La luna hunde entre escombros insensatos,
> Tras los mares de mármol su agonía.
> En constricción de enormes pugilatos,
> Raya los lurtes acerada estría,
> Cuajando largas venas de agua umbría
> Inyectadas de lívidos sulfatos.[66]
>
> [The moon sinks into unfeeling ashes,
> Behind seas of marble hides its agony.
> Compressed in enormous caches,
> With avalanches of steel in grooved harmony,
> Injected from the veins of a metal anatomy
> Dark waters curdle under livid sulfur splashes.]

These images of steel, sulfur, agony, and ruin recall scientific experiments and scenes of apocalyptic destruction that are defining characteristics of the collections of short stories, *Las fuerzas extrañas* (*Strange Forces*) (1906) and *Los cuentos fatales* (Fatal stories) (1926); in these collections, Lugones explores spiritism, alchemy, insanity, mythological apocalypse, and the limits, dangers, and failures of scientific exploration. Such themes were not far removed in the writing from the reality of a hard-charging development in Buenos Aires at the turn of the century where, with the wealth generated from the prosperous cattle industry, the city began a construction boom completing subway lines, theaters, race tracks, most of the buildings that still stand in the city's downtown, and housing, both grand and humble, for the huge immigrant population drawn to the city by its economic prosperity. The wrenching experience of explosive growth and striking demographic alterations had Lugones (and other *modernista* writers) disturbed by the way in which the culture was changing.

As I mention in the first few paragraphs from the introduction to this chapter, these and other political concerns led Lugones to become an advocate for fascist political theory, opposed to liberal programs of development, suffrage, and immigration.[67] Thus, while many of his stories can be read from an ecocritical perspective for the ecological devastation occasioned by the hyper-industrialization of Buenos Aires, it can be difficult to reconcile the nascent environmental sensibilities in Lugones's work with the tragedy of fascism from the first half of the twentieth century. To countenance Nazism in any form is troubling, and even more so here because it means that the argument for ecological sensitivities inferred from these short story collections means constructing an ecocritical account of Lugones's work that must originate with the same ideas that eventually lead the Argentine nation toward Nazi ideology.

An example of this ambivalence comes in "El espejo negro" (The black mirror) where a learned doctor has discovered that highly polished carbon can provoke visions that vary according to the state of mind of the one who stares into it. The doctor has fashioned a kind of black mirror, but the narrator, when he stares into the reflection, contemplates "un mundo de generaciones ardiendo abajo" ["a world of burning generations down below"].[68] A similar apocalyptic story—"La lluvia de fuego" (Rain of fire)—narrates the destruction of the mythical city of Gomorrah as the product of a storm that rains molten copper upon humanity. The description of the city's destruction invokes ecological imagery that would not seem out of place as a description of the ultimate apocalypse for any highly industrialized and highly contaminated city of the present:

> Esa tarde y toda la noche fue horrendo el espectáculo de la ciudad. . . . El derrumbe de los edificios, la combustión de tantas mercancías y efectos diversos, y más que todo la quemazón de tantos cuerpos acabaron por agregar al cataclismo el tormento de su hedor infernal. Al declinar el sol, el aire estaba casi negro de humo y polvaredas. . . . Empezó a soplar un viento ardentísimo, denso, como alquitrán caliente. Parecía que se estuviese en un inmenso horno sombrío. Cielo, tierra, aire: todo acababa. (Lugones, *Espejo negro*, 133–34)

> [That afternoon and night, the spectacle in the city was unbearable. . . . The collapse of buildings, the inferno of goods and other products, and more than anything the hellish stench of so many burned corpses made the cataclysm a true torment. As the sun set, the air was nearly black with smoke and dust. . . . Then a dense, burning wind began to blow like hot tar. It was like being in a huge, dark oven. Earth, air, heaven: all coming to an end.]

Images of flaming destruction may invite criticism for the support of a ideology of racial purity: visions of the immolation of previous generations can be understood as a reference to the disappearance of some kind of an idealized, pure "Original" while the moral failings of Sodomites are held up for divine punishment. Lugones's later political writings certainly make such associations more credible. Yet our present fears about an ever-warmer future climate can also recast such images as ecological catastrophe: the representation of unbearable heat, air polluted with tar, and the fiery destruction of Sodom and Gomorrah has much in common with concerns about warming, pollution, and a coming ecological cataclysm that characterize present environmentalist advocacy.

In one other story by Lugones, "Viola acherontia," a similarly ambivalent interpretation presents itself. The story deals with a botanist who is attempting to breed the flower of death. He succeeds in developing a strain of black violets by exposing them to the pollen of carnivorous and narcotic plants and by recreating in front of them "escenas crueles que las impresionaran más" ["scenes of cruelty so as to make more of a profound effect"]. The violets can be heard to audibly produce sighs of distress and moans "muy semejantes a los de un niño" ["similar to those that might be made by a child"].[69] In the end, the narrator is disgusted by the achievements of the botanist:

> Pero mi convicción de ahora es que se trata de un verdadero bandido, de un perfecto hechicero de otros tiempos, con sus venenos y sus flores de crimen. ¿Llegará a producir la violeta mortífera que se propone? ¿Debo entregar su nombre maldito a la publicidad? (Lugones, *Espejo negro*, 84)

> [But I am convinced now that with his poisons and criminal flowers, he was a true miscreant, an all-too-real wizard from another time. Will he be able to produce the flower of death that he has envisioned? Should I expose him for the danger that he represents to society?]

In the context of the rest of Lugones's stories, the fearful reflections by the narrator at the end of the story are familiar enough. And in light of Lugones's later political development, the genetic manipulation of species takes on somewhat disturbing undertones after the horrors of Nazi ideas about purity, ethnicity, and the terrible experiment in eugenics that was the Holocaust. But from an ecological perspective, issues relevant to the current debate about the manipulation of nature are also broached in "Viola acherontia." While the moaning violets are a lot more dramatically diabolical, the revulsion expressed by the narrator at the close of the story represents a fairly accurate description of environmentalist aversion toward so-called "scientific" manipulations of nature: cloned reproduction, genetic engineering, agro-industrial mono-culture, etc.

It might have been more convenient to simply ignore the ecological elements of Lugones's texts and leave him out of the discussion entirely; his late political theory makes for difficult conclusions about his early work. If Lugones embraced fascist thinking after the texts considered here, then the seed of that thinking may be found in those same stories and similarly stain any ecological conclusions that may be garnered therefrom. Such a line of reasoning may seem plausible given the way in which environmentalism as a movement has sometimes strayed toward militant (if not necessarily fascist) models. But that is precisely why Lugones is in: I believe a critical engagement with potentially problematic ecocritical approaches must also be a concern for this project and its aim to assemble a history of ecology and environmentalism in Spanish American literature. As Greg Garrard documents in a broader context, "The Nazis not only appealed to small farmers and georgic philosophers, but also to conservationists, enacting the world's first comprehensive nature conservation and animal welfare laws."[70] And even earlier than that, Richard Grove finds that "in the second half of the nineteenth century, . . . forest conservation and associated forced resettlement methods were frequently the cause of a fierce oppression of indigenous peoples and became a highly convenient form of social control."[71] Thus, it is to be expected that the development of an ecological sensibility in Latin America must be comprised of disparate elements, no matter how distasteful some of them may be. In fact, I am especially keen to document early literary manifestations of autocratic or authoritarian environmental positions because, as shall be seen in chapter 10, it becomes a central concern for some recent environmentalist fictions that challenge seemingly sacrosanct principles and advocate for the development of local, not globalized, models of conservationism, environmental justice, and political ecology.

Finally, the development of the *modernista* oeuvre led to the emergence of a subsequent literary sub-category described as *decadentismo*; Gabriela Mora approximates a definition for works of this stripe by making reference to "los últimos años del imperio romano para aludir a declinación, fin, muerte próxima" ["the last years of the Roman empire to allude to decline, an end, and near death"].[72] Thus, these writers, especially Clemente Palma, son of noted nineteenth century author Ricardo Palma, explore the final boundaries within which *modernista* literary expression would range: an interest in spiritualism and the occult, a lifestyle of conspicuous consumption, and the ingestion of narcotics and other hallucinatory drugs.[73] All of it seemed to take the *modernista* dissatisfaction with the progress of modern living to the absolute extremes of rejection. Yet Clemente Palma's texts find a place here for the way in which they contain an imagery of extermination similar to the expression of concern about species extinction so common in current environmentalist thinking. A good example of these tendencies in his work comes from the story "Leyendas de haschischs" ("Hashish legends") from *Cuentos malévolos* (Malevolent stories) (1904, 1913) where the narrator records impressions from his drug-fueled trips. One of the visions is a nightmare of plant and animal extinction:

> Por todas partes se veían las enmarañadas copas de árboles extraños, cuyos troncos estaban llenos de pústulas. . . . Las flores, apenas abiertas, caían moribundas sobre el césped raquítico y gris; sus pétalos ardían en violenta fiebre, y sus estambres se estremecían y retorcían en las convulsiones de intenso dolor.[74]

> [The tangled trees, whose trunks were filled with pustules and whose tops were all so strange, could be seen all around. . . . The hardly-opened flowers, fell dead upon the withered, gray grass; their petals seemed to burn in a kind of violent fever, their stamens shook and writhed in intensely painful convulsions.]

The theme of natural extinction and disappearance continues in "La última rubia" (The last blonde woman) where a narrator seeks to find his distant cousin, the last blond-haired woman. Set in the year 3025, the world faces ecological crisis in the form of complete exhaustion of gold as a resource: "Era tal la escasez de este mineral que solo uno que otro erudito tenía noticias de que hubiera existido" ["The mineral was so scarce that it was rare to find even well-educated people who remembered that it had ever existed"]. To reformulate the metal, alchemists have hit on a formula that requires human hair from blonde females, but inter-racial sexual coupling has left the planet dominated by "los mongoles y los tártaros, esas malditas razas amarillas, habían inundado el mundo y malogrado las razas europeas y americanas" ["Mongols and Tartars, these cursed yellow races had overrun the world and polluted the European and American races"].[75] The implicit racism of these lines is troubling and makes it difficult to see anything else here: Palma's work seems to express the influence of Lugones's philo-fascist sentiments. But the idea that certain "resources" could be exhausted—gold, blond-haired women—allows for the observation that, as early as the publication of Clemente Palma's work in the first decades of the twentieth century, the threat of resource scarcity due to over-consumption was again being imagined in the literature as it had in the nineteenth century with orange trees and honeysuckle in Argentine Vicente F. López's *La novia del hereje* (The heretic's lover) (1854); with river fisheries in Uruguayan Eduardo Acevedo Díaz's *Nativa* (Native) (1890); with forests in Ecuadoran Luis Martínez's novel *A la costa*

(To the coast) (1904), and with guano in *Los amigos de Elena* (Elena's friends) (1874) by Peruvian Fernando Casós.

The *modernista* writers of the late nineteenth and early twentieth century represent a highly imaginative precursor to the critical regional and *indigenista* works that would, with few exceptions, characterize Spanish American literature until the "Boom." While the *modernista* aversion to the culture and values of the modern world took the form of seemingly escapist/fantastical (Darío), apocalyptic/extremist (Lugones), or narcotic (Palma) narrative, the political ideology implicit in these rejections of modernization prefigure many of the criticisms that were central to the regional and *indigenista* novels that I consider in part 2. The industrialization of the coasts, mountains, plains, and jungles of Latin America was often represented in the regional novels with an unblinking awareness of the false myth of resource inexhaustibility. The large-scale agricultural, mining, and extractive industries, and the infrastructure construction that developed to support them are represented in *indigenista* literature with particular attention for the injustice such activities would visit upon indigenous populations. In both cases, the criticism comes from similar critical perspectives toward the modern world as that expressed by the *modernistas*. In the second part of this book, I explore how such developments have had a deleterious effect on lands and indigenous populations and how this is represented in Spanish American regional and *indigenista* literature.

NOTES

1. Jennifer L. French, *Nature, Neo-Colonialism, and the Spanish American Regional Writers* (Hanover, NH: Dartmouth University Press, 2005), 13.
2. Michael Lowy and Robert Sayre, *Romanticism Against the Tide of Modernity* (Durham, NC: Duke University Press, 2001), 17.
3. Rubén Darío, *Prosas profanas* (Madrid: Editorial Mundo Latino, 1917), 9.
4. Luis Vitale, *Hacia una historia del ambiente en America Latina* (Mexico City: Editorial Nueva Imagen, 1983), 79–80.
5. Fermín Adrián Rodríguez, "Nacionalidad forzada. Leopoldo Lugones, entre la tradición del espíritu y la herencia del monstruo," *Revista Iberoamericana* 75, no. 227 (April–June 2009): 389.
6. Mariana Amato, "Force Majeure: Leopoldo Lugones toward a Vitalist Fascism," *Dissidences: Hispanic Journal of Theory and Criticism* 3, no. 6 (Spring 2010): 3.
7. Shawn William Miller, *An Environmental History of Latin America* (New York: Cambridge University Press, 2007), 207.
8. Lowy and Sayre, *Romanticism*, 6.
9. José Martí, *Nuestra América* (Caracas: Biblioteca Ayacucho, 1977), 63.
10. José Enrique Rodó, *Ariel* (Buenos Aires: Kapelusz, 1966), 35, 36.
11. Ibid., 47–48.
12. Diego Alonso, "José Enrique Rodó: una retórica para la democracia," *Revista canadiense de estudios hispánicos* 34, no. 2 (Winter 2001): 187.
13. Cathy L. Jrade, *Modernismo, Modernity, and the Development of Spanish American Literature* (Austin: University of Texas Press, 1998), 2.
14. Vitale, *Hacia una historia*, 78–80.

15. Rubén Darío, *El modernismo y otros ensayos* (Madrid: Alianza, 1989), 161, 165.
16. Darío, *Prosas profanas*, 43.
17. Rubén Darío, *Azúl* (San José, Costa Rica: Editorial Universitaria Centroamericana, 1981), 50.
18. Ibid., 14.
19. Darío, *Modernismo,* 61.
20. Rubén Darío, *Obra poética* (Madrid: Biblioteca Corona, 1914–1916), 147.
21. William Cronon, ed., *Uncommon Ground: Rethinking the Human Place in Nature* (New York: Norton, 1995). See chap. 1, note 51.
22. Darío, *Modernismo,* 168.
23. Beatriz González-Stefan, "Martí y la alta modernidad," in *Escribiendo la Independencia: Perspectivas postcoloniales sobre la literatura hispanoamericana del siglo XIX,* ed. Robert Folger and Stephan Leopold (Madrid: Iberoamericana, 2010), 343–71. In her article, González-Stefan identifies a few moments from Martí's journalistic writing that seem to indicate his ambivalent fascination with modernization, particularly the use of electricity.
24. José Martí, *Ismaelillo. Versos sencillos. Versos libres* (Habana: Impr. De Rambla, Bouza, 1913), 125.
25. José Martí, *Versos de José Martí* (Buenos Aires: Ediciones Mínimas, 1919), 46.
26. Guillermo Castro Herrera, "Naturaleza, sociedades y culturas en José Martí," *Cuadernos Americanos*, 2nd series, 51 (May-June 1995): 95–96.
27. Kay Milton, *Loving Nature* (New York: Routledge, 2002), 105, 101.
28. Julio Herrera y Reissig, *Los éxtasis de la montaña, eglogánimas* (Buenos Aires: Ediciones Selectas América, 1920), 13, 34.
29. Jrade, *Modernismo*, 131–32.
30. Julio Herrera y Reissig, *Las pascuas del tiempo (poesías y otros poemas)* (Montevideo: O. M. Bertani, 1913), 113.
31. Herrera y Reissig, *Los éxtasis*, 38.
32. Milton, *Loving Nature*, 75.
33. Karen J. Warren, *Ecofeminist Philosophy: A Western Perspective on What It Is and Why It Matters* (Lanham, MD: Rowman & Littlefield Publishers Inc., 2000), 103–04.
34. Amado Nervo, "Éxtasis," in *Antología poética hispano-americana; con notas biográficas y críticas*, ed. Calixto Oyuela (Buenos Aires: A. Estrada y cía., 1919), 163.
35. Amado Nervo, "La nieve misteriosa de la montaña," in *Los cien mejores poemas de Amado Nervo*, ed. Enrique González Martínez (Mexico: Tip. Murguia, 1920), 183.
36. Manuel José Othón, *Poemas escogidos; selección de Agustín Loera y Chávez* (Mexico City: Impr. Victoria, 1917), 15.
37. Luis G. Urbina, *Puestas de sol* (Paris: Librería de la Vda de Ch. Bouret, 1910), 75.
38. Ricardo Jaimes Freyre, *Los sueños son vida. Anadiomena. Las víctimas* (Buenos Aires: "Buenos Aires" Sociedad Cooperativa Editorial Limitada, 1917), 107–8.
39. Eugenio Chang-Rodríguez, *La literatura política de González Prada, Mariátegui y Haya de la Torre* (Mexico City: Ediciones de Andrea, 1957), 62.
40. Ibid., 89.
41. Juan Guillermo Gómez García, "La significación continental de Manuel González Prada: sobre la génesis del anarquismo en Hispanoamérica," *Estudios Políticos* 37 (July-December 2010): 226.
42. Manuel González Prada, *Exóticas* (Lima: Tipografía De "El Lucero," 1911), 106.
43. Manuel González Prada, *Antologia poética* (Mexico City: Editorial Cultura, 1940), 37, 180. The images are from *Minúsculas* (*Lower-cased*) (1901) and *Trozos de vida* (Pieces of life) (1933), poem 12.
44. González Prada, *Exóticas*, 59.
45. Jrade, *Modernismo*, 1.
46. Rolando Pérez, "Irony, Love, and Political Economy in José Asunción Silva's *De sobremesa*," *Hispanófila* 150 (May 2007): 88.
47. José Asunción Silva, *Obras completas* (Caracas: Ediciones Ayacucho, 1977), 132, 163, 165, 305.
48. Pérez, "Irony, Love, and Political Economy," 95.

49. Silva, *Obras completas*, 195, 261. There are also references to the loss of nature in the face of an ever more destructive pattern of development in poetry by José Clemente Silva; the most explicit comes from the poem "Obra humana" (Human work) that decries the fact that a bucolic scene, "En lo profundo de la selva añosa" ["In the deepest, oldest part of the jungle"], is supplanted by "el paso de la audaz locomotora, / Que en el carril durísimo cruzaba" ["the crossing of the haughty locomotive / Passing over the hard rails of its route"]. See *Los mejores poemas de José Asunción Silva* (Mexico City: Cultura, 1917), 25.

50. Silva, *Obras completas*, 264–65.

51. Manuel Díaz Rodríguez, *Narrativa y ensayo* (Caracas: Ayacucho, 1982), 46.

52. Ibid., 60.

53. Eduardo García Piedrahita, "Dos novelas colombianas olvidadas," *Boletin Cultural y Bibliográfico* 17(1980): 217–28; James J. Alstrum, "*La Diana cazadora* de Clímaco Soto Borda: Un discurso disonante del fin de siglo," *Estudios de Literatura Colombiana* 5 (July–December 1999): 9–22.

54. Clímaco Soto Borda, *Diana cazadora* (Bogotá: Editorial ABC, 1942), 9.

55. Manuel Latorre, *Zurzulita* (Santiago de Chile: Editorial Chilena, 1920), 13.

56. Ibid., 180.

57. Jonathan Tittler, "Una relectura del canon *criollista*," *Tabula Rasa* 7 (July-December, 2007): 206–5.

58. Gabriela Mora, *El cuento modernista hispanoamericano* (Lima: Latinoamericana editores, 1996), 22, 23, 32.

59. Ibid., 24.

60. Manuel Gutiérrez Najera, *Cuentos y cuaresmas del duque Job* (Mexico City: Porrua, 1966), 6.

61. Ibid., 58.

62. Ibid., 60.

63. Jrade, *Modernismo*, 118.

64. Leopoldo Lugones, *Montañas de oro* (Montevideo: Editorial Rioplatense, 1919), 18.

65. Lawrence Buell, *Writing for an Endangered World* (Cambridge, MA: Harvard University Press, 2001), 2.

66. Leopoldo Lugones, *El libro de paisajes* (Buenos Aires: Otero y García, 1917), 20.

67. For a political analysis of Lugones's ideas, see Alberto Spektorowski, "The making of an Argentine fascist. Leopoldo Lugones: from revolutionary left to radical nationalism," *History of Political Thought* 17, no. 1 (1996): 79–108.

68. Leopoldo Lugones, *El espejo negro y otros cuentos* (Buenos Aires: Editorial Abril, 1988), 59.

69. Ibid., 83–84.

70. Greg Garrard, *Ecocriticism* (New York: Routledge, 2004), 112.

71. Richard Grove, *Green Imperialism* (New York: Cambridge University Press, 1995), 12.

72. Mora, *El cuento modernista hispanoamericano*, 143.

73. For a more detailed analysis and definition of *decadentismo*, see Ibid., 141–200.

74. Clemente Palma, *Cuentos malévolos* (Barcelona: Imp. Salvat y Compañía, 1904), 152.

75. Ibid., 91, 94.

Part Two

Land, People, Ecology

Chapter Four

Swallowed: Environmentalism and the Spanish American *novela de la selva*

In Chilean Luis Sepúlveda's *Un viejo que leía novelas de amor* (*The Old Man who Read Love Stories*) (1989), the narrator describes how in Ecuador, gold panners waited for the end of the rainy season before setting off to look for more of that precious metal. Every year several would set out early and some would disappear, "quién sabe si tragados por la corriente o por la voracidad de la selva" ["maybe swallowed up by the river or the voracious jungle; no one knows"].[1] The "devouring" image of *selva* (jungle or rainforest) is memorable for the closing line of Colombian José Eustasio Rivera's *La vorágine* (*The Vortex*) (1924). In Rivera's text, Arturo Cova and companions disappear into the dangerous jungle; the cable from the consul declares: "Los devoró la selva" ["The jungle has swallowed them up"].[2] The intertextuality of the image suggests a consideration of the similarities between the two texts. First, *La vorágine,* one of the earlier examples of the *novela de la selva*[3] genre, and *Un viejo que leía novelas de amor,* one of the more recent, feature the familiar romantic representation of the educated man or urban dweller who leaves the city to engage in what Jorge Marcone has characterized as a "return to nature."[4] Yet, despite these similarities, the two novels differ markedly in each one's representation of the ecological ideas of their time: in *La vorágine* opposition to the human injustice of rubber extraction is primary, with concern about deforestation an insignificant element of the text's political ecology; in Sepúlveda's novel, however, the language of environmentalism is used explicitly. The difference is what will inform the central question of this chapter: what precedent existed in previous *novelas de la selva* for the environmentalist rhetoric that emerges in Sepúlveda's novel?

I begin with the assumption that there are certain fixtures within the "canon" of the *novela de la selva:* the short stories of Uruguayan Horacio

Quiroga, *La vorágine*, *Canaima* (1935) by Venezuelan Rómulo Gallegos, *La serpiente de oro* (*The Golden Serpent*) (1935) by Peruvian Ciro Alegría, *Los pasos perdidos* (*The Lost Steps*) (1953) by Cuban Alejo Carpentier, and *La casa verde* (*The Green House*) (1966) by Peruvian Mario Vargas Llosa. These are the most-often cited in scholarly analyses of the genre and typically contain expressions of political ecology specific to their temporal and political context. However, several other texts, contemporary to those in the list above but neglected in the "canon," take political ecology in different directions and even pre-figure aspects of current environmentalist principles: deep ecology, conservationism, sustainable development, and notions about the rights of nonhuman animals. Among these are *Toá* (1933) and *Andágueda* (1947) by Colombians César Uribe Piedrahita and Jesús Botero Restrepo, *Una mujer en la selva* (A woman in the jungle) (1936) by Nicaraguan Hernán Robleto, *Su nombre era muerte* (His name was death) (1947) by Mexican Rafael Bernal, the trilogy comprised of *Sangama* (1942), *Selva trágica* (Tragic jungle) (1956), and *Bubinzana* (1960) by Peruvian Arturo D. Hernández, *El mensú que triunfó en la selva* (The peasant who triumphed in the jungle) (1951) by Argentine Valentín Barrios, and *Llanura, soledad y viento* (Plains, solitude, and wind) (1960) by Colombian Manuel González Martínez. I call attention to these seemingly forgotten titles not simply to expand the canon, but to demonstrate that they express many of the aspects of environmentalism eventually articulated in Sepúlveda's novel.

LITERARY ENVIRONMENTALISM: LUIS SEPÚLVEDA'S *UN VIEJO QUE LEÍA NOVELAS DE AMOR*

Several moments in Sepúlveda's novel articulate an environmentalist ideology from the perspective of the Spanish American *novela de la selva*. Even before the novel opens, Sepúlveda explicitly frames the environmentalism of the narrative; in the Tusquets edition of *Un viejo que leía novelas de amor* (1993), he notes that Chico Mendes was killed only a couple of days before the novel was awarded the Tigre Juan prize, but that the prize would belong to his memory and to "todos los que continuarán tu camino, nuestro camino colectivo en defensa de éste, el único mundo que tenemos" ["all those who will follow your path, our collective path in defense of this, the only world we have"].[5] In the novel proper, the narrator characterizes the interaction in the jungle between nature and its non-indigenous inhabitants as a series of errors:

> Tanto los colonos como los buscadores de oro cometían toda clase de errores estúpidos en la selva. La depredaban sin consideración, y esto conseguía que algunas bestias se volvieran feroces. . . . Y estaban también los gringos venidos desde las instalaciones petroleras.

> Llegaban en grupos bulliciosos portando armas suficientes para equipar a un batallón, y se lanzaban monte adentro dispuestos a acabar con todo lo que se moviera. Se ensañaban con los tigrillos sin diferenciar crías o hembras preñadas, y, más tarde, antes de largarse, se fotografiaban junto a las docenas de pieles estacadas. (Sepúlveda, *Un viejo*, 59–60)

> [The settlers along with the gold-seekers made all kinds of stupid errors in the jungle. They sacked it without a second thought and as a result, some animals turned fierce. . . . Then there were the *gringos* come down from the oil installations.
>
> They arrived in noisy groups carrying enough guns to arm a battalion and they would venture up the mountain ready to shoot anything that moved. They would vent their rage at ocelots, young and pregnant females included, and then, before leaving, they would take pictures of themselves holding up dozens of bullet-ridden skins.]

These errors are contrasted to the wisdom that the protagonist José Antonio Proaño acquires living among the Shuar in Amazonian Ecuador. He learns their customs concerning hunting: not killing "un cachorro, ni de tigrillo ni de otra especie. Sólo ejemplares adultos, como indica la ley shuar" ["neither cubs of ocelots nor any other species. Only adults, as dictated by Shuar law"].[6] The principles of what is called "Shuar law" are characterized by a way of living that rejects excessive use of resources and deforestation in favor of a lifestyle of conservation and sustainable consumption: "Los shuar no cazan tigrillos. La carne no es comestible y la piel de uno sólo alcanza para hacer cientos de adornos que duran generaciones" ["The Shuar do not hunt ocelots. Their meat is inedible and the skin of just one is enough to make hundreds of adornments that will last for generations"]. The natural setting is written as a limitless supplier of nutrition and a restorer of vitality: "Comía en cuanto sentía hambre. Seleccionaba los frutos más sabrosos, rechazaba ciertos peces por parecerle lentos" ["He ate when hungry. He chose the tastiest fruits, didn't eat certain fish because they seemed overly slow to him"]; its healthful effect on Proaño is described in the following terms: "La vida en la selva templó cada detalle de su cuerpo. Adquirió músculos felinos que con el paso de los años se volvieron correosos" ["Life in the jungle tempered his body. He acquired cat-like musculature and became remarkably flexible over the course of the years"].[7] The idealized vision of the jungle or rainforest as a nourishing and invigorating place complements the critique of development in the novel and many moments in Sepúlveda's novel describe several of the most central tenets of environmentalism: conservation, sustainability, condemnation of deforestation, the idealization of nature, the wisdom of indigenous groups.

Kay Milton's definition of environmentalism as the "concern that the environment should be protected, particularly from the harmful effects of human activities" and as the emulation of non-industrial societies, perceived as "models for a sustainable or conserver society" that maintain "spiritual ties"[8] to their land are at the heart of the ideology of *Un viejo que leía novelas de amor*. In the novel, harmful human activities include those of *gringo* hunters and others that have settled the jungle to search for gold. But the Shuar community's knowledge of the land is held up as the preferred non-commercial model, the one followed by Proaño, and one that hews closely to the precepts of deep ecology. His transformation from a character who first wants to destroy the natural world—"quería vengarse de aquella región maldita, de ese infierno verde. . . . Soñaba con un gran fuego convirtiendo la amazonía entera en una pira" ["The old man wanted to take revenge upon that accursed region, upon that green Hell that stole his love and his dreams. . . . He dreamed of seeing the whole Amazon region converted into a pyre"]—to one who identifies with it to such an extent that he curses "todos los que emputecían la virginidad de su amazonía" ["all those who pimped out his virginal Amazon"].[9] The use of the possessive effectively communicates Proaño's total identification with the *selva*; he thinks of it as "*su* amazonía." Each of these elements from the novel illustrates the way in which the environmentalism in *Un viejo que leía novelas de amor* represents an explicit element. As shall be seen in the following pages, Sepúlveda's text represents something of a culmination of the environmentalist tendencies in the *novela de la selva.*

THE FIRST TO THE CANON: THE LEGACY OF ROMANTICISM IN THE *NOVELA DE LA SELVA*

In *La novela de la selva hispanoamericana*, Lydia León de Hazera begins her survey with an analysis of the romantic tendency to personify nature in what are considered two antecedents of the tradition: *María* (1867) by Colombian Jorge Isaacs and *Cumandá* (1879) by Ecuadoran Juan León Mera; both are also foundational nineteenth century texts to which I have dedicated some space in chapter 2. In *María* the land ambivalently embodies the protagonists' attitudes and emotions while in *Cumandá* the jungle is idealized as helpful friend or mother. In Isaacs's novel, as the narrator falls in love with María, the forests, plains, and rivers are metaphors for her beauty, but in the closing chapters of the book, when they are apart and María is dying, nature is transformed into a cruel obstacle. In Mera's text, the unfailingly idealized forests facilitate the protagonist Cumandá's escape from her pursuers: "Toda la naturaleza convida a acompañarla en sus magníficas armonías matinales. Hay gratísima frescura en el ambiente, dulces susurros en las hojas, suave fragancia en las flores" ["All of nature conspires to accompany her in those magnificent morning melodies. There is a wonderful refreshment to the environment, sweet sighs in the leaves, a soft fragrance to the flowers"]. Cumandá and her indigenous kin are called "hijos de las selvas" and "hijos del desierto" ["children of the jungle and of nature"].[10] These manifestations of the romantic idealization of nature led Hazera to describe *María* as "la manifestación más lograda del romanticismo hispanoamericano" ["the most well-effected manifestation of Hispano-American romanticism"], while regarding *Cumandá,* she comments that "en los párrafos dedicados a la descripción de la selva, Mera persiste en su interpretación romántico-católica" characterized by "una actitud idealizante hacia la naturaleza a lo largo de la novela" ["in the paragraphs dedicated to the jungle, Mera persists in a Catholic/romantic interpretation [characterized] throughout the novel by an idealizing attitude toward nature"].[11] The proto-environmentalist ideas of romantic writers are well-documented[12] and eventually, the twentieth century *novela de la selva* that evolved from these early Spanish American romantic representations of nature similarly develops environmentalist tendencies, but only through a literary process in which the genre alternately diverges from, but also readily returns to its origins in European romanticism.

The divergence can be traced to the fact that, as Hazera observes: "lo que modifica el aparato romántico es la naturaleza americana, de cuyas voces autóctonas y de cuyo realismo nunca se han despegado del todo los autores hispanoamericanos" ["that which modifies these romantic tendencies is Latin American nature, tendencies toward native voices and characterized by a kind of literary realism that Spanish American authors never really got away from"].[13] Though influenced by the idea of nature as an escape from overwhelming industrialization common to the writings of European romantics, certain ecological "facts on the ground" also conditioned Spanish American writing about the jungle. Thus the *novela de la selva* both maintains but often subsequently rejects or modifies canonical elements of romanticism: its metaphors of nature, its representation of the awe of human interactions with sublime nature, and especially, as Raymond Williams put it in *The Country and the City,* "a feeling for unaltered nature, for wild land."[14] As was seen in chapters 1–3, the influence of the European romantics on nineteenth and early twentieth century Spanish American writers was considerable, but the examples of the *novela de la selva* that emerge in the twentieth century articulate uniquely Spanish American political ecologies and even proto-environmentalist ideas: radical adaptation to nature as in deep ecology, the value of both human and non-human nature, the emulation of non-industrial societies, etc.

With regard to the actual environmental contexts within Latin America concurrent to the development of the *novela de la selva* beginning in the nineteenth century through to Sepúlveda in the 1990s, the ecological elements that were at play for the texts considered in the first three chapters only become more urgent as the twentieth century wore on. Urban migrations beget the explosive growth of cities with serious demographic and geographic consequences: forests were now being cut down not only to support agricultural development but to make room for cities. In 1983, Luis Vitale calculated that "la deforestación continúa a un ritmo galopante en América Latina: entre 5 y 10 millones de hectáreas anuales" ["deforestation continues at an alarming rate in Latin America: from 5 to 10 hectares annually"].[15] By the second decade of the twenty-first century, that number has not much changed, despite the growing awareness of environmental degradation.[16] Vitale also emphasizes the contamination of air and water caused by the growth of cities:

> Según un estudio de la CEPAL [Comisión Económico para América Latina], en las ciudades de América Latina sólo las aguas servidas de 5.7 millones de habitantes no generaban contaminación. Otros 55.6 millones de habitantes urbanos tampoco generaban contaminación 'hídrica' pues sencillamente carecían de alcantarillado. Y unos 51.7 millones disponían de alcantarillado pero sus aguas servidas constituían fuente de contaminación de ríos, lagos o mares por no ser tratadas.[17]

> [According to a study by CEPAL [the Economic Commission of Latin America], only 5.7 million urban residents in Latin American cities utilize water that does not eventually cause environmental contamination. Another 55.6 million inhabitants also do not contribute to the direct pollution of waterways for the simple fact that their neighborhoods simply do not have sewers. But another 51.7 million people who do have sewers generate contaminated water that is deposited directly and untreated into nearby rivers, lakes, and oceans.]

Finally, several of the export commodities upon which the economies of the region relied were either dwindling, exhausted, or responsible for widespread environmental harm. Rubber extraction, so central to Rivera's *La vorágine*, had by the early twentieth century been farmed out to British possessions in Malaysia. And while the automobile and bicycle industries in Europe and North America powered the demand for rubber, it was growth in agro-industry throughout the world that did the same for the miracle-fertilizer guano. The only problem was that guano had been extracted to unsustainable levels and ceased to function as an economic stimulus by 1906, which marked the occasion of the closing of South Chincha Island off the coast of Peru by that country's government.[18] Similarly, the discovery of oil in Mexico, Venezuela, and Ecuador had been an economic boon during the middle decades of the twentieth century, and while Venezuelan reserves are likely the largest in the world, the supply of Mexican oil has more recently begun to run out. In 2010, New York Times reporters covering the political difficulties in Mexico for seeking foreign companies to extract oil from deep-water sites in the Gulf write that "Mexico's readily accessible oil is used up."[19] And elsewhere in

Latin America, the oil company Chevron, parent company of well-operator Texaco, continues a legal battle to defend itself against multiple suits seeking damages for the huge volume of toxic water that contaminated the ground around the Lago Agrio oil field in the Ecuadoran Amazon. These points of environmental conflict are representative of the ecological state of affairs in twentieth century Latin America; and no area was more profoundly affected than the Amazon river basin and its surrounding forests, a region that includes parts of every South American nation except for Chile and Uruguay. The jungles of Central America were similarly afflicted with severe environmental degradation, targeted as they were for the same kinds of development that was undertaken in various parts of the Amazon.

Therefore, it is not surprising that the literary development of the *novela de la selva* should mature and develop into explicitly environmentalist forms as it did in *Un viejo que leía novelas de amor*. The pervasive air and water pollution, deforestation, and exhaustion of natural resources in the Amazon had become an environmentalist *cause célèbre* by the time Sepúlveda's novel was published in 1989. However, the argument in what follows is that environmental awareness concerning these issues did not simply materialize from the memorable visits by *Police* frontman Sting and his founding of the Rainforest Foundation that same year. Rather, I should like to identify the ecological aspects of a Spanish American literary heritage prior to the publication of *Un viejo que leía novelas de amor* in several lesser-known titles in the tradition. And even canonical examples of the *novela de la selva* represent something quite different from their romantic forebears in that they confound the representation of nature as a place of refuge for those who wish to flee the seething crowds of over-industrialized cities; instead, these subsequent fictional works propagate political ecologies critical of neo-colonial political economics. But even what Raymond Williams called the romantic yearning for "unaltered nature" and "wild land" matures in the *novela de la selva* to become an ever more explicit ecological, even environmentalist, ideology, as much in several novels that fit the profile of the genre, but also in those that for one reason or another, no longer are or never have been considered canonical.

THE EARLY CANON: POLITICAL ECOLOGIES IN QUIROGA, *LA VORÁGINE, CANAIMA,* AND *LA SERPIENTE DE ORO*

Uruguayan Horacio Quiroga's short stories feature representations of the jungle that, for the first time in the literature, have it construed, not as sublime refuge, but as consumable resource. Jennifer L. French has documented how, as manifest in Quiroga's work, the forces of British neo-colonialism in

Latin America were a primary cause for the problematization of people's relationship to the land. She indicates that "nature is represented in its productive capacity, specifically as the locus of a political contest among local workers, the national elite, and foreign capitalists."[20] These economic factors and negotiations for control and consumption of the forests are characteristic of conflicts that drive the plots of many novels in the established tradition of the *novela de la selva* genre. As a significant departure from the romantic literary models of European writers after Chateaubriand via Spanish American novels like *María* and *Cumandá*, French finds that Quiroga's ecologism "oscillates between careful attention to the natural world and the crucial fact that control of nature is almost always a question of social, political, and economic power."[21] An analysis of stories like "Las fieras cómplices" (Complicit beasts) (1908) and "Los mensú" (Peasants) (1914) from that perspective bears this out and scholars have also focused on the tension between nature and "social, political, and economic power" in Quiroga's work.[22]

I return to Rivera's *La vorágine* which was published just after most of Quiroga's stories first appeared and features a political ecology critical of the laxity of state bureaucracies concerning the exploitation of rubber tappers. French finds that Rivera's novel proceeds largely "in terms of the relationship between the law and the land, arguing that the government's faulty administration was visible in the landscape itself."[23] Others identify authority and land use, social injustice, and political economy as central elements of the novel's political ecology.[24] But while the plight of the workers is the issue with which the novel most explicitly concerns itself, the environmental damage of the rubber extraction industry is not always ignored. In one moment from the novel, the rubber tapper Clemente Silva expresses regret about the effects of the industry on a certain species of tree: "Los caucheros que hay en Colombia destruyen anualmente millones de árboles. En los territorios de Venezuela el *balatá* desapareció. De esta suerte ejercen el fraude contra las generaciones del porvenir" ["The rubber tappers in Colombia destroy millions of trees annually. In the territories of Venezuela the *balatá* has disappeared. In this way, they commit fraud against the generations of the future"].[25] French notes that the sentiment—something approaching an advocacy of sustainable development—represents a subtle but important anomaly: "*La vorágine* begins to articulate—tentatively, intermittently—a relationship to nature that is based on the desire to understand rather than to dominate and subdue."[26] Inevitably, the relocation of the rubber-producing *Hevea brasilensis* tree to plantations in Malaysia where the *Microcyclus ulei* fungus endemic to South America did not threaten the trees was the death knell for the South American rubber extraction industry early in the twentieth century: "By 1913, British plantations matched South America's entire productions. . . . Only six years later, Asian plantations out-produced the Amazon by a factor of seven."[27] *La vorágine*'s appeals to the state for reform of the latex industry were rendered moot by the British, French, and Dutch plantations in Asia. Even so, the political ecology in *La vorágine* is associated almost exclusively with statist entities, an ideology that characterizes most canonical examples in the *novela de la selva* tradition. But as I detail in the analysis below, there are anomalies within both the canonical tradition and a lesser-known counter-tradition that depart from this model of literary political ecology. I identify a small corpus of novels not characterized by advocacy for reform under the auspices of governmental agencies but by appeals to conservation for its own sake, for the love of nonhuman nature, and out of concern for the way in which the destruction of the jungle can have fearful consequences as much for human as nonhuman fauna.

Such an analysis can begin even among those titles that have long been considered canonical: there are many exceptional moments that depart from those political ecologies that are typical of the genre. For example, Rómulo Gallegos's *Canaima*, the coming-of-age story of protagonist Marcos Vargas's intense experience in the jungle, ultimately ends in failure. Such a fate is shared by many other of the protagonists from the *novela de la selva*: "the alternative modern life or the alternative path for modernization is defeated by natural forces and local factors or by the contradictions of modernity."[28] Marcone has commented on these defeats as the failure of the "return to nature" in several other novels including *La serpiente de oro*, a novel about the *balseros* (raft-pilots) of the Peruvian Upper Maranón river. But, Vargas's return features several episodes where the protagonist proves his "manhood" through gambling, duelling, working, and most significantly, by surviving the dangers of the jungle as a hurricane blew through. The experience is referenced several times in the concluding chapters of the novel as the definitive moment in the development of Vargas from boyish youth to fully realized man:

> Lo que Marcos Vargas traía de la selva no era para narrarlo. Sus experiencias de allá se habían fundido todas en la emoción de la tormenta, la más intensa y plena emoción de sí mismo que jamás había sentido y esto ni cabía en la memoria ni podía serle comunicado a otro.[29]
>
> [No story could do justice to what Marcos Vargas brought with him from the jungle. His experiences there had been forged from the emotions of torment, from the most intense and authentic experience of his own self that he had ever felt. Something like that could not be retained as a memory or communicated to someone else.]

As Marcone observes, texts that feature the failure of a "return to nature" invoke criticism of the political ecologies associated with present ideas about sustainable development. But as Vargas's intense experience construes the *selva* as a destination for proving one's full worth in a one-on-one encounter, such moments also express something other than political ecology, something more akin to the romantic ideal of nature as refuge: sublime moments like Marcos Vargas's experience in the hurricane or that pang of regret for the disappearance of the *balatá* in *La vorágine* or when the protagonists of a few of Quiroga's stories retreat to the *selva* to escape the bewildering excesses of modernization. The Uruguayan's "La miel silvestre" (Wild honey) (1912), for example, features a certain Gabriel Benincasa who in lieu of a bachelor party goes to spend a few days in the jungle "con su libertad como fuente de dicha y sus peligros como encanto" ["attracted by its danger and the happiness of its freedom"].[30] Although things do not turn out very well for Benincasa (he eats some fermented wild honey, passes out, and is devoured by a horde of ants), the liberating joy of jungle danger functions as a powerful magnet for human desire, even if the consequences of poorly planned environmental stewardship ultimately result in suffering or even death. Moments such as these are more a central motif in several other non-"canonical" *novelas de la selva*; some even feature returns to nature that do not end in failure; it is to their consideration that I now turn.

LOST TO THE CANON: POLITICAL ECOLOGY IN NEGLECTED *NOVELAS DE LA SELVA*

Up until this point in the literary history of the *novela de la selva*, scholars have largely concurred with regard to canon: *María* and *Cumandá*, then Quiroga, *La vorágine*, *Canaima*, *La serpiente de oro*, and the leap forward to *Los pasos perdidos* and *La casa verde*. However, several other novels, written contemporaneously with or in gaps between these better known titles contain alternatives to the political ecologies of the "canonical" texts, something more akin to the literary environmentalism of Sepúlveda's *Un viejo que leía novelas de amor*. The earliest of these novels but one that clearly represents an early example of current ecological ideas is *Andágueda* by Jesús Botero Restrepo. Specifically, Restrepo's text makes pessimistic reference to the way in which the trees, fauna and flora, and the general environment of the *selva* are being laid to waste. The images, while they do not dominate the narrative, are strongly evocative of the fears about climate change, species extinction, deforestation, and other concerns about ecosystem alteration that are so central to current dialogues within ecology and environmentalism. In the novel, there are warnings about deforestation and permanent loss of the jungle: "la selva, tierra nueva y antigua, su nuevo mundo que pronto se iba a extinguir" ["the jungle, both a new and old land, a new world that was to soon disappear"]; "esa era la selva moribunda que . . . estaba tan cambiada que si hubiese vuelto al tiempo no lo habría reconocido" ["this was like a dying jungle that . . . had been so altered such that even if it were possible to travel back in time it would not be recognizable"]; the remaining trees performed genuflections to the "genio extinto del bosque, allí bostezan los últimos árboles" ["spirit of the jungle, now extinct, where the last of the trees stood and yawned"].[31] While the novel does not make these concerns central to the narrative, it provocatively concludes with the third of the lines cited above and a reference to extinctions and the last of the trees: concern for the disappearance of trees and the destruction of the jungle are important enough that the Colombian author chooses to conclude his novel with a dramatic representation of what that would be like.

Another of these early authors, César Uribe Piedrahita, was a Colombian contemporary of José Eustasio Rivera, but the 1933 publication of *Toá* has fallen by the canonical wayside. It may be due to the fact that despite the presence of many of the same themes of the abuse of rubber tappers as in *La vorágine*, *Toá* does not feature a narrator anywhere near as intriguing as Rivera's Cova. Instead, the emphasis in *Toá* is on indigenous labor and on the love between a doctor from Bogotá and an indigenized *mestiza* (woman of mixed Spanish/indigenous heritage) in the Putumayo region of Colombia. The integration of these two stories allows for an exploration of the way in which the rubber industry affects indigenous Colombians. As Lesley Wylie argues, "By foregrounding the Putumayo's distinctive geography and culture, Uribe Piedrahita's novel complicated Colombian efforts to assimilate the region and its inhabitants, and marked an important step in the development of an indigenous-inflected poetics in Spanish American literature."[32] So even though the political ecology in *Toá* remains characteristic of an appeal to the state for mediation of conflicts concerning the natural world, the premises of the advocacy also invoke indigenous elements: "Ese informe . . . tiene que decidir al gobierno . . . o si no, nos llevan todos los diablos, y al fin se acaban las tribus, se agota el caucho y esto se vuelve un desierto" ["That report . . . must make the government act . . . because if not, it's Hell to pay: the tribes will disappear, the rubber will run out, and this will become a desert"].[33] In Uribe Piedrahita's novel there is an awareness, often implicit, but occasionally explicit (as in the above passage), that the baleful effects of rubber extraction may not simply be a question of reforming an industry for its human trafficking, its abuse of workers, and its manipulation of prices, but that these elements have to do with intimate connections between indigenous communities and nature itself: cultural extinction, resource exhaustion, and desertification are intimately linked. These have all become targets of the ethical ideology of current conservationism, "the concern that the environment should be protected, particularly from the harmful effects of human activities"[34] and environmental justice, concerned "not only with animals, plants, and the purity of the atmosphere, but with the joint and global relations of human beings and nature."[35] What surprises is the expression of a clearly environmentalist concern in a literary moment from the early 1930s, particularly when a novel (*La vorágine)* written by an author from the same country, about the same jungle, less than ten years before has almost nothing like that.

If *Toá* has faded from the scholarly spotlight, *Una mujer en la selva* by Hernán Robleto can be characterized by its absolute non-status, entirely neglected by scholars as a whole. The novel recounts the tale of Emilia Rivera kidnapped to the jungle by a gorilla that she subsequently grows to love. Finally, when the gorilla dies, she is given the chance to reenter "civilization" but does not return. She visits a nearby village only long enough to steal a pad of paper which becomes her diary, the literary frame for the novel. Apparently, the story was inspired by legends told in Robleto's hometown of Camoapa in Nicaragua "sobre monos que robaban mujeres que se descuidaban en los lavaderos" ["about monkeys that kidnapped women washing clothes by the river"][36] and likely also by the simian/human love story from the first film version of *King Kong* released in 1933, just three years before Robleto's novel. Also, the story about a woman taken from her home to the jungle, the female protagonist's surname—Rivera—and the novel structured as found manuscript are clear references to *La vorágine.* And like the criticism of the rubber industry in Rivera's novel, the text is not without its own political ecology of the kind described by Peet and Watts that seeks to make "the causal connections between the logics and dynamics of capitalist growth and specific environmental outcomes rigorous and explicit."[37] In the closing chapter, Emilia hears the approach of loggers who are clearing the forest for farming and what follows is an unequivocal condemnation of development: logging is "un mordisco a la montaña" ["a chewing-up of the mountain"], that results in "la muerte del bosque" ["the death of the forest"] by loggers "destruyendo para la civilización" ["destroying for the sake of civilization"].[38] But, there is another side to it in Robleto's novel, something that makes *Una mujer en la selva* quite different from *La vorágine.*

In Rivera's novel, the perception of nature and the representation of his lover Alicia are both realized from the perspective of a male narrator, the infamous Arturo Cova. As Sharon Magnarelli observes, "la representación de mujer y naturaleza en *La vorágine* depende directamente de la percepción del personaje principal, Arturo Cova, y ha sido moldeada o desfigurada por él" ["the representation of both nature and women in *La vorágine* is directly dependent upon the perceptions of the central character, Arturo Cova, and has thus been molded or even disfigured by him"].[39] However, in *Una mujer en la selva* the found manuscript is construed as that of a woman (although the author—Hernán Robleto—was a man). And the representation of nature and the "autobiographical" descriptions by Emilia are quite different from what is in *La vorágine*: in the latter, there is fear and dread and madness in the jungle, but the former has the experience of nature and the forest as thoroughly positive, a wilderness more Edenic than savage, more healthful than deadly.

First, as the Emilia of Robleto's novel spends more and more time among the trees, she grows to love them: "Al principio corrían escalofríos de incomprensión por mi espinazo; pero ahora veo como la cosa más natural que los árboles tengan alma" ["In the beginning, I had shivers of total incomprehension running down my spine; but now, it seems quite natural to me that trees have souls"].[40] The expression of love for the forest is what makes Robleto's novel a counterexample to the failure of the "return to nature" that Marcone identifies as so characteristic of the texts considered in the first part of this chapter: unlike the protagonists from those novels, here is a non-indigenous visitor who finds the romantic ideal of nature: peaceful refuge, even love. In her journal, Emilia describes the favors of Nature—"perfumes de copel, del liquidámbar y del bálsamo; frutas deliciosas y la música . . . semejante a un concierto maravilloso" ["resinous perfumes of the sweetgum and the balsa tree; delicious fruits and music . . . like a marvelous concert"]—and receives embraces from "los árboles amigos" ["my tree-friends"].[41] She has literally become a tree-hugger and nature-lover, which as Kay Milton argues in her anthropological study of environmentalists comprises "a perfectly respectable credential for nature protectionists and is more or less taken for granted in many contexts."[42] In the novel, Emilia falls in love with the gorilla that has kidnapped her (like Ann Darrow from *King Kong*), names him, shares a nest with him, grooms him, receives food from him, finds water for him, and eventually grieves his death as if he were truly her mate. Emilia's relationship with the animal she calls Jongo and other moments of identification with nature from her journal in the novel also manifest many of the key elements of ecofeminism: the near-coupling between the woman and the ape offers "a perspective that complicates cultural conceptions of human identity and human relationships with nonhuman nature"[43] and the reference to trees as friends with souls aptly describes a disposition "to conceiving of oneself as fundamentally being in relationships with others including the nonhuman world."[44] The representation of a literal love for nature in the human/gorilla couple comprises these aspects of feminist ecocriticism inasmuch as human identity is reconfigured as relational rather than isolated, unaffected by, and not accountable to the nonhuman world.

Una mujer en la selva is also remarkable for the way that it pre-figures the intelligent sentience of animals that is central to Sepúlveda's work. In *Un viejo que leía novelas de amor*, the ocelot cunningly hunts and sets traps for humans, but also grievously mourns the loss of her mate at the hands of the *gringo* hunter. When Proaño is forced to kill her, it represents something akin to killing a part of himself so intimately does he experience an identification with nature. The fictionalization of the principle of identification from

Naess's deep ecology in Sepúlveda clearly has precedent in Robleto's novel with its representation of human/animal companionship and the unwillingness of Emilia to abandon her natural home and return to civilization. But in Rafael Bernal's *Su nombre era muerte*, the exile to the forest produces a more ecologically nightmarish outcome. In Bernal's novel, an unnamed narrator who goes by the name of Tecolote Sabio settles among a small Lacandón indigenous community where he eventually comes down with malaria. As he suffers the paludic fevers typical of the disease, he kills time by studying the life cycles and behavior of the Anopheles mosquitoes that invade his hut as he convalesces. It occurs to him that the insects' buzzing is some kind of language and eventually he learns their form of communication and constructs a small flute to imitate it thus becoming the first human to communicate with the mosquitoes. He then hatches a plot with the insects to take over the world through the release of fatal and as yet unknown strains of mosquito-borne disease. While the fantastical apocalyptic nature of the story has its own appeal as a prophecy for the demise of humanity, Bernal's novel is remarkably prescient as a critical gesture toward extremist eco-activism.

The ecological aspects of *Su nombre era muerte* seem precocious for their articulation in a novel from the 1940s:

> La selva, cobijadora de desgracias, ocultadora de amarguras y de odios, la selva bendita. Quien quiere vivir en ella puede hacerlo si se sujeta a sus leyes y se conforma con ser un adorador miserable entre tanta magnificencia, si se conforma con dejar a un lado su orgullo de hombre, para ser tan sólo un arrimado, un advenedizo sin derechos, que vegeta a la sombra de la bondad de la selva.[45]
>
> [The jungle: welcoming host of misfortune and concealer of bitterness and hatred: blessed jungle. Whoever may want to live there can do so on the condition that she or he obey its laws and agree to bow down in worship among so much magnificence, agree to leave aside pride, and agree to think of oneself as a novice, a newcomer with no rights but whose only purpose is to give thanks to the grace of the jungle.]

There is no rhetoric of development here nor terror like in *La vorágine*, only an expression of respect for nature in which human rights are secondary to the magnificence of the *selva*. These seem strikingly responsible and progressive attitudes toward the habitation of the jungle, but the context of the novel makes such statements seem ominously naïve. That is because "*el Gran Consejo*," the hive mind of the mosquito species with which Tecolote can now communicate, has developed a blueprint for the almost total extermination of humankind with the narrator as their pawn: the plan is to unleash as-yet-unknown insect-borne disease, many times more fatally virulent than malaria, dengue, or yellow fever. While a mosquito-directed apocalypse seems farcical from a literal reading of the novel, the advent of a rebellion against an elite class from among the worker class of insects in the hive inspired by Tecolote's human culture of liberty, fraternity, and democracy suggests an allegorical reading for the novel: the underclasses must revolt against those that control them. Notwithstanding the obviously leftist allegory as the novel's central conceit, what I find intriguing about *Su nombre era muerte* is the way in which the conditions required by the insects to allow human survival foreshadow some of the very sacrifices that may be required to avert future environmental catastrophe.

The mosquito vision is for total destruction of all human culture, religion, and social organization unrelated to the production of sugar cane (for insect subsistence) and the engendering of humans to ensure a continuing supply of blood (for insect reproduction):

> A los hombres se les permitiría tener ganados, casas, algo de ropa hecha de pieles y telas burdas, fuego y útiles de labranza rudimentarios, pero se cerrarían y destruirían todas las fábricas, se acabarían los ferrocarriles, se prohibiría el uso del petróleo, de la luz eléctrica y de toda fuente de armas. (Bernal, *Su nombre*, 139)
>
> [Humans are permitted livestock, shelter, simple clothing made from skins and coarse cloths, the use of fire, and simple tools; but all factories must be destroyed, as well as all railroads, and the use of gasoline, electricity, and all forms of weapons will be strictly prohibited.]

These elements—the phasing out of factories, means of transport, and other toxic sources of fuel—represent a central aspect of current environmentalist thought. If it means that current human populations will be unsustainable, some extremist groups may allow for that fate.[46] But such a sentiment is eventually criticized in the text when the narrator imagines the painful death of a woman that he has grown to love:

> Pero ¿era hacia el bien el cambio que yo premeditaba? Sin sentirlo me había yo formado la teoría de que tan sólo mediante este cambio podría vivir la humanidad. . . . Con estas teorías había yo logrado que durmiera en mi cerebro la duda, pero ahora, ante la realidad la duda renacía . . . porque una cosa es pensar en el dolor en general y otra es verlo palpable, en los pies de una mujer que se arrastran por el polvo. (Bernal, *Su nombre*, 204-5)
>
> [But, was the path I might tread the right one, the good one? Without realizing it, I had come up with the idea that this was the only way for humankind to survive. . . . I managed to set aside my doubts with these theories, but now, in the face of what was actually happening, the doubt came back . . . because, it was one thing to think of sacrifice in general terms, but quite another to see it close up, in the feet of a dead woman, dragged through the dust.]

Reductions to human population and the elimination of technology as a vision of the future where nature would be protected are ultimately discarded in the novel as viable solutions: the narrator eventually leads the underclass mosquitoes in a revolt against the ruling council. The revolt fails but the casualties on both sides force a one hundred year delay in the implementation of any definitive mosquito attack against humans. Thus ideas about catastrophic human decline from uncontrollable environmental factors are mulled by Bernal in his novel long before groups like Earth First!, Greenpeace, and the Environmental Liberation Front began to consider direct action against the purveyors of environmentally damaging technologies and developmental policies. As Donald Worster points out in his historical analysis of nature, "if people once trod the earth more lightly than we do today, they did so by practicing a brutal infanticide or by suffering warfare, hunger, and disease that restrained their populations."[47] Like a restoration of the ecological and industrial conditions of the distant past, the pursuit of primitive ecological models (of which the apocalyptic mosquito vision for the future enslavement or abolition of humankind is one possible version) could occasion some or all of the unbearably negative consequences mentioned by Worster.

Similar to *Su nombre era muerte*, a trilogy of novels by Peruvian Arturo D. Hernández has protagonists with apocalyptic utopian visions that degenerate into dystopian catastrophes. The author himself was from among a generation of writers in Peru that included Ciro Alegría and José María Arguedas, with the former a canonical author in the *novela de la selva* tradition and the latter, one of the most well-known *indigenista* writers. While perhaps eclipsed by his countrymen in literary reputation and current scholarly esteem, Hernández's first novel in the trilogy, *Sangama*, is, according to Peruvian scholar Ulises Juan Zevallos Aguilar, "una de las novelas más leídas en el Perú" and as such "ha contribuido a la formación de una visión de la amazonía en varias generaciones de ciudadanos peruanos" ["is one of the most widely read novels in Peru" and as such "has contributed to the way in which various generations of Peruvian citizens view the Amazonian region"].[48] If the trilogy (or at least, *Sangama*) has been as widely read as Zevallos Aguilar documents, then serious reconsideration of Hernández's work seems appropriate: with novels this familiar to so many Peruvians, their inclusion in the wider canon of Spanish American literature makes sense numerically. But as the following ecocritical analysis demonstrates, the trilogy's dystopian elements suggest that *Sangama*, *Selva trágica*, and *Bubinanza* are also important as forebears to the tradition of ecological apocalypse that characterizes recent environmentalist fiction in the region.

In *Sangama*, the indigenous protagonist foresees an apocalyptical future jungle civilization, but when he fails to bring it about, commits ritual suicide. *Selva trágica* narrates the bitter experience of a non-indigenous captive among the Capanahua of Peru. And, *Bubinzana* features a Roman Catholic priest who tries and fails to create an idealized community of former rubber tappers. The ecologically utopian aspects of life in a rainforest that are characteristic of Sepúlveda's novel and of the romantic idealization of nature are central aspects in Hernández's trilogy as well. In *Sangama*, the jungle is contrasted favorably to the city, "aquellos que arroja como desperdicios la ciudad, aquí se regeneran" ["those thrown out from the cities like so much garbage, gradually get better here"], and is seen as the center from which a new society can emerge: "¿por qué no adaptarse a los dictados de la selva y crear en ella un nuevo tipo de civilización? . . . [E]l verdadero hombre libre está aquí; sólo que se trata del hombre adaptado al medio" ["why not adapt to the law of the jungle and create here a new type of civilization? . . . Here is the place for a truly free humankind: a new type of humanity totally adapted to nature"].[49] In the other two novels, a similarly utopian inclination permeates the narrative. In *Selva trágica*, an anthropologist taken prisoner by the Capanhua believes that within the *selva* "radica, a no dudarlo, la esperanza de la humanidad angustiada" ["can be found, no doubt about it, some hope in the face of humanity's troubles"].[50] The priest from *Bubinzana* asserts a similar ideal: "No era cuestión de rehuir a la civilización para formar una comunidad utópica, extrahumana; sino de venir con ella para conquistar la selva y descubrir sus secretos en bien de la humanidad" ["It wasn't a question of escaping from civilization in order to form a utopian community; rather, the idea was to come here to conquer the jungle and discover its secrets for the good of humankind"].[51] Each of the protagonists represents an example of the failure of the "return to Nature" like those from the canonical novels considered in the first part of this chapter: their utopian ideals are never realized. However in these three novels, the idea of utopian nature, "expressed in specific locations outside industrial society,"[52] is articulated as the only authentic possibility for hope, freedom, and the future well-being of humanity. The requirement that in order to enjoy these benefits, humans must be "adapted to nature," expresses a central principle of "deep ecology" instead of any political ecology. Hernández's trilogy does not feature the negotiation of access to resources, but advocates an adaptation to the natural world so that the question of access to resources becomes moot. In the trilogy, the representation of such radically ecological adaptation in the form of characters that seek a form of transcendent belonging to the forest—spiritual, regenerative, based on indigenous models—comprises an example of Rozelle's rhetorical ecosublime and on a grand scale given *Sangama*'s large domestic audience: in each novel, the reader is fully immersed in stories that produce "heightened awareness of real natural environments"[53] and is challenged to adapt to the natural world just as the characters in the Hernández trilogy do. Despite the obvious inspirational deficiencies of suicide, disillusionment, and captivity, a belief in the power of nature for the benefit of humankind in Hernández's novels effectively articulates principles that anchor current environmentalist thought.

In the Introduction, I offered a working definition of environmentalism that included political ecologies whose emphases were almost entirely upon entities of the state: appeals for governmental reform of land management, the expectation that properly informed representatives will halt certain abuses, the role of neo-colonialism in resource extraction, etc. However, political ecology may also highlight "the plurality of perceptions and definitions of environmental and resource problems."[54] In *El mensú que triunfó en la selva* by Valentín Barrios, a tale of trials and triumphs of the *campesino* (peasant) Tristán in the *yerba mate*[55] groves of the upper Paraná region of Paraguay, both of these manifestations of political ecology are present: an articulation of the ambivalence of perceptions of the *selva* and an examination of local practices. In the first case, the novel contains several passages that attest to its sublime qualities but which are tempered by context. Statements like "el machete . . . parecía entonar cánticos a la sordina en honor de la sublimidad de la naturaleza" ["the machete seemed to intone muted songs in honor of the sublimity of nature"] or "el muchacho . . . bebía su inspiración en la sublime fuente de la Naturaleza" ["the boy . . . drank at the sublime fountain of Nature's inspiration"][56] are often followed by an expression of the desire to abandon these sublime places. The vacillation between a transcendent experience of nature and the ever-present desire to get away from it represents that ambivalence in the *novela de la selva* with its heritage of European romanticism in conflict with the realities of the natural world on the American continent. From the stories of Quiroga to the misadventures of Cova and even the initial experiences of Proaño within the jungle, the genre is filled with protagonists who are attracted by the allure of rainforest, but suffer gravely once they actually get there. What is remarkable about *El mensú que triunfó en la selva* is that these equivocations occur so close together and from the perspective of the same character. In fact, that is what constitutes the uniqueness of the novel's political ecology: Tristán's plurality of perception is illustrative of the challenges for a definitive representation of Latin American *selva* which in turn is allegorical of the very real conflicts that arise when negotiations of access to resources rest upon definitions and perceptions of value, sacredness, rights, authority, and ownership of the natural world.

Finally, Manuel González Martínez's *Llanura, soledad y viento* represents another unfortunate example of the way in which more well-known authors can blunt the awareness of alternate tendencies within certain traditions as it did with Arturo Hernández in Peru. In this case, González Martínez's novel was eclipsed by literary developments in Colombia and elsewhere in Latin America: most notably, the publication of Gabriel García Márquez's *Cien años de soledad* (*One Hundred Years of Solitude*) (1967).

Readers' fascination with magical realism, and before that with the originality of Alejo Carpentier's extensive use of ideology, intertextuality, and culture in the representation of the jungle in *Los pasos perdidos*, does a lot to explain why González Martínez's novel published exactly seven years before and after these other two has had much less of an impact on the Spanish American literary scene. *Llanura, soledad y viento* features the same humanization of animals as in Quiroga's short stories, but the dominance of statist political ecology inaugurated in the call for development in Sarmiento's *Facundo* comes in for sharp criticism here. The effects of development on the natural world are dramatized in the book as the animals gradually lose their habitat to dynamite blasting used by petroleum prospecting engineers employed by foreign corporations with the support of the Colombian state. In one exchange between a vulture and a reptile, the bird reports how, from its privileged aerial perspective, the land has been poisoned:

> Si vieras ahora esas tierras como las he visto yo. Hasta el color les han cambiado. Aquél no es un color de tierra, es un color negro, más fúnebre que el color de mi plumaje; y parece que a aquellas tierras les cae muy bien ese color de luto porque están muertas, no producen nada, ni siquiera hierbas venenosas.[57]
>
> [If you could see those lands like I have. Even their color has changed. It's not a color of earth, but a blackish, funereal hue like the feathers of my plumage; and it seems that these colors are fitting because the land is dead, nothing grows there, not even poisonous plants.]

This makes a marked departure from the representation of animal perspectives in Quiroga; here, the conflict between humans and animals is not a struggle for survival that takes the form of the hunting and killing of one species by another, but one (the humans) poisoning the habitat of another (the animals). Yet it continues a trend in the representation of the sentience of the natural world and its animals seen in Quiroga, *Una mujer en la selva, Su nombre era muerte*, and eventually in Sepúlveda's novel.

Not all humans in *Llanura, soledad y viento* are as obtuse as the careless oil prospectors. In fact, the novel features an exchange between the rancher Victor Ramón and his representatives in the house of deputies in Bogotá. Ramón proposes several methods of sustainable agriculture that will ensure the land's fecundity for generations to come, but realizes he will get nowhere with the deputies who are concerned with votes and not sustainability: "Y sus tierras, que son las reservas del futuro . . . ; esas tierras—pensaba—, continúan siendo quemadas, destruidas y deforestadas" ["And his lands, lands that should have been conserved for future generations . . . ; those lands—he thought—will continue to be burned, destroyed, and deforested"].[58] Victor Ramón's "demand for changes in land use"[59] expresses an ecological protest as strong and unique as had yet been seen in Latin

American letters. However, the evolution of nature representation in Spanish American literature from the prevalence of a statist political ecology to a more diverse kind of environmental writing was not to be explored in the following years. García Márquez, Mexican Carlos Fuentes, Argentine Julio Cortázar, Mario Vargas Llosa and others were about to embark on a revolution of the more formal aspects of literature—structure, narrative voice, language, etc.—in what was to be called the "Boom" of Latin American literature.

LAST TO THE CANON: THE USURPATION OF POLITICAL ECOLOGY FROM THE BOOM AND BEYOND

As the "Boom" explodes then draws to a close, Mario Vargas Llosa's unique literary style evolves from a decidedly implicit to a somewhat unreservedly explicit emphasis on ecological ideas. *La casa verde* and *Pantaleón y las visitadoras* (*Captain Pantoja and the Special Service*) (1973) represent typical examples of Vargas Llosas's totalizing texts that attempt to represent all facets of a given reality through the use of various literary techniques: non-chronological narration, interior monologues, the uninflected use of dialogue in an indirect style, multiple narrators, and the creation of myth. These meta-narrative techniques have the effect of removing any kind of message from the locus of narrator to the dialogue of character. The peculiar narrative style may allow for the occasional expression of ecological ideas, but only as reactions, recognitions, expressions, or platitudes. Because the narrative exists almost exclusively to frame character dialogue, the only access to "message" occurs in the context of what the characters say. It is difficult therefore to come away from a reading of some of Vargas Llosa's earlier work with the feeling that the narrative delivers an ethical message *per se*, much less one as specific as that of current environmentalism.

In Vargas Llosa's later novels, however, things are quite different. In *El hablador* (*The Storyteller*) (1987) (published roughly contemporaneously with Sepúlveda's *Un viejo que leía novelas de amor*) these same literary techniques—multiple narrators, non-chronological narration, the incorporation of myth—are widely used, but the dialogues among the characters and the values expressed by the indigenous Machiguengan storyteller convey environmentalist ideas more explicitly. One narrator mentions that the Machiguengua tribe was able to "preservar aquella naturaleza, aparentemente tan exuberante, y, en realidad tan frágil y perecedera, de la que dependía para subsistir" ["preserve that nature, seemingly so abundant, and, in reality so fragile and delicate, on which they all depended to survive"].[60] Later in the text, Vargas Llosa has the *hablador* or "storyteller" describe the land in terms of injury and mistreatment:

> Si un daño ocurre en la tierra, es porque la gente ya no le presta atención, porque no la cuida como hay que cuidarla. ¿Puede la tierra hablar como nosotros? . . . No se olviden de mí, diciendo. Yo también vivo, diciendo. No quiero que me maltraten. . . . Tal vez los Padres Blancos . . . querían hacerle daño a la tierra. (Vargas Llosa, *El hablador*, 217)
>
> [If something happens to the land, it's because people don't pay any attention to it, because they don't take care of it like they should. Can the land talk like we can? . . . Don't forget about me, it is saying. I am also alive, it is saying. I don't want you to mistreat me. . . . Maybe the white folks . . . wanted to harm the land.]

Both of these statements communicate ideas familiar to environmentalist positions concerning consumption, conservation, the fragility of the natural world, and alternatives to development. Marcone observes that Vargas Llosa here imposes an "environmentalist agenda onto the storyteller's function";[61] considering the late date of publication in 1987 (two years before Sepúleveda's text), the explicitly ecological language does not surprise, even if the novel is not by an author usually associated with that ideological position.

Most of the texts that I have surveyed above, some well-known, others not so much, are among the most promising as evidence to support my argument that Sepúlveda's novel was preceded by a corpus of texts that implicitly express many of the same environmentalist ideas that are so explicit in *Un viejo que leía novelas de amor.* Furthermore, the preceding survey of some titles that I have characterized as "lost to the canon" incorporates a counter-tradition to the canonical works of the *novela de la selva*, one that communicates principles of conservation, deforestation, alternatives to development, sustainability, and deep ecology. *Toá*, *Una mujer en la selva*, *Su nombre era muerte,* and *Llanura, soledad y viento*, in particular, merit further review. The primacy of *La vorágine* in the *novela de la selva* tradition has overshadowed the contribution of *Toá* to conceptions of environmental justice as related to indigenous communities just as the novelty of the "Boom" left *Llanura, soledad y viento* out of the critical spotlight; likewise, the literary obscurity of Hernán Robleto has condemned *Una mujer en la selva* to a similar fate, but the recent interest in animal studies begs for a revaluation of both Robleto's novel and *Su nombre era muerte*. Only recently has scholarship on animals been formalized with the founding of the Institute for Critical Animal Studies in 2007, yet animal characters have been mainstays of the *novela de la selva* since Quiroga and with an implicitly environmentalist emphasis since Robleto, Bernal, and González Martínez.[62] In the novels by these three authors, as well as in *Toá,* the literary representation of biodiversity, deforestation, and environmental justice comprises conservationist tendencies and aspects of deep ecology well before such terms became conventions of current environmentalist terminology. So discussions of literary environmentalism in *Un viejo que leía novelas de amor* should begin with an awareness that Sepúlveda's novel, as a late entry in the *novela de la selva* genre, can be thought of as a culmination and not an innovation given the long history and wide range of ecological ideas within that tradition.

NOTES

1. Luis Sepúlveda, *Un viejo que leía novelas de amor* (Barcelona: Tusquets, 1993), 78.
2. José Eustasio Rivera, *La vorágine* (Buenos Aires: Losada, 1952), 250.
3. See Introduction, note 43.
4. Jorge Marcone, "Del retorno a lo natural: *La serpiente de oro*, la 'novela de la selva' y la crítica ecológica," *Hispania: A Journal Devoted to the Teaching of Spanish and Portuguese* 81, no. 2 (May 1998): 299–308.
5. Sepúlveda, *Un viejo*, 9.
6. Ibid., 122.
7. Ibid., 123, 45, 50.
8. Kay Milton, *Environmentalism and Cultural Theory: Exploring the Role of Anthropology in Environmental Discourse* (New York: Routledge, 1996), 27.
9. Sepúlveda, *Un viejo*, 44, 137.
10. Juan León Mera, *Cumandá* (Boston: DC Heath, 1932), 116, 39, 68.
11. Lydia de León Hazera, *La novela de la selva hispanoamericana: Nacimiento, desarrollo y transformación* (Bogotá: Caro y Cuervo, 1971), 25, 48.
12. Laurence Coupe, ed. *The Green Studies Reader: From Romanticism to Ecocriticism* (New York: Routledge, 2000). Note the reference to a progression from romanticism to "ecocriticism" in the phrase Coupe employs as a subtitle.
13. Hazera, *La novela de la selva*, 23–24.
14. Raymond Williams, "The Green Language," in *Green Studies Reader: From Romanticism to Ecocriticism*, ed. Laurence Coupe (New York: Routledge, 2000), 51.
15. Luis Vitale, *Hacia una historia del ambiente en América Latina* (Mexico City: Editorial Nueva Imagen, 1983), 90.
16. I use figures from the University of Michigan's Global Change Curriculum to make the calculation. http://www.globalchange.umich.edu (accessed November 27, 2012)
17. Vitale, *Hacia una historia*, 94.
18. Shawn William Miller, *An Environmental History of Latin America* (New York: Cambridge University Press, 2007), 150.
19. Clifford Kraus and Elizabeth Malkin, "Mexico Oil Politics Keeps Riches Just Out of Reach," *New York Times*, March 8, 2010.
20. Jennifer L. French, *Nature, Neo-Colonialism, and the Spanish American Regional Writers* (Hanover, NH: Dartmouth University Press, 2005), 36.
21. Ibid., 63–64.
22. Jorge Marcone, "Cultural Criticism and Sustainable Development in Amazonia: A Reading from the Spanish American Romance of the Jungle," *Hispanic Journal* 19, no. 2 (Fall 1998): 281–94; Bridgett W. Gunnels, "An Ecocritical Approach to Horacio Quiroga's 'Anaconda' and 'Regreso de Anaconda,'" *Mosaic* 39, no. 4 (December 2006): 93–110; and Beatriz Rivera-Barnes, "Yuyos Are Not Weeds: An Ecocritical Approach to Horacio Quiroga," *ISLE: Interdisciplinary Studies in Literature and Environment* 16, no. 1 (Winter 2009): 35–52.
23. French, *Nature, Neo-Colonialism*, 130.
24. Carlos J. Alonso, *The Spanish American Regional Novel* (New York: Cambridge University Press, 1990); Montserrat Ordoñez, ed., *La vorágine: Textos críticos* (Bogotá: Alianza, 1987); Montserrat Ordoñez, introduction to *La vorágine*, by José Eustasio Rivera (Madrid: Cátedra, 1995).
25. Rivera, *La vorágine*, 177.

26. French, *Nature, Neo-Colonialism*, 147.
27. Miller, *Environmental History*, 127.
28. Marcone, "Cultural Criticism," 288.
29. Rómulo Gallegos, *Canaima* (Madrid: Austral, 1982), 201.
30. Horacio Quiroga, *Nuevos cuentos de la selva* (Buenos Aires: Solaris, 1997), 1:105.
31. Jesús Botero Restrepo, *Andágueda* (Bogotá: Librería y Ediciones Teoría, nd), 89, 100, 208.
32. Lesley Wylie, "Frontier Fictions: The Place of Amazonia in César Uribe Piedrahita's *Toá*," *Bulletin of Spanish Studies* 87, no. 7 (2010): 961.
33. César Uribe Piedrahita, *Toá* (Manizales, Colombia: Arturo Zapata, 1933), 74.
34. Milton, *Environmentalism*, 27.
35. Leonardo Boff, *Ecology & Liberation: A New Paradigm* (Maryknoll, NY: Orbis, 1995), 14.
36. Nicasio Urbina,"Estilo y estructura de 'Sangre en el Trópico' de Hernán Robleto," *El nuevo diario*, September 1, 2001.
37. Richard Peet and Michael Watts, "Liberation Ecology: Development, sustainability, and environment in an age of market triumphalism," in *Liberation Ecologies: Environment, Development, Social Movements*, ed. Richard Peet and Michael Watts, (London: Routledge, 1996), 9.
38. Hernán Robleto, *Una mujer en la selva* (Santiago de Chile: Editorial Ercilla, 1936), 158.
39. Sharon Magnarelli, "La mujer y la naturaleza en *La vorágine*: A imagen y semejanza del hombre," in *La vorágine: Textos críticos*, ed. Montserrat Ordoñez (Bogotá: Alianza, 1987), 336–37.
40. Robleto, *Una mujer en la selva*, 111.
41. Ibid., 153, 138.
42. Kay Milton, *Loving Nature* (New York: Routledge, 2002), 24.
43. Karla Armbruster, "A Poststructuralist Approach to Ecofeminist Criticism," in *Green Studies Reader: From Romanticism to Ecocriticism*, ed. Laurence Coupe (New York: Routledge, 2000), 199.
44. Karen J. Warren, *Ecofeminist Philosophy: A Western Perspective on What It Is and Why It Matters* (New York: Rowman & Littlefield Publishers Inc., 2000), 103.
45. Rafael Bernal, *Su nombre era muerte* (Mexico City: Editorial Jus, 1947), 11–12.
46. The Voluntary Human Extinction Movement, http://www.vhemt.org, (accessed December 2, 2011).
47. Donald Worster, "Nature and the Disorder of History," in *Reinventing Nature: Responses to Postmodern Deconstruction*, ed. Michael E. Soulé and Gary Lease (Washington, DC: Island, 1995), 67.
48. Ulises Juan Zevallos Aguilar, *Las provincias contraatacan: Regionalismo y anticentralismo en la literatura peruana del siglo XX* (Lima: Vicerrectorado Académico UNMSM, 2009), 68.
49. Arturo D. Hernández, *Sangama* (Lima: Juan Mejía Vaca, 1958), 52–53, 95.
50. Arturo D. Hernández, *Selva trágica* (Lima: Juan Mejía Vaca & P.L. Villanueva, 1956), 227.
51. Arturo D. Hernández, *Bubinzana* (Lima: Luis E. Valcarcel, 1960), 137.
52. Milton, *Environmentalism*, 28.
53. Lee Rozelle, *Ecosublime: Environmental Awe and Terror from New World to Oddworld* (Tuscaloosa: University of Alabama Press, 2006), 3.
54. Peet and Watts, "Liberation Ecology," 11.
55. A bush-like tree whose leaves are harvested to make *mate* tea, popular in Uruguay and Argentina.
56. Valentín Barrios, *El mensú que triunfó en la selva* (Buenos Aires: Librería Perlado, 1951), 205, 241.
57. Manuel González Martínez, *Llanura, soledad y viento* (Buenos Aires: Sudamericana, 1965), 217.
58. Ibid., 85.
59. Milton, *Environmentalism*, 27.
60. Mario Vargas Llosa, *El hablador* (Barcelona: Seix Barral, 1987), 29.

61. Jorge Marcone, "Jungle Fever: Primitivism in Environmentalism: Rómulo Gallegos's *Canaima* and the Romance of the Jungle," in *Primitivism and Identity in Latin America: Essays on Art, Literature, and Culture*, ed. Erik Camayd-Freixas and José Eduardo González (Tucson: University of Arizona Press, 2000), 165.

62. And before that in the nineteenth century novel *El moro* (Blackie) (1897) by Colombian José Manuel Marroquín which I analyzed above in chap. 2.

Chapter Five

Other Lands: Ecology in the Spanish American *novela de la tierra*

Like the tension between a more-or-less established canon of the *novela de la selva* and the ecological counter-tradition that I analyze in chapter 4, there is a similar duality when it comes to the *novela de la tierra* genre.[1] In chapters 6 and 7, I develop an analysis of political ecologies in *indigenista* texts where identity, ownership, access to, and authority over the land comprises a central element; the same analysis can also be undertaken in texts where the representation of land is central, but where protagonists and antagonists are not necessarily associated with indigenous peoples and their abusers. Profound demographic alterations and unrestrained growth of cities brought deforestation, resource exhaustion, and contamination of air, land, and water and set the backdrop for the ecological tendencies in the *novela de la selva* tradition where such phenomena occurred in cities and other developed areas within or near the Amazonian region and other forested regions of Latin America. But the plains, mountains, and coasts also suffered similar environmental degradation along with some other ill effects not usually seen in rainforests and jungles. For example, from early in the nineteenth century, "la propiedad territorial, concentrada en grandes latifundios, fue dedicada a la crianza masiva de ganado o al cultivo de determinados cereales y plantaciones, consolidándose un subsistema agrícola de escasa diversificación que acentuaba el deterior de los ecosistemas" ["the land was divided into territories, organized as huge plantations dedicated to the production of cattle or the cultivation of a small number of grains and in such a way was limited to agricultural production of limited horticultural diversity, a limitation that only accelerated the decline of those ecosystems"].[2] By the twentieth century, the borders of these huge plantations and ranches had begun to employ wire-fencing to mark their limits, to keep the cattle in, and invasive predators out. While these fences in themselves did not have any pronounced ecological effect, their impact on the way in which the livestock industries operated was profound: with the round-up rendered quaint, an entire population of culturally symbolic men—the gauchos—were suddenly out of a job.

And on the plant-based agricultural front, the impact of the monoculture crop model introduced in the nineteenth century and industrialized in the twentieth was in equal measure cultural and ecological: various tree infestations, fungal epidemics, and other crop blights spurred economic policies and commercial investment in remedies that led to land exhaustion, desertification, and the widespread use of harmful pesticides. Finally, the discovery of oil in Mexico, Ecuador, Patagonia, and especially in Venezuela also occasioned serious ecological contamination of the lands in and around the oilfields; as the industries in these nations and locales matured, the environmental impacts of oil drilling, extraction, and transport became more and more obvious. Just in the region around Venezuela's Lake Maracaibo, "the uncontrolled search for oil also proved damaging to the environment and the health of the people who inhabited the production sites. Wells often spewed their contents over lakes, land, and foliage. The smell of oil and leaking gas wafted over the communities, and the once pristine Lake Maracaibo, the largest body of freshwater in South America, became quickly contaminated."[3] These profound ecological changes in the landscapes of Latin America represent the historical context against which the regional novels that are the focus of this chapter are set.

THE ENTRENCHED ECOLOGIES OF *DON SEGUNDO SOMBRA* AND *DOÑA BÁRBARA*

The literary moment of 1926 when Argentine Ricardo Güiraldes published *Don Segundo Sombra* features a formerly wide-open pampa (the plains, also "*el llano*") which had by then been closed through the widespread use of barbed wire fencing to enclose lands acquired by large-holders. There was by now no need for the gaucho rodeos to separate, sell, and trade livestock whose owners were identified by branded cattle; the fences took care of that. Thus, the story in Güiraldes's text is an exercise in nostalgia, pining for a past way of life, idealized in the novel as a hostile but invigorating landscape, nothing like residence in Argentina's urban centers. In fact, the (now-disappeared) gaucho lifestyle was far removed from the urban lives of the book's urbane Argentine readership and representations of the wild pampas marked dramatic contrasts to the highly structured parameters of city life. The novel creates a city/country dialectic where the boundless plains, although threatening and dangerous, could be overcome and, as a result, make the city "man," whose urban existence was made to seem artificial, more of a "real man." So it seems that an obvious ecocritical move would be to read *Don Segundo Sombra* and look for signs of "nature."

But for almost every narrative case in Güiraldes's novel, the representation of the pampa is absent. Although it is the sole physical setting in *Don Segundo Sombra*, descriptions of the pampa are only effected in order to flesh out human values: the importance of nature for the "education" of the city-dweller, the importance of topography in the determination of character, the value of the land for self-improvement. Again, the attitude toward the natural world as in *La vorágine* and Uruguayan Horacio Quiroga's stories cannot yet properly be called environmentalist; the representation of nature in these texts is limited to the concerns of political ecology where the central issue is the justification or condemnation of certain ways of using the Argentine plains. There is still no expression of the inherent value of nature as is common to the present ecological debate, and concern for the future of the land is only articulated in the context of its utility. Yet the idea that a vast wilderness such as the pampa had value not given it from the perspective of industrialization leads inevitably to the conclusion that the land as it was ought not to have been disturbed and thus expresses something like the conservationist ideal of wilderness preservation.

It is also worth noting that the narrative in *Don Segundo Sombra*, as in all other texts in the gauchesque tradition, proceeds from the perspective of the city. Nature is and must be evoked and described so that city-folk can understand it. The novel sets up a mediator—the narrator Fabio—who has participated in the classic Dantesque voyage of descent and return, but the endless plains were not, indeed could not, be represented for what they truly were but for how they might be experienced from an urban perspective. The notion raises serious questions in terms of the feasibility of the representation of the natural world as it reveals how the tension between reality and representation is a constant pressure on the writing of texts that deal with nature. In Güiraldes's novel, the pampa, as textually represented, was evoked not as an independent entity, but as a function of the imagery that folks from the city could understand. In the wider context of Spanish American fiction, the tendency to represent the land by privileging an urban perspective had been widespread and the influence of the city on literature pervasive: a legacy from the nineteenth century writings of Argentine Domingo Faustino Sarmiento.

Doña Bárbara (1929) by Venezuelan Rómulo Gallegos—the classic tale of the Venezuelan plain—is another of the *novelas de la tierra* that portrays the natural world from an urban perspective. It is a text that prominently features the categories of urban civilization and rural barbarism in the mode of Sarmiento's *Facundo* (1845) to describe Latin American reality. In Gallegos's novel, the inhabitants of the Venezuelan plain include the diametrically opposed Doña Bárbara and Santos Luzardo. The former was the incarnation of barbarism, a life-long inhabitant of the plain, and imbued with all the "savage" qualities that such an existence bestows. Her antithesis, Luzardo, educated in Caracas but descendant from former ranchers in that part of the country, brings a program of progress to the property he has inherited and ultimately reclaims the land from Doña Bárbara, thus vanquishing barbarism and replacing it with civilization.

In the chapter of Carlos J. Alonso's *The Spanish American Regional Novel* devoted to *Doña Bárbara,* Alonso notes that it seemed inevitable that a reader would imagine Sarmiento's categories in the reading of Gallego's text:

> The novel has consistently been interpreted as yet another document depicting the stale contest between civilization and barbarism that has had such a long history as a metaphor for Latin American reality. . . . The text has, in a sense, incorporated into itself its own reading, so that the reader is faced with what could be described as an already-read work, one that conceives of the act of reading as corroboration rather than as interpretation.[4]

Following the civilization/barbarism dialectic, one finds that the rural, undeveloped land represents a corruption of humanity while progress and civilization are portrayed as an extension of the city; the wild, savage *llano* require redemption: a baptism of urban progress. When read from that perspective, *Doña Bárbara* seems to legitimize the authority of a kind of enlightened landownership by associating city-educated landowners with the ideals of civilization such that they are perceived as the rightful stewards of the land. Alonso considers the allegory present in all parts of Gallego's text in much the same way that *Hamlet*'s Gertrude thought that the "lady doth protest too much": the text's insistence on reiterating the same allegorical message from beginning to end—the conflict between civilization and barbarism—is actually evidence that another more complex reading might reside underneath the allegorical one.

Released by Gallegos in Spain, it was received by critics there and subsequently by readers in Venezuela as an authentic example of Latin American reality. Gallegos's novel expressed in literary form the Venezuelan and Latin American nationalist criticism of European racial theory; Julie Skurski observes that "the prevailing currents of nationalist thought in Latin America after World War I involved a critique of determinist theories of racial evolution and the assertion of mestizaje as a uniquely Latin American contribution to the advancement of world civilization."[5] Thus, as a rejection of Eurocentric thought, traditionally associated with not only a racial theory of purity and elites but also with the economic superiority of modernization, *Doña Bárbara* suggests an alternative to these economic systems. Additionally, Skurski notes that "changes brought about by the impact of the oil export industry helped create a desire among the conservative elite for a constitutional regime, while demands for democracy and reform arose primarily from the nascent middle class."[6] According to Skurski's reading, the image of Santos Luzardo, representative of an enlightened elite, was a model for the political struggle against the twentieth century Venezuelan dictatorship of General Juan Vicente Gómez about whom critics have said that "the novel was a not-so-veiled indictment of his dictatorial regime."[7] The alternative to the backwardness of the barbarity of the Doña Bárbara/General Gómez figure, while still based on the ideals of civilization as they proceed from the city to the country, does not propose the city as the site of a new foundation for Venezuelan identity nor does it mention the discovery of oil or an economy based around the petroleum industry as possible alternatives; instead, a third reading should be considered.

If one considers the date and historical circumstances surrounding the publication of Gallegos's novel, a reading from the perspective of economic theory is more appropriate. By the late 1920s, Latin American modernization was well underway, but its effects on Latin American people were far from positive. With the crash of the New York stock exchange in 1929 (the same year that *Doña Bárbara* was published), regional economies across the developing world were thrown into a decade-long depression. The Great Depression as it was called in the United States took an economic toll on Latin America as modernization and industrialization began to fail as models for development in the face of a large-scale financial slowdown. Read from that perspective, *Doña Bárbara* can be understood as not just a literary representation of the categories of civilization/barbarism as expressed on the Venezuelan plains, but the proposal for a rural alternative to urban-centered models of industrialization and development.

The linguistic properties in the narrative seem to indicate that that would be a valid approach to the novel as José Castro-Urioste notes in his analysis of the textual notation that accompanies the use of regional colloquialisms:

> Un conjunto de términos provenientes de la llanura son recopilados y "marcados" por medio de las comillas. El empleo de este signo indica la utilización de una palabra "extraña," su uso fuera del contexto habitual, el distanciamiento entre el sujeto "civilizado" y la voz del otro.[8]
>
> [A group of plains-terms are recopied and "marked" by quotation marks. The use of such punctuation indicates the utilization of a "strange" word, its occurrence out of context, and the distance between the "civilized" subject and the voice of the other.]

The idea that Gallegos should have to use quotation marks, italics, or other textual devices to signal terminology outside the ken of the novel's probable readership demonstrates that the text itself refers to an alternative from the predominant mode of thinking upon the novel's release. Castro-Urioste makes reference to the urban center that was the foundation for such thinking when he observes that:

> la novela de Gallegos contiene un glosario bajo el título de "Vocabulario de venezonalismos que no figuran en los últimos diccionarios de la Lengua Española." Este "vocabulario" se construye porque *Doña Bárbara* es un texto dirigido a un lector citadino que, obviamente, desconoce el lenguaje usado en la llanura que Gallegos incorpora en su novela. (Castro-Urioste, "Utopía y transgression," 339)
>
> [Gallegos's novel contains a glossary entitled "Vocabulary of Venezuelanisms that do not Appear in the Latest Dictionaries of the Spanish Language." This "vocabulary" is necessary because *Doña Bárbara* is a text meant to be read by an urban reader who, obviously, does not know the language of the plains that Gallegos incorporates in his novel.]

The fact that Gallegos does not translate or explain these "*venezolanismos*" evidences the conclusion that the author advocated for the viability of the plains as a rural alternative to the hegemony of the city. Notwithstanding the overwhelming tendency to consider its representation under the lens of the civilization/barbarism dialectic, the Venezuelan plain as its own entity, not contained by urban categories, represents a response to the failures of modernization and development as they were practiced in the cities. And despite the constant reiteration of these urban categories, an alternative reading seems possible in light of the textual evidence: that *Doña Bárbara*—where the plains are portrayed as an unindustrialized, unmodernized locale—suggests the viability of alternatives to the polluting, industrialized city and to the potential for environmental destruction represented by the various extractive industries that rapidly developed in early twentieth century Latin America. Such were the concerns registered in the novels by Güiraldes and Gallegos, but also in several far less well-known *novelas de la tierra* considered below.

RURAL ECOLOGIES: PROVINCIAL *NOVELAS DE LA TIERRA*

El inglés de los güesos (The Englishman of the bones) (1924) by Argentine Benito Lynch had been largely forgotten until Jennifer L. French's insightful contextualization of the novel within a Spanish American tradition of literary response to the economic fallout of nineteenth century British neo-colonialism.[9] Like the nostalgia for the disappearance of the gaucho in *Don Segundo Sombra*, Lynch's novel features a similar longing for a disappeared natural past (or a longing for one that never existed). The story is about James Gray, an English archaeologist conducting field work in the Argentine pampa, and about the love that blossoms between him and Balbina, a *campesina* (female peasant) living on the ranch where Gray is conducting his research. French's analysis emphasizes the neo-colonial relationship among British investment, local elites, and the underclass often victimized by these economic arrangements:

> The seemingly limitless land of the pampa was almost entirely divided up into private holdings, not as small family farms but in an anachronistically feudal arrangement that concentrated enormous expanses . . . in the hands of a few wealthy proprietors and foreign corporations, while families like Balbina's earned a humble living as tenant farmers or *puesteros.* [10]

The ecological implications of such arrangements have everything to do with a political agenda of land consolidation which seems to be targeted for criticism, however subtly, in Lynch's novel:

> La sombra de los bosquecillos misteriosos, el agua cristalina de los arroyuelos, el canto de los pájaros y el perfume de las flores, todo eso era y sería más bien para los incapaces, para los abúlicos, para los decadentes, para los poetas y para los locos. . . .
>
> ¡Caramba! Bueno era trabajar, sin duda; bueno era ceñirse a una severa línea de conducta si se quería llegar a ser algo en la vida; pero aquello de convertirse en eterno esclavo del propio perfeccionamiento individual, también era al cabo cosa triste. [11]

> [The shadow of mysterious forests, the crystal-clear water of brooks, birdsong, and the perfumed scent of flowers, all of this was and should be for the useless, the weak-willed, the decadents, the poets, and the insane. . . .
>
> But shoot! It was all well and good to toe the line and work hard in order to be somebody; but it was also kind of sad to become a slave to self-improvement.]

The foil to the life of hard work (read as a sign for the British presence in the economic affairs of nineteenth and early twentieth century Argentina) is poignantly represented here as appreciation for forests, streams, birds, and flowers, domains typically reserved for the shiftless, the unmotivated, the decadents, and the mentally unstable. Thus, the criticism in Lynch's novel of British neo-colonialism and the loss of an idyllic past is an ecological one: if commercial pursuits come to sadness in the end, the alternative proposed in the novel takes the form of appreciation for nature and for the unspoiled natural phenomena which had long since been fenced in, cut down, dammed up, and developed out of existence.

In the novels that most quintessentially qualify as *novelas de la tierra*, stewardship of the land, political ecology, and notions of progress and civilization proceed from the principles of development applied to the earth and access to its soil for farming. In chapters 6 and 7, I compile a long list of titles in the *indigenista* tradition where the relationship between indigenous groups and their ancestral land is explored from various angles; but in most cases, the conflict revolves around a tenuous "legal" right guaranteed to, or more usually taken from, these groups. A similar conflict is at the heart of *Ranquil: novela de la tierra* (1942) by Chilean Reinaldo Lomboy. The novel deals with a *campesino* insurrection in 1934 violently put down by government forces. According to historian Olga Uliánova, Lomboy's work was "considerada por décadas lectura obligatoria para todo militante de izquierda, en cuanto fuente de inspiración de sensibilidad social" ["for its inspiring sense of social responsibility, considered required reading for decades by any leftist militant"][12] and several instances in the novel illustrate the way in which such social sensibilities were possessed of a political ecology characterized by protectionism for the rights of small-scale land cultivators lacking legal title to the lands they worked. The labor is hard and the landowners absent, but the specter of violent loss of traditional plots hangs over the text from the beginning: "muchos años defendiendo la tierra contra los elementos antes que hubieran de defenderla contra los hombres" ["so many years trying to protect the land from the elements, and now, they had to protect it from other men"].[13] The novel is highly critical of a legal process that does not recognize *de facto* ownership established over many years and generations of labor on parcels uncultivated by any but the *campesinos* who had been there all their lives:

> Pero, entonces, ¿no había ley para el pobre? El pobre no tiene más que su pobreza y su hambre. Si un día se creyó dueño de la tierra que cultivaron sus manos y las de sus padres, la ley le pide documentos de posesión; llega y le quita la tierra y se la da al rico que ahí la deja, porque no alcanza a cultivarla toda. Pero le pertenece aunque no la cultive, aunque nada sino maleza haga brotar en ella. El pobre no sería pobre si tuviese tierra. (Lomboy, *Ranquil*, 148)
>
> [But weren't there any laws to protect the poor? The poor seem to have nothing more than hunger and their own poverty. If one day they think that some land that had been cultivated by their fathers and mothers might one day belong to them, the legal processes insist on proper title; the powers-that-be come along, take the land away, and give it to the rich who just let it sit because they don't have the time to plant all the fields they own. But it belongs to them even if they do not cultivate it, even if nothing but weeds grow up on the parcel. The poor would not be poor if they had land.]

Thus, the ideas in the novels by Lomboy and Gallegos can be said to support competing political ecologies: the progress promised by ideas brought from the city implies, on the flip side, the possible dislocation of long-standing residents removed from traditional lands when "legal" requirements are enforced. While Gallegos's text deals with ideas about the development of ranching and Lomboy's is more concerned with small-scale agriculture, both contain competing notions of access to and authority over the land as the central conflicts with which their novels are concerned.

Colombian Jaime Ibáñez's *Cada voz lleva su angustia* (Each voice its anguish) (1944) features a somewhat more bizarre approach to these issues by prophesying future ecological effects as explicitly causal consequences of present practices. The novel describes a small peasant community's battle with soil exhaustion and unproductive crop yields—a pervasive ecological concern of the present and represented with surprisingly foresighted awareness of the potential environmental harm that can befall a formerly fertile parcel: "Ya no es aquella cosa tierna, perfumada, henchida, firme. Es un polvillo endiablado como el que le queda a úno cuando ha cogido una mariposa de las alas" ["It's not soft, perfumed, full, and firm like it used to be. The land is nothing but a cursedly fine dust like what you get after picking up a butterfly by its wings"].[14] What is unique about the novel is that it features the voice of the land itself, portrayed as suffering the effects of overuse and pointedly blaming human activity for those effects: "Por tu obra! . . . Por tu obra! . . . Por tu obra! [¿] . . . Es una maldición? . . . Estéril será por tu obra" ["By your hand! By your hand! By your hand! . . . Is it a curse? The land left sterile by your hand"].[15] The novel recognizes the importance of the issue of land infertility, but the narrative becomes ecologically ideological when the representation of the relationship between human and earth blames the former for the destruction of the latter. In a final textual flourish that highlights the political ecology in the novel, the narrator makes an explicit connection between the destruction of the land and the eventual demise of humanity as a whole: "Esto era lo que la tierra hacía en su proceso de destrucción. Quitarles el alma a los hombres" ["This is what the land was doing as it was being destroyed: stealing the souls of men"].[16] The novel is not satisfied to merely describe the destruction of the natural world wrought by humans; it also insists on the way in which the destruction will eventually afflict humans themselves.

While the sense of nostalgia for a disappeared way of life out on the Argentine pampa was implicit in *Don Segundo Sombra* and criticism of the neo-colonial weight of English economic interests was similarly muted in *El inglés de los güesos*, these ideas are expressed quite a bit more explicitly in *Campo arado* (Ploughed land) (1953) by Argentine Ernesto L. Castro. The novel deals with competing perceptions of such changes to the Argentine land including the question of what to do with its indigenous inhabitants, the undesirability of using fences to bound it, and the disruptions to rural life brought by the railroad. In all cases, *Campo arado* adopts a conservative position, but one that can also be read as an environmentally sensitive posture from an ecocritical perspective. First, the novel criticizes the widespread use of barbed wire to enclose property titled to individual holders: "Todo se golvió surcos y alambraos. ¡Arruinar ansí el campo!" ["Everything was posts and barbed wire. What a way to ruin the country!"].[17] The argument that progress represented by the introduction of these boundaries brought ruin to the land is repeated several times throughout the novel, particularly with regard to the appearance of the railroad: "El ferrocarril al llegar le mata el caballo. . . . Comienzan los males. Tras ése vendrán otros. Es la maldición de la tierra ensangrentada"; "la vida fue siempre más simple y libre. Y tal hubiera continuado de no trastrocar el ferrocarril la quietud de tantos años" ["The railroad killed many horses. . . . This is the first sign of bad times, and there will be more. It's the curse of a blood-stained land;" "life was always simpler and more free. And it would have stayed like that, but the railroad came to upset the peace of so many years"]. The invocation of the appearance of the railroad invariably calls to mind the influence of English economic interests in Argentina and the novel makes it plain: "porque nos tiene por haraganes, el gobierno quiere quitarnos la tierra y dárselas a los gringos" ["because they think we're lazy, the government wants to take away our land and give it to the *gringos*"].[18] While the complaint about the loudness of trains could be that of any crotchety old crank bothered by the noise of modern life, the novel is specific to associate these ills with "la tierra ensangrentada" and with a governmental policy of land concessions to foreign "*gringo*" entities. The ill-ease with the way that progress is often bloody, imposed from without, more complicated, and less free is at the root of many of the regional novels that deal in nostalgia—for disappeared plains, jungles, or mountains—and challenge political ecologies that result in the permanent alteration of lands and the cultures adapted to them. The ideas are often implicit in novels like *Don Segundo Sombra* and *El inglés de los güesos*, but *Campo Arado* and several others of these somewhat less well-known novels bring them out into the open.

While the *novelas de la tierra* from the 1920s to the 1950s that I have considered in this section thus far are all fictionally located in rural settings, it was as early as the publication date in 1934 of *Madre América* by Argentine Max Dickman that several challenges to the prevailing political ecologies of urban land development appear. In Dickman's novel, an attempt is made to reckon the emigrant experience of displacement from rural areas to more urbanized small towns and cities in the Argentine interior. As an Argentine writer of Jewish heritage, Dickman is thought to be among a wave of writers arriving in Argentina that "focused upon the genre of the American Utopia. . . . It is not surprising that it was Jews who pioneered the American Utopia; this space-other, refuge, and land of asylum for Europe's persecuted, Jewish emigrants who dreamed of finding a new possibility."[19] As was so characteristic of the Jewish experience from the first half of the twentieth century, the narrative deals with the migration of people from one place to another; here, it has to do with those who arrive from a farming community around an unnamed river and begin to inhabit the small city of San Itatí. While the geographical accidents in the novel seem somewhat contrived,[20] they serve more as literary conceits to explore the experience of urban influx in the twentieth century. In the novel, it takes many forms, but there are several moments where the ecologically psychological impacts of such movements are made clear. In one such moment, Lafuente, a former inhabitant of the lands around the river, doubts whether the urbanization of the Argentine populace is really beneficial after all:

> Soy de esos hombres para los que un bello amanecer, el color de una nube, los cambios de la naturaleza, tienen un valor, un encanto superior a todas las complejidades y misterios del alma humana. . . . El progreso con su artificio ha cambiado un poco la naturaleza humana, hasta hacerla un poco insensible a las emociones directas. Hay multitudes que sienten y reaccionan mejor ante lo ficticio que ante lo verdadero. A veces impresionan más los paisajes pintados o entrevistos en el cinematógrafo, que los reales, por lo menos ante aquéllos, muchos hombres ejercen mejor su facultad de sentir que ante la naturaleza.[21]
>
> [I'm one of those men that find true pleasure and great value in a beautiful sunrise, the color of a cloud or the change of seasons, more so than in all the mysteries of the human soul. . . . The artificiality of progress has changed human nature such that we have become numb to these kinds of raw emotions. There are so many people that more readily identify with fiction than with real life. Sometimes, they are more impressed by painted landscapes or by those seen in the movies than by the real thing. Or at least it seems like they react much more profoundly to such things than to an actual experience of nature.]

The separation from the land because of exile to urbanized landscapes here embodies the idea that a landscape or a similar shot in a movie are often much more likely to resonate with the "modern" individual than the real thing. Such a sentiment has been confirmed by anthropologist Eric Hirsch in *The Anthropology of Landscape*. In the introduction to his book, Hirsch traces the history of landscapes to demonstrate the importance of painted representations in the formation of expectations about what nature ought to look like:

> What came to be seen as landscape was recognized as such because it reminded the viewer of a painted landscape, often of European origin. . . .
>
> . . . This ideal or imagined world as depicted in various genres of landscape painting . . . was linked to the perception of countryside scenery and its subsequent improvement (through landscape gardening, estate management, etc.): the goal was to achieve a correspondence between the pictorial ideal and the countryside itself.[22]

However, *Madre América* explicitly rejects such sentiments with its lament for the difference between textualized reproductions and "real" nature where the former suffers in comparison and where appreciation for the latter is slowly being eliminated by the march of progress. In the novel, characters unambiguously express their disgust with the new urban living arrangements: "¡Qué asco todo esto! ¡Qué felicidad la del aire fuerte y perfumado del campo, sin olores a fenol, sin el quejido de los enfermos y las protestas de los parientes!" ["It is all so disgusting! It is so much better to enjoy the strong and aromatic country air without the smell of phenol nor the moans of the sick nor the nagging of relatives!"].[23] Complaining in-laws seem as inescapable in the country as in the city, but it is notable that complaints about city-living are so ecological in the text: city folk are sickly, the noxious odors of the plastic-additive phenol (or carbolic acid) pollute the air and are far less desirable than the purity of the country. While Dickman's principal concern in *Madre América* was an exploration of urban relocation and its social consequences, the narrative was also somewhat prophetic in those brief moments when the ecological consequences of such relocations (although highly undeveloped in these types of novels) nevertheless were expressed explicitly, albeit briefly and almost as an afterthought.

FICTIONS OF RESOURCE EXTRACTION

Another of the issues taken up by the *novelas de la tierra* and particularly relevant to the analysis here are the numerous texts related to mining, the discovery and exploitation of oil, the industrialization of agriculture, and other forms of resource extraction. While I narrow the focus of analysis to the literary representation of the effects that these industries had specifically upon indigenous communities in chapters 6 and 7, the emphasis here will be more upon the industries themselves, the effects on nature from these industries' extractive practices, and the effects of such practices upon local populations whether *criollo*, *mestizo*,[24] or indigenous. While there are various novels in the sub-genre, those examples with the most compelling ecological elements are *Mancha de aceite* (Oil stain) (1935) by Colombian César Uribe Piedrahita (author of *Toá*, discussed in chapter 4), *Paralelo 53 sur* (Latitude 53, south) (1936) by Chilean Juan Marín, *Mene* (1936) by Venezuelan Ramón Díaz Sánchez, Honduran Ramón Amaya-Amador's *Prisión verde* (Green prison) (1945), and *Mi camarada padre* (My father, my comrade) (1958) by Chilean Baltazar Castro. These novels explore the cultural, sociological, and ecological effects of extractive and large-scale agricultural industries in Spanish America.

The first three texts all deal with the petroleum industry and were published within one year of each other: *Mancha de aceite* and *Mene* about the discovery of oil under and around Lake Maracaibo in Venezuela and *Paralelo 53 sur* about the petroleum industry in the Chilean *Tierra de fuego.* As shall be seen in the imagery of ecological contamination quoted below, these novels are sensitive to the environmental destruction that was the product of the oil industry in Venezuela and Chile and as Cósimo Mandrillo observes,

> Puede considerarse entonces que este contexto de creación de partidos, extensa difusión del pensamiento marxista, rápida aparición de organizaciones de trabajadores, especialmente en las compañías petroleras, y una visión crítica del rol que juega el gobierno nacional con relación a los intereses extranjeros, rodea y condiciona la aparición de la narrativa del petróleo escrita en las décadas de 1930 y 1940.[25]
>
> [The creation of political parties, extensive diffusion of Marxist thought, rapid unionization of workers, especially in the petroleum industry, and a critical vision of the role of the national government with regard to foreign interests can be thought of as the historical context within which these petroleum narratives appear in the 1930s and 1940s.]

The political atmosphere into which these works are inserted represent one of the most salient examples of the way in which the environmental sensibilities of a literary work can be said to have had something more than a literary effect. That is, if these works were part of a dialogue that included political organization, unionization, and political ecology, then the representation of the toxic contamination of the oil industry would have had as much of a real impact for change as a political tract or a union-organizing petition.

The first, *Mancha de aceite,* features a loosely connected "story," but its most impactful characteristic for consideration here is its use of "toxic discourse,"[26] particularly the gothic description of damage unleashed by the petroleum industry on Lake Maracaibo. The men employed by the various foreign companies are described as "hombres esclavos de una industria devoradora" ["men enslaved by a devouring industry"][27] and the most persistent ideological discourse in the novel deals with the way in which human labor is exploited to extract the oil from the ground. However, in many instances, the effects on the land are also extensively, graphically, and gothically explored. There are references to "un arroyo sulfuroso eructando vapores fétidos y calientes" ["a sulfurous river, belching steaming, fetid vapors"]; the land is fouled "bajo la lluvia espesa y pestilente del aceite maldito, que cubrió el paisaje, marchitó las hojas, ahogó los árboles, y sumió . . . los millares de insectos, gusanos y sabandijas que poblaban el suelo" ["under a thick, stinking rain of the accursed oil that covered the landscape, turned the leaves brown, swamped the trees, and drowned . . . the thousands of insects, worms, and other bugs that lived in the soil"]; "el aceite se insinuaba a través de los poros de las rocas mismas" ["the oil even got into the very pores of the rocks"]; "parecía que hasta el sol llegaban las escupas del aceite asfáltico" ["and it seemed that the oil even got on the sun, like a gob of asphalt spit"].[28] These images invoke the utter destruction of water, fauna, land, and sky; nothing can escape the contamination of the oil. The resource which is thought to bring wealth, prosperity, and development to a nation is portrayed in Piedrahita's novel as a curse: the companies that make up the petroleum industry and the way in which the land itself is so thoroughly polluted "han hecho de este pueblo y de todos los que tienen el infortunio de poseer petróleo, unos pueblos esclavos" ["has made slaves out of this, and every other society unfortunate enough to have oil in their land"].[29] The double curse of petroleum is categorized in *Mancha de aceite* as a harbinger of economic dependence and ecological destruction.

Mene, for its part, is most concerned with the Trinidadian immigrant experience on the oil fields as a source of cheap labor: the novel is sympathetic with "the ostracized Trinidadian, without family and friends, facing racism because of the color of his skin."[30] But Díaz Sánchez's novel also simultaneously laments the disappearance of lacustrine cultures through a strong expression for lost communities in boom cities like Cabimas around Lake Maracaibo, forever altered by the overwhelming changes brought by the petroleum industry: "Ésta no es Cabimas, la Cabimas de nosotros" ["This isn't Cabimas, our Cabimas"].[31] There are intense feelings that condemn the unsettling nature of change and the novel is at its most ecologically aware when the alterations have to do with the way in which natural places are utterly destroyed by the inevitable contamination of spilled oil. *Mene* decries the petro-pollution: "Una laguna de betún anegaba las casas y las plantas, ennegreciéndolo todo" ["A sea of asphalt flooded buildings and plants, blackening them both"]; condemns the invasion of foreign industries and their "maquinarias fornidas, saturadas, diríase, de un espíritu de odio contra todo lo verde" ["huge machines that saturated the land with a kind of spirit of hatred for everything green"]; and laments the way in which the industry has affected the watershed and local flora: "El petróleo, el petróleo que llena todo esto por debajo, es lo que no deja brotar el agua dulce y crecer las matas" ["It's petroleum; petroleum is what is getting into everything down below, polluting the water and stunting crop growth"].[32] But the poetic prose in the introductory pages to the second part of the novel is where the criticism is most intense for the connections it makes between the appearance of the petroleum industry in Venezuela and a litany of ecologically ill effects:

> Negras proas rasgan la feliz virginidad.
>
> La linfa gris se escinde, himen roto de América en su latitud himenal.
>
> Hace cuatrocientos años dolió por vez primera este desgarramiento. Sin embargo, era más lento, más parsimonioso entonces. Ni esta impetuosidad de tajo que acuchilla ni este voltijear que cava, ahonda y remueve las entrañas

líquidas sobresaltando el cieno y levantando la protesta de las espumas. Entonces, ni este humo negro, maculador de cielos, ni este grito bronco que llena de estupefacción el alma amodorrada del paisaje. (Díaz Sánchez, *Mene*, 16)

[The black bows of ships penetrate the virgin happiness.

These tropical latitudes now so much like that torn feminine membrane, the broken hymen of the Americas.

It's been 400 years of suffering the pain of this violation. At first, the injury was not so brutal, not like the present slashing or incessant probing of the insides, burrowing ever more deeply until the raging black eruption of slime and foam. Then: black smoke and a brutishly stupid scream that echoes above the sleepy landscape.]

The images of deflowering a virgin often associated with the original Spanish conquest seem to pale in comparison with the sexualized images of ecological catastrophe in the quoted paragraph: polluting smoke, the probing of entrails, the eagerness to violate, the foamy protest of contamination. While the novel does not offer any alternative political ecology to the economic concessions from Dutch Royal Shell and the other foreign oil companies that were behind the development of the oil industry in the Maracaibo region and responsible for the resulting toxic environment described above, the conclusion of the novel makes the implicit argument. On the final pages, a remnant of local inhabitants, tired of the contamination, the disappearance of flowers, and the lack of potable water decide to flee from the lake and make for the mountains. What had been done to their homeland was so bad that the only solution was escape.

Many of these issues are similar to an entirely different geographic latitude represented in Juan Marín's *Paralelo 53 sur*—the Chilean *Tierra de fuego*—and include several other of the extractive industries including logging, the fur trade, and mining. The parameters of the critical posture in Marín's novel were just as ecological as in *Mancha de aceite* and *Mene*, but while the issues of deforestation, toxic metal exposure, and species endangerment do not quite make an explicit expression in the novel, two passages concerning petroleum extraction reveal the stirrings of disquiet associated with ecological effects from the industry. The first deals with the noxious byproducts of oil: "De las honduras de la tierra ascienden náuseas aceitosas que rompen la costra amarilla por uno u otro lado y los gases ignescentes envenenan el aliento de los hombres y el viento de los cajones montañosos" ["Noxious petrol vapors ascend from the depths of the earth and spill out of the yellow crust here and there; fiery gases poison the breathable air and the breezes from mountain ravines"].[33] The contamination is offensive to human existence and the moral depravity of these extractive industries for the harm they have unleashed upon the lands of the Chilean Tierra del Fuego is portrayed with an imagery of oil extraction that is most ecologically ominous.

After a strike and the subsequent concession for a safer, more controlled plan to exploit an oil field in a region called Tres Puentes, engineers in the employ of foreign economic interests insist upon exploding a larger opening to the reserve. When it leads to a cataclysmic inferno of spilled oil over a vast tract of neighboring land, the engineers escape into exile in disgrace and the novel concludes with a particularly troubling image:

> Sólo quedó igual que antes, un inmenso pozo tapado de ruinas, como un ojo ciego, al cual le fue negado la luz. Inmenso ojo turbio, cuya inmóvil mirada acusadora se cierne sobre los hombres, las tierras y los mares del sur. (Marín, *Paralelo,* 203)

> [Now, all that remained was a huge pit, ringed with ruins and lightless, like a blind man's eye, a turbid eye whose sightless accusing stare ponders humankind, the land, and the southern seas.]

The threat of the oil industry to the humans and lands of the Tierra del Fuego which has been such a pressing environmental concern in recent environmentalist efforts is foretold in this brief passage.[34] The narrative in Marín's novel does not yet use the language of environmental catastrophe to which we have become all too accustomed after the Exxon Valdez and Deepwater Horizon petroleum catastrophes of 1989 and 2010, but the ominous image of a blind and accusatory eye fixed upon humankind and the lands and seas we inhabit indicates that *Paralelo 53 sur* was sadly prophetic in 1936 about the dangers and potential damage of the petroleum extraction industry.

In a similar vein to the petro-novels of the previous few paragraphs, Ramón Amaya-Amador's *Prisión verde* deals with irresponsible behavior by corporate and foreign interests that abuse local workers and unleash destructive ecological effects. Here, the object of criticism is the industrial fruit industry in the Central American nation of Honduras and attempts by workers at massive North American-owned banana plantations to organize and resist the slave-like conditions in which they labored. In *Prisión verde*, the importance of worker health, labor organization, and condemnation of abuses contains certain ecological elements. The most explicit moment occurs when the narrative describes the use of a liquid copper sulfate fungicide which is sprayed indiscriminately on the banana trees to battle the leaf-spot disease Black Sigatoka. Inevitably, some of the poisonous treatment winds up ingested or otherwise absorbed by workers and gives rise to disease:

> Se fundía la vida de los peones con la vida de los bananeros y la fuerza de las máquinas, sobre aquella tierra que pedía dolor para su fecundación. Sangre, azul de sulfato por las tuberías largas. Sangre, rojinegra sangre de hombres con los bacilos de Koch en impulsión de muerte.[35]

[The life of the peasants seemed fused to the banana groves and the powerful machines that worked them; the land seemed to require human pain in exchange for its fruit. A sulfurous, blue-tinted blood in the pipes and men, at death's door, with black-tinged red blood in their veins from Koch's bacilli.]

"Koch's bacilli" refer to the bacteria that cause tuberculosis (TB); the fatal combination of a weakened respiratory system due to exposure to atomized and blue-colored copper sulfate along with the presence of these bacteria explain the prevalence of TB infections among the workers. The reference to fungicides remains secondary in the narrative compared to the imperatives of labor organization, but presently it represents a significant current environmental health concern among farm workers.[36]

The novel does not just implicitly advocate for the relief of environmental health concerns among agricultural laborers, but it also expresses a sense of regret for the deforestation of the Honduran countryside to provide land for the sprawling plantations:

> Los boscajes incultos y las haciendas desaparecieron ante el machete y el hacha de las cuadrillas desmonteras. . . .
>
> . . . Siguieron las cuadrillas de hacheros jóvenes haciendo temblar la naturaleza con el eco de sus hachazos, en los troncos gemidores de las arboledas que iban cayendo, como vencidos monstruos sobre el dorso de la tierra humedecida; los guayacanes y caobas . . . le fueron cedidos a mister Still como madera inútil . . . y el gringo, organizó un "benque" magnífico cuya producción en pies cúbicos de madera exportada fue fabulosa ganancia. (Amaya-Amador, *Prisión verde*, 88–90)

> [Wild stands of trees along with nearby haciendas disappeared, hacked down by squads of clear-cutters with their machetes and axes. . . .
>
> . . . These squads of young loggers just kept cutting as the whole of nature trembled at the sound of axes against the moaning bark of trees falling upon the face of the dewy earth like slain monsters; evergreens and mahogany trees . . . were given to Mister Still like so much useless wood . . . and the *gringo* organized a magnificently profitable operation whose countless square feet of exported wood was turned into an incredible profit.]

The exportation of vast Honduran wealth to foreign corporations and individuals and the transformation of the countryside into agricultural tracts carry certain ecological implications that are perhaps clearer now than they may have been in 1945 when Amaya-Amador's novel was published. This is especially true after the difficulties in controlling Black Sigatoka and Panama disease, a similar fungal infection that afflicts the banana plant. The large agro-industries battled the plagues with anti-fungal treatments, but could find no chemical solution to Panama disease. The solution was to "adoptar un sistema de plantación móvil, que consistía en abandonar las plantaciones invadidas por la enfermedad, desmontar una nueva selva y establecer una nueva plantación" ["to adopt a kind of mobile plantation system which consisted of the abandonment of blighted plantations, clear-cutting a new tract of land in some other jungle, and the establishment of a new plantation"].[37] The effect of the strategy was as predictable as it was ecologically disastrous: the fungus followed the banana plantations wherever they went and the changes to the landscape were permanent: "Its transformation of the landscape was so sudden and so ephemeral that banana planting left not even the shadow of a ghost town but only deforested tracts littered with sickly banana stumps between which poor farmers tried to eke out a bare subsistence."[38] *Prisión verde* documents all of these social, economic, and ecological effects and the importance of the possession of local land by local individuals is emphasized in the text when the small-holder Lucho López refuses to sell to the North American company and declares a genetic connection to the land: "¿No comprenden ustedes que esa es mi heredad, que estoy ligado a ella con todas las fuerzas de mi vida? Soy como un árbol, tengo mis raíces muy adentro de esa tierra" ["Don't you understand that this is my inheritance, that I am attached to this place by my heart, soul, strength, and mind? I am like a tree with roots that go down deep in the land"].[39] After the expression of such a sentiment, the deforestation and large scale transformations later in the novel take on something more than an implicit argument against such practices.

Finally, (and while there are dozens of novels from the middle part of the twentieth century that deal with themes related to the Chilean mining industry, very few explore ecological issues related to that industry), I consider the exceptional *Mi camarada padre* by Baltazar Castro whose novel is principally concerned with the organization of a miners' union, but in which an ecological consciousness also clearly emerges. One moment describes the pollution over neighboring populations that is the result of the various metal refining processes in a city near the larger Sewell called Caletones,

> cuya presencia podía advertirse desde treinta kilómetros antes de arribar por la gigantesca chimenea que repartía los humos de los hornos al cajón cordillerano: a la postre serían los sewelinos los que respirarían las negras nubes con que cubrían el cielo los caletoninos.[40]
>
> [whose presence could be perceived up to 30 kilometers away because of the huge smoke stack that spewed smoke from the furnaces up over the mountain ravines: in the end, it would be the residents of Sewell that would breathe in the black, smoggy clouds sent up to the sky from Caletones.]

Another moment expresses frustration at a political ecology of corrupt concessions to North American companies: "Las minas están en Chile; los obreros que se meten a trabajarlas son chilenos; los cementerios en que enterramos a los mineros silicosos son chilenos, pero nada tenemos que ver con nuestro cobre" ["The mines are in Chile; the workers that work in them are Chilean; the cemeteries where we bury the miners who die from silicosis are in Chile, but Chileans don't receive one cent from the copper that we ourselves mine from the mountain"].[41] The way in which foreign companies abuse the Chilean labor force and the efforts of the protagonists in Castro's novel make up the bulk of the narrative, but the questions of access to resources and the condemnation of industrial mining for its creation of toxic environments that make diseases like silicosis commonplace are ecological corollaries to the critical perspective in *Mi camarada padre*. Castro's novel is typical of others like it that deal with industrialized agriculture and extractive operations: they emphasize the effects upon workers of toxic oil spills, of the use of dangerous pesticides, of the contamination of air by smokestacks, of the economic fallout from deforestation, of the harmful pulverization of silica in copper mines. Yet as each of these industrial practices are condemned for the way in which they produce unbreathable air and unleash afflictions like tuberculosis and silicosis, the implicit corollaries to the explicit ethical ideology of worker welfare represent ecological ideas that remain central to current environmentalist doctrine: concern about the production and consumption of oil for its contamination of land, water and air; advocacy for bans on the use of pesticides; efforts to resist deforestation; and the condemnation of industrial mining practices for its noxious by-products.

In the end, these works of regional literature, the *novelas de la selva* and *de la tierra* are the literary forebears to the more recent texts like Chilean Luis Sepúlveda's *Un viejo que leía novelas de amor* (*The Old Man who Read Love Stories*) (1989) where the ecological elements are explicitly couched in terms familiar to the environmentalist advocacy of the present. Yet, Sepúlveda's novel, with its condemnation of deforestation, oil installations, and the commodification of precious metals and furs, deals with many of the very same ecological issues first raised in the novels considered above. These earlier works deserve a more prominent place in the canon for the way in which they anticipate the environmentalism of Spanish American literature before it became more explicitly thus. In the next two chapters, I select a corpus of texts generally considered examples of the *indigenista* genre, but many of the concerns that I have discussed in this chapter concerning mining, industrialized agriculture, deforestation, and the petroleum industry are common to those works as well. The ethical emphases in the *novelas de la tierra* are similar to those of the Spanish American *indigenista* tradition: condemnation of the abuse of labor, of the accumulation of vast tracts of land by ever fewer individuals, and of the industrial practices of local, but more usually, foreign corporate entities. In the *indigenista* context, this means criticism of the appropriation of traditional tribal lands and of the slave-like labor conditions in mines, fields, and quarries as well as the representation of the validity of indigenous economic models. Such are the ideological elements that undergird the analysis in chapters 6 and 7.

NOTES

1. See Introduction, note 43.
2. Luis Vitale, *Hacia una historia del ambiente en América Latina* (Mexico City: Editorial Nueva Imagen, 1983), 77.
3. Miguel Tinker Salas, *The Enduring Legacy: Oil, Culture, and Society in Venezuela* (Durham, NC: Duke University Press, 2009), 7.
4. Carlos J. Alonso, *The Spanish American Regional Novel: Modernity and Autochtony* (New York: Cambridge University Press, 1990), 116.
5. Julie Skurski, "The Ambiguities of Authenticity in Latin America: *Doña Bárbara* and the Construction of National Identity," *Poetics Today* 15, no. 4 (Winter 1994): 624–25.
6. Ibid., 633–34.
7. Alonso, *Spanish American Regional Novel*, 109.
8. José Castro-Urioste, "Utopía y transgresión: El imaginario nacional en la obra de Rómulo Gallegos," *Hispanic Journal* 17, no. 2 (Fall 1996): 336.
9. Jennifer L. French, *Nature, Neo-Colonialism, and the Spanish American Regional Writers* (Hanover, NH: Dartmouth University Press, 2005), 71–111.
10. Ibid., 79–80.
11. Benito Lynch, *El inglés de los güesos* (Mexico City: El Libro Popular, 1955), 129.
12. Olga Uliánova, "Levantamiento campesino de Lonquimay y la Internacional Comunista," *Estudios Públicos* 89 (Summer 2003): 175.
13. Reinaldo Lomboy, *Ranquil: Novela de la tierra* (Santiago de Chile: Editorial Orbe, 1966), 32.
14. Jaime Ibáñez, *Cada voz lleva su angustia* (Bogotá: Ediciones Gamma, 1961), 24.
15. Ibid., 74.
16. Ibid., 235.
17. Ernesto L. Castro, *Campo arado* (Buenos Aires: Losada, 1953), 48.
18. Ibid., 57, 98, 161.
19. Leonardo Senkman, "Jewish Latin American Writers and Collective Memory," in *Tradition and Innovation: Reflections on Latin American Jewish Writers*, ed. Robert E. DiAntonio and Nora Glickman (Albany: State University of New York Press, 1993), 33.
20. Itatí is a real small town in Argentina and the river referenced in the novel must be the Paraná which borders Itatí to the north.
21. Max Dickman, *Madre América* (Buenos Aires: Santiago Rueda, 1960), 107–8.
22. Eric Hirsch, "Introduction: Landscape: Between Place and Space," in *The Anthropology of Landscape: Perspectives on Place and Space*, ed. Eric Hirsch and Michael O'Hanlon (Oxford: Clarendon, 1995), 2.
23. Dickman, *Madre América*, 126–27.
24. *Mestizo* refers to individuals of mixed Spanish/indigenous heritage while *criollo* means an individual born in the Americas, but of Spanish (as opposed to indigenous or genetically mixed) heritage.
25. Cósimo Mandrillo, "Antiimperialismo, tecnología y petróleo en la novela venezolana," *Enl@ce: Revista Venezolana de Información, Tecnología y Conocimiento* 9, no. 1 (January–April, 2012): 54.
26. That is, the language of environmental toxicity, as detailed in chap. 2. According to Lawrence Buell, *Writing for an Endangered World* (Cambridge, MA: Harvard University Press, 2001), 30–54, toxic discourse contains four different types of defining characteristics: the promotion of awareness about toxicity, the representation of a world without refuge from toxic penetration, the expression of outrage against the hegemonic centers of power perceived as responsible for the toxicity, and the gothic representation of the consequences of toxicity.
27. César Uribe Piedrahita, *Mancha de aceite* (Bogotá: Oveja Negra, 1985), 80.

28. Ibid., 11, 26, 27.
29. Ibid., 50.
30. Tinker Salas, *Enduring Legacy*, 130.
31. Ramón Díaz Sánchez, *Mene* (Buenos Aires: Editorial Universitaria de Buenos Aires, 1966), 55.
32. Ibid., 80, 38, 147
33. Juan Marín, *Paralelo 53 sur* (Santiago de Chile: Editorial Nascimiento, 1955), 12.
34. The Wildlife Conservation Society in particular has targeted the region as part of its activist efforts throughout the world. See the Tierra de Fuego place page on their website http://www.wcs.org/where-we-work/latin-america/argentina.aspx (accessed January 7, 2013).
35. Ramón Amaya-Amador, *Prisión verde* (Buenos Aires: Agepe, 1957), 100.
36. Dixie E. Snider et al., "Tuberculosis and Migrant Farm Workers," *JAMA* 265, no. 13 (April 3, 1991):1732; J. E. Poss, "The Meanings of Tuberculosis for Mexican Migrant Farmworkers in the United States," *Social Science Medicine* 47, no. 2 (July 1998): 195–202; Centers for Disease Control and Prevention, Morbidity and Mortality Weekly Report, "Prevention and Control of Tuberculosis in Migrant Farm Workers Recommendations of the Advisory Council for the Elimination of Tuberculosis," Centers for Disease Control and Prevention, http://www.cdc.gov/mmwr/preview/mmwrhtml/00032773.htm (accessed December 6, 2011).
37. Santiago Muñoz Arbelaez, review of *Banana Cultures: Agriculture, Consumption, and Environmental Change in Honduras and the United States* by John Soluri, *Historia Crítica* 38 (May-August, 2009): 210.
38. Shawn William Miller, *An Environmental History of Latin America* (New York: Cambridge University Press, 2007), 132.
39. Amaya-Amador, *Prisión verde*, 8.
40. Baltazar Castro, *Mi camarada padre* (Santiago de Chile: Zig-Zag, 1958, 1967), 35.
41. Ibid., 40.

Chapter Six

Ruin: The Precedents of Ecological Destruction in Early and Canonical *indigenista* Novels

The ethical gesture of the *indigenista* text has had a long history. Writing in the 1540s and 1550s, Spanish Dominican monk Bartolomé de las Casas decries "las matanzas y estragos de gentes inocentes, y despoblaciones de pueblos" ["the killing and ruin of innocent peoples and the general depopulation of their towns"].[1] More recently, *Los caminos de Paxil* (The ways of Paxil) by Guatemalan Arturo Arias represents a relatively recent entry in the *indigenista* tradition published in 1990, but features similar condemnations of the killing of native peoples and destruction of the places where they live. The latter novel features a conflict between the employees of a U.S. oil company and a Mayan community in Guatemala that comes down to different perceptions of the land, but with ecological considerations now an explicit aspect of the advocacy. *Indigenista* texts often go to extremes in their condemnation of practices damaging to indigenous peoples by portraying their victimizers as unequivocally evil; Arias's text is not different in this respect. The novel lampoons the hostility of the corporate world toward nature by having Emerson, the comically evil foreman of the oil company, say "—Los árboles y matorrales van cayendo; los sustituye el asfalto y el acero. Yo aplaudo la caída de cada árbol" ["—The trees and thickets are coming down substituted by asphalt and steel. I applaud the felling of each tree"].[2] Here is obviously wry irony—no one would actually say something like this out loud—and while the larger critique of the novel employs standard *indigenista* rhetoric to condemn the destruction of a Mayan community, the argument in *Los caminos de Paxil* rests on premises that include an ecological idea about deforestation as part of its ethical position.

In *Los caminos de Paxil*, the conflict is consummated with the massacre of the indigenous resistors by the national militia. While the eventual outcome is tragic, the text attempts to force a recognition of the legitimacy of Mayan claims to traditional lands according to local conceptions of place as opposed to a valuations imposed by external entities. In the novel, the invocation of Mayan mythology to make a case for the condemnation of the oil extraction enterprise invokes another of the classic ideological conventions of the *indigenista* genre: the role of spiritual belief in a native community's stewardship of their land. The criticism is not articulated as a legal argument about property rights nor as a social argument about civil rights; rather, religious convictions about how supernatural entities expect humans to behave are what counts for indigenous communities and their interactions with the land. And, as in many other texts from the tradition, Arias's novel explores the question of social injustice in the context of religious belief, labor, land use, and indigenous rights. But *Los caminos de Paxil* also expands the boundaries of the genre because the ethical rhetoric that condemns the oil drilling endeavor includes condemnation of the contaminating tendencies of the petroleum industry, support for a program of conservation of traditional lands, and environmental justice for the Mayan community. Because ecological concerns are so central to the novel's ethical ideology, *Los caminos de Paxil* suggests that an ecocritical reconsideration for the whole of the *indigenista* genre is in order: what I document here and in chapter 7 is the extent to which these same elements were at play in earlier titles.

Arias's novel was among the first to explicitly integrate ecological concerns with the ideology of an *indigenista* text, but it was not the first. In *Me llamo Rigoberta Menchú* (*I, Rigoberta Menchú*) (1983) (edited and arranged for publication by Elizabeth Burgos-Dubray), the testimonial novel that lead to the awarding of the Nobel Prize for Peace to Rigoberta herself, similar concerns are also expressed. However, the controversy generated by anthropologist David Stoll's investigation of the factual elements of the story seems to have eclipsed the analysis of other elements in her testimony.[3] Not only has the debate over Menchú's claims garnered more of the critical attention, but the absolutely appalling episodes of abuse, torture, assassination, and terrorization of Guatemala's indigenous populations are so wrenching that it is hard to pay attention to anything else.[4] Nevertheless, the inclusion of K'iche' cultural attitudes toward nature in the text deserves further exploration and corroborates my argument that ecological concerns represent an important element from previous *indigenista* literature.

In her account, Rigoberta communicates the K'iche' love for nature, for trees, animals, and plants in terms of familial affection, spiritual belief, and mythological identity: "La tierra es como una madre que multiplica la vida del hombre" ["The land is like a mother that multiplies human life"]; "hay que saber respetar a la naturaleza. Saber respetar a los árboles, a la tierra, al agua, al sol y saber respetar al hermano" ["it is important to know how to respect nature. To know how to respect trees, the land, the water, the sun, and to know how to respect other human beings"].[5] One of the central elements of the *indigenista* tradition has been to challenge the characterization of indigenous individuals as somehow less than human; in Menchú's testimony, that means respect for humankind but also for the natural world upon which humans depend. According to the narrative, the attitude of respect for the natural world is instilled in members of the K'iche' community from before they were born. In their tradition, the seventh month of pregnancy marks an important moment in the ecological education of the yet-to-be-born fetus:

> Es cuando la señora se pone en contacto con toda la naturaleza, según nuestra cultura. Saldrá al campo, irá a caminar en el monte. Así el niño está encariñándose con toda la naturaleza. . . . Empieza a platicar constantemente con su hijo, desde cuando está en su vientre. . . . Es como si estuviera acompañada de un turista, donde le explica las cosas. Por ejemplo; "De esta naturaleza nunca tienes que abusar. . . ." (Burgos-Debray, *Rigoberta*, 40)

> [This is when the mother-to-be seeks intimate contact with all of nature, according to our culture. She will go out into the country and walk around in the wild. This way, the child begins to cherish all nature. . . . The mother begins to constantly speak with her child, even from when he or she is in the womb. . . . It's like going around with a tourist, explaining things. For example: "You must never abuse nature. . . ."]

Throughout her testimony, there are similar prohibitions: against the over-exploitation of nature, against the wasting of water, against the careless use of the land. She refers to her community of Chimel as a paradise because of "todo lo lindo que es la naturaleza en esos lugares ya que no hay carreteras, no hay vehículos" ["how beautifully natural it is there because there are no highways and no vehicles"].[6] The rejection of modernization is manifest in these lines for the pastoral quality that the lack of development lends to her home town, but later in the narrative, such sentiments are incorporated into the ideological critique that is at the heart of the *indigenista* character of her testimony: the toxic effects of crop fumigation, corporate deforestation, and the decrease in life expectancy among indigenous communities for the use of chemical fertilizers and other synthetic agents that have replaced traditional agricultural techniques.

In an account of the first of the Menchú family to die from contact with the non-indigenous world, Rigoberta describes how an older brother succumbs to toxic poisoning on a coffee plantation: "Habían fumigado el café, con el avión como acostumbran hacerlo mientras nosotros trabajamos, entonces mi hermanito no aguantó el olor de la fumigación y se murió intoxicado" ["They were fumigating the coffee by air which they always did, even if we were working in the fields. My brother could not take the stench of the fumigation and he died from its poisonous chemicals"].[7] The facts of her brother's death represent an indictment of chemical-based agricultural practices with which all readers can identify. The injustice of the death is plain; but we all also drink coffee or know someone who does and are thus involved personally as potential consumers of poisonous products (or supporters of inhumane industries). Later in the narrative, Rigoberta refers to her grandfather who blames the use of such chemicals for the decrease in life expectancy among the K'iche':

> Él dice que durante su niñez él vio a gentes con ciento diecisiete, ciento quince años que estaban vivos. Mujeres con ciento diez años. Él decía, ustedes hijos, ustedes no son los culpables, los culpables son los aparatos modernos que vinieron a nuestra tierra. . . . Decía, es que ahora ustedes comen cosas químicas y eso no les permite la larga vida que tienen que tener. (Burgos-Dubray, *Rigoberta*, 299–300)
>
> [He says that when he was younger, he remembers people living to one hundred seventeen, one hundred fifteen years old. Women who lived to one hundred ten. He would say, you young folks, it's not your fault, it's those modern machines on our land. . . . He would say that you all eat things with chemicals in them, chemicals that cut short the long lives that we should all enjoy.]

Again, the reference to chemical products intimately involves the reader: all people living in modernized societies are exposed to food chemically aided in some way and wish for the same long life that the K'iche' apparently used to have but no longer do. Even though most readers are not a part of an indigenous community, Rigoberta's testimony provides an intimate glimpse through the incredible power of her witness, but also through the invocation of ecologically harmful industrial practices to which we are all subject. Throughout her narrative, there are several other similar statements of political ecology, conservationism, condemnation of deforestation, and other ecological concerns. And despite the fact that these are overshadowed by the ideological message of her narrative concerning political corruption, military brutality, and tactics for indigenous resistance, they seem to inform Rigoberta's critique at a basic level. Although the testimonial novel as a genre was a relatively recent arrival on the Spanish American literary scene, *Me llamo Rigoberta Menchú* follows many of the conventions of the traditional *indigenista* novel. The articulation of ecological ideas as central to the defense of indigenous communities in *Los caminos de Paxil* and *Me llamo Rigoberta Menchú* through the well-established form of the *indigenista* novel suggest that such ideas were present all along, even back to the earliest examples of the genre.

THE SACK OF THE LAND: ECOLOGY AND EARLY *INDIGENISMO*

In *Los caminos de Paxil*, Justin, a junior manager in the oil company, begins to develop an intimate, facts-on-the-ground relationship with the realities of what dislocation of native peoples will mean for the physical and cultural survival of the nearby Mayan community and of the ecological harm that will befall the sites targeted for oil exploration around where they live. In *Gray World, Green Heart*, Robert L. Thayer Jr. discusses the sense of regret that arises from growing awareness of the negative consequences of modernized life: "It is less and less possible to conduct one's personal life innocently—to live quietly within one's own realm without some degree of awareness and an increasing portion of shared responsibility for environmental problems." Thayer calls it "environmental guilt," the tendency to "*feel guilty about what technological development has done to the landscape, to nature, and to the earth*" (emphasis given).[8] The problem with environmental guilt however, unlike feelings of guilt that one might experience for having committed a single bad act, is that humans must experience it as an ongoing sense of regret. The environmentally damaging products and by-products of modernization must continue to exist in a society that has incorporated them so fully into its culture; it is difficult to imagine life without electricity, gas engines, and industry. Justin's awareness of the harm that the company's oil drilling project will do to the indigenous community's traditional lands contributes to his own strong sense of environmental guilt and cognitive dissonance: he cannot square his role as an employee of the company with a moral sense that the dislocation of indigenous communities represents undeniable injustice. To reconcile the dissonance and eliminate his feelings of guilt, Justin acts radically to bring the dissonant action into consonance with his beliefs and his transformation represents a kind of spiritual identification with the Mayan perspective.

He tells the indigenous community of his affinity for their culture: "Nos dijo que venía a unirse a nuestro mundo. Que él ya estaba cansado del desperdicio de la civilización" ["He told us that he was coming over to our side, to our world. That he had grown tired of the excesses of civilization"]. As the novel progresses, he moves further and further toward the Mayan perspective as he exclaims that he would never consent to "este insensato robo, nunca botar un árbol sin pedirle permiso al guardián de la selva y al guardián de la estrella, nunca, no, nunca" ["this senseless robbery nor fell a tree without first asking permission of the guardian of the jungle and the guardian of the stars, never, no, never"].[9] The fundamental motivation in the development of a character like Justin in Arias's novel is similar to many of the *indigenista* novels of the nineteenth century; it began with a dawning awareness of the fate of indigenous communities in the Americas as their communities were disrupted, their lands surrendered, and the forests cut down on a much larger scale.

If the promotion of awareness about injustices suffered by indigenous communities is accepted as a primordial definition of the genre, then *indigenista* literature can trace its roots in Spanish American literary history to the moment that the very first texts were produced in and about the Americas. In general, these contain representations of abuse of native peoples by landowners, priests, government officials, judges, and other representatives of the elite. The extensive history of *indigenismo* as a literary genre owes its long record to a tradition of injustice established from the very first steps Europeans took upon the lands that were later called the Americas. After meeting a community of native Taino inhabitants on the first island to which his expedition had arrived somewhere in what are now the islands of the Bahamas, Christopher Columbus disembarks and, in the first act of imperial acquisition in the western hemisphere, plants the royal Spanish flag on the beach:

> El Almirante llamó a los dos capitanes y a los demás que saltaron en tierra, y a Rodrigo d'Escobedo escribano de toda el armada, y a Rodrigo Sánches de Segovia, y dixo que le diesen por fe y testimonio cómo él por ante todos tomava, como de hecho tomó, possessión de la dicha isla por el Rey e por la Reina sus señores, haziendo las protestaçiones que se requirían, como más largo se contiene en los testimonios que allí se hizieron por escripto.[10]
>
> [The admiral gathered the two captains and the others who had come ashore, as well as the scribe Rodrigo d'Escobedo and also Rodrigo Sánches de Segovia, and swore an oath before the whole gathering and pronounced with all the required, official declarations that the lands upon which they were all standing were at that point, and henceforth, possessions of their Royal Majesties, the King and Queen of Spain, declarations that were later written down, properly signed, and duly witnessed so as to make the pronouncement official.]

Thus, with the stroke of a pen and a proclamation neither heard by nor remotely comprehensible to the local residents, indigenous lands become Spanish and all native rights to them are automatically forfeit. Later, these lands are not only possessed by Spanish conquistadors and settlers, but all indigenous inhabitants that were then living within the boundaries of each parcel are likewise and irrevocably enslaved by the individual granted "legal" title; the system was called the *encomienda* and was the principle process by which the property of the "new" world was transferred to new "owners" from the old. These processes which were unendingly repeated throughout the sixteenth century in Latin America had long-lasting and profound impacts upon notions of race, access to resources, concepts of property, and economic power: "The development of racial ideologies and systems has been driven by the appropriation and exploitation by White/European-identified people of land and other natural resources (mineral, forests, human bodies) that had been lived in and managed by other people."[11] These land-grabs affected ideas about racial identities, but they also established a kind of political and ideological prominence enjoyed by Spanish and other European invaders for the control that they could exert upon the land and over access to its resources. Inevitably, this state of affairs led to profound environmental change in the land itself.

As Shawn William Miller notes, the first catastrophic ecological alteration was human: facilitated by humans and nearly universally fatal, it meant a reduction in indigenous human population from somewhere near 50 million to only 12 million by 1750.[12] The native inhabitants were decimated by warfare and Spanish imperial policies, but far more extensively by European and African diseases brought by conquistadors, missionaries, slaves, and other settlers. However, the importation of European nutritive flora and fauna thrived in the Americas as did the production of food crops from native plants after the introduction of European agricultural techniques: cattle, sheep, swine, bananas, sugar cane, rice, and the indigenous potato, maize, cacao, and pineapple. This newly abundant food supply led to a demographic recovery among the original inhabitants, but the socio-economic landscape was profoundly altered: because of the earlier depopulation, "many Indian groups lost access to their personal lands due to Spanish usurpations."[13] Thus, with the exception of the first example from the *indigenista* corpus by Bartolomé de las Casas who was writing just a few decades after the first contact between native communities and European conquistadors, the *indigenista* genre appears in the context of an already socially disadvantaged indigenous population.

The novels themselves adopt various slants to expose and condemn different aspects of abuse relating to the victimization of the indigenous character, but the narrative was often developed from a concerned, Spanish or *criollo*[14] perspective. Later texts, in the mode of a kind of *neo-indigenismo*, began to fictionally adopt an indigenous narrative voice to tell these tales, even though the authors could still not be said to share that identity. Antonio Cornejo-Polar characterizes earlier manifestations of the genre as a "literary reproduction of the structure and history of disintegrated societies"[15] while Mabel Moraña defines it as "un esfuerzo de diferenciación . . . como respuesta a los modelos de racionalidad iluminista y modernización" ["an effort at differentiation . . . as a response to Enlightenment thinking and modernization"].[16] And Jean Franco calls neo-*indigenismo* an "attempt to comprehend the Indian mind through his mythology, poetry and legend."[17] Defined as such, *indigenismo* has a long history indeed.

As I mention above, the very first example of the genre appeared in the mid-sixteenth century, by Bartolomé de las Casas and only fifty years after

Columbus's arrival in 1492 to what began to be called the "new world" in the 1500s. Las Casas wrote *Brevísima relación de la destrucción de las Indias* (*A Short Account of the Destruction of the Indies*) in 1542 and published it in 1552 as a summary addressed to Prince Felipe II of Spain recounting the violent sacking of the newly discovered lands and the genocidal extermination of indigenous groups wherever the Spanish went, from Florida to the Southern Cone. Thus, about 300 years before the nations of Spanish America declared their independence from Spain, the contours of *indigenista* ideology was to emerge from las Casas's accounts of the conquest of the original inhabitants from present-day North, Central, and South America and the Caribbean. While it would be anachronistic to think that las Casas articulated as explicitly an environmentalist agenda as in *Los caminos de Paxil* or even as implicitly an ecological position as in the other *indigenista* texts to be considered in this chapter, there are many aspects of the denunciation of the extermination of indigenous communities in the *Brevísima relación* with which ecological concerns are inherently conflated.

The first of these concerns—ecosystem disruption—appears as an implicit element of las Casas's account of the enslavement of native peoples on the islands around Hispaniola—Cuba, Puerto Rico, Jamaica, and the lesser islands of the Bahamas: "La más sana tierra del mundo, en las cuales había más de quinientas mil ánimas, no hay hoy una sola criatura. Todas las mataron trayéndolas y por traellas a la isla Española, después que veían que se les acababan los naturales della" ["The most healthful land in the world in which there were more than five hundred thousand souls. They were all killed either on the way to or after arriving at the island of Hispaniola, an island whose own indigenous population had already been finished off."][18] The Spanish lust for gold and for slave labor to help them get it result, according to las Casas, in the extinction of human populations on entire islands; the elimination of the top link in an ecosystem's food chain meant widespread ecological alterations as documented above. The reality of microbial infection and the human, livestock, and mosquito vectors that made viral and bacterial epidemics so extensive were poorly understood when las Casas was writing. That the dramatic military conflicts and massacres were not the only cause of the demise of indigenous populations does not lessen the validity of las Casas's criticism of a Spanish conquest as often accompanied by bloodshed as by a methodical pillaging of the land through trampling, fire, and senseless consumption.

Although the Dominican monk is always politically careful to avoid naming names, the Cátedra edition of las Casas's text that I consulted identifies Captains Pedro de Alvarado and Cristóbal de Olid[19] as the perpetrators of systematic desolation of land and population in Guatemala: "haciendo matanzas y robos, quemando y robando y destruyendo donde llegaba toda la tierra con el título susodicho" ["killing, robbing, burning, and destroying wherever they went on the above-mentioned island"].[20] In Venezuela, las Casas laments the concession of government and legal jurisdiction to German economic interests for the barbarous destruction visited upon the land by governors like Ambrosio Alfinger, Jorge Spira (Spier), and Nicolás Federmann.[21] One of these three ventures inland and leaves such a wake of destruction that even the later-arriving Spanish "tyrants" (las Casas's term for most Spanish captains, commanders and other authorities) are impressed by the carnage:

> Hallaron toda la tierra más de doscientas leguas tan quemada y despoblada y desierta, siendo poblatísima y felicísima como es dicho, que ellos mesmos, aunque tiranos y crueles, se admiraron y espantaron de ver el rastro por donde aquél había ido, de tan lamentable perdición. (las Casas, *Brevísima relación*, 150)

> [They found more than 200 leagues of land which had previously been happily inhabited by many peoples, so burned, depopulated, and deserted that they themselves, though they were also cruel and tyrannical, couldn't help but be impressed and somewhat shocked to see the path of ruin and destruction left behind by that German.]

In each of these moments from *Brevísima relación* (and in many others not cited above), las Casas connects indigenous victimization and depopulation to the Spanish desire for gold and slaves and particularly to the destruction of the land. His account is not only titled for the genocide of indigenous groups, but for the destruction of an entire super-region from the southern coasts of present-day United States to the territories surrounding the Río de la Plata.

As discussed in chapter 1, the nineteenth century marked the inauguration of Spanish American literature as a separate body of work from that produced by Spaniards and Spanish *criollos*, but it also came in the midst of an indigenous population recovery after the Conquest. Thus, the development of the *indigenista* genre in the nineteenth century occurs during a demographic increase among poor, landless, and largely cultureless ethnic people groups in the context of a political contest for the articulation of national identities for newly independent American states. From there, the form eventually matures as a literary genre in the twentieth century,[22] but it develops at a time when there were both antipathetic and sympathetic ideas with reference to indigenous groups. Opposition to Argentine Domingo Faustino Sarmiento's characterization of indigenous elements in Argentina as manifestations of barbarity begin to appear even in early works of *indigenismo* such as the historical novels *El último rei de los muiscas* (The last king of the Muiscas) (1864) by Colombian Jesús S. Roza and *Enriquillo* (1882) by Dominican Manuel de Jesús Galván. The representation of the harmonious relationship between indigenous communities and nature in Ecuadoran Juan León Mera's *Melodías indígenas* (Indigenous melodies) (1858) and *Cumandá* (1879) and in *Tabaré* (1888) by Uruguayan Juan Zorilla de San Martín, the articulation of indigenous perspectives in the context of the *cautiva*[23] story *Huincahual* (1888) by Chilean Alberto del Solar, and the representation of abuse and victimization in Peruvian Clorinda Matto de Turner's seminal novel *Aves sin nido* (*Birds without a Nest*) (1889) all represent a counterpoint to Sarmiento's dismissal of indigenous groups and effect positive representations meant to defend and protect the rights of such communities.

Among other things, *indigenista* literature offers a representation of native communities in such a way so as to actively combat the negative associations perpetuated about them in various expressions of Spanish American culture. Jesús S. Roza's *El último rei de los muiscas* does not contain any of the classic elements of the genre such as a defense of indigenous rights, advocacy for land reform, or a general plea for justice and retribution for wrongs suffered; but it does attempt a vindication of Muisca culture. With Roza's fairly detailed narrative that recreates the demise of their last king, several ecological principles implicitly and inevitably emerge. The novel concerns events that unfold largely before the conquest of their land and culture by the Spanish, but details several noteworthy moments from Muisca mythology. One of these is an episode from the people's ancient history:

> Hubo un dia en que un hombre de instintos perversos se rebeló contra el órden establecido, rompió los lazos que los unian a sus semejantes, i, hollando todo sentimiento de razon i de justicia, echó una cerca en derredor de un campo i dijo: "ESTO ES MIO." Fijado el primer eslabon de robo, la cadena se aumenta dia por dia, hasta que las jeneraciones desheredadas se convocan, se juntan, i se dicen: "La tierra ha sido criada por Chiminigagua para todos sus hijos, él no ha esceptuado de esa herencia preciosa a ninguna jeneracion, i si ha habido hombres que se la hayan usurpado, derribemos sus cercas i castiguemos su detentacion."[24]
>
> [One day, a perverse man with base instincts decided to rebel against the established order, broke off all relationships with his fellow humans, and thinking it just and right, put up a fence around a field and said, "THIS IS MINE." After the first stone was set in the foundation of injustice, it had daily gone on in this way until future generations that lost their land gathered together and said amongst themselves, "The land has been created by the god Chiminigagua for all his children, he has not excluded any generations and if some men have gone against his will, then we should knock down their fences and punish their haughtiness."]

After the people's complaint reaches heaven, the god Chiminigagua sends his son to reinstruct the community concerning the arts, science, the sun as the proper object of worship, and, most critically for the analysis here, the sacred command to respect "el derecho ajeno" ["the rights of others"].[25] In the passage, Roza recreates Muisca mythology where putting a boundary around, disinheriting future generations from, and concentrating ownership of the land are portrayed as highly objectionable actions, the opposite of civilization, and meriting divine punishment. In this way, the novel addresses one of the fundamental polemical elements of the *indigenista* genre: the appropriation of ancestral indigenous lands. As shall be seen in the rest of the chapter, the central issue comes down to ecologically competing notions of place; as in *El último rei de los muiscas*, a defense of the validity of perspectives among the original inhabitants toward their own traditional lands is a central component of the *indigenista* critique.

Manuel de Jesús Galván's historical novel *Enriquillo* retells the story of a revolt against Spanish rule in the context of colonial intrigues involving romantic entanglements, indigenous uprisings, and faithless Spanish colonizers and conquistadors. The majority of the novel focuses upon various romantic pairings, unrequited passion, political marriage, and other amorous frustrations associated with love in the time of colonial rule. And as Doris Sommer argues, Galván's novel represents a gloss or "a kind of *midrash* on the earliest chronicles of Bartolomé de las Casas and Gonzalo Fernández de Oviedo" that functions as an "ideological shelter against the imputation of African (that is, Haitian) identity in a country of dark people."[26] The political purposes to which Galván put his novel in the second half of the nineteenth century are convenient for my analysis because it allows for an early glimpse of textually represented nature, concerned as the novel was with Enriquillo's revolt headquartered in the rugged sierra of Bahoruco in the southwest of what is now the Dominican Republic. The representation of wild nature in the context of a political revolt against Spanish rule inspired principally upon the unjust nature of the *encomienda* strongly associates the novel both with a dual narrative of sublime nature and the problems with a political ecology of conquest and domination.

There are several moments where the sublime role of nature as an idealized site for political, religious, and recreational purposes are recorded throughout the novel. The beauty of the interior is compared to paradise: "Esa linda *sabana* . . . es de las que hacen creer al padre Las Casas que en nuestra hermosa tierra estaba el paraíso de Adán" ["This beautiful plain . . . is what made Father Las Casas believe that our beautiful land might have been the Garden of Eden"].[27] Later, when Enriquillo returns to rule his people in an uprising against Spanish rule, he arrives to the sierra from the more urban confines of Santo Domingo and the agriculturally oriented provinces surrounding it and finds the mountainous environment much more healthful: "Los puros aires de la sierra devolvían la salud y el vigor a sus miembros" ["The purity of the mountain air returned health and vigor to his limbs"]. One of his captains welcomes him back by pointing to the peaks and valleys to declare, "¡Allí está la libertad!" ["Here is where you can find true freedom!"].[28] In these examples, the undeveloped, natural terrain of the interior and its forests, plains, and sierras are associated with solace, ceremony, beauty, health, and liberty. Throughout the novel, the representation of sublime nature contrasts with the corruption, contamination, and moral evil of the urbanized cities and towns founded and inhabited by the Spaniards.

In *Enriquillo*, the critique also takes the form of a denunciation of the injustice of the *encomienda* system. Galván has the famous and original *indigenista* author Barolomé de las Casas appear in his novel with "la convicción íntima, inquebrantable, profundamente arraigada en su conciencia, de que para salvar a la raza india de la opresión que diariamente la diezmaba no había otro medio que acabar de una vez con el sistema fatal de las encomiendas" ["the internalized, unbreakable, and profoundly foundational conviction that in order to save the endangered indigenous communities of the Americas, there was no other choice than to end the *encomienda* system"].[29] This aspect of the *indigenista* critique involves questions of political ecology, land management, and power; as such, it is not just a question of access to resources, but one intimately tied to human rights; that is, the right not to be exploited or enslaved merely on the basis of one's origins and place of residence.

Another of the concerns that early *indigenista* texts dealt with was perhaps the most profoundly destructive foreign ambition: the Spanish lust for gold. I have already referenced the representation of indigenous groups as "hijos de la selva" (children of the jungle) in Juan León Mera's *Cumandá* in chapters 2 and 4; and although Mera's *Melodías indígenas* continues with similarly sympathetic representations, it also emphasizes the way in which Spanish conquistadors' and colonists' economic objectives clash with indigenous values concerning the conservation of place. In "La tola volcada" (The overturned idol), the conflict comes to a head with a Spaniard's desire to knock over a holy fetish (*la tola*) that marks the grave of a woman that the indigenous poetic speaker had once loved:

> –¿Qué haces? Mal extranjero, ¡tente! ¡tente!
> No esa tola derroques;
> No la ira de los dioses, imprudente,
> Con un crimen provoques.
> ¿Juzgas que encierra en sus entrañas oro?
> ¡Falsísima noticia!
> Oculta el montezuelo gran Tesoro,
> Más vano á tu codicia:
> Huesos de una que fue rara belleza
> Que mi alma idolatraba
> Polvo de un corazón cuya terneza
> Mi vida deleitaba
> . . .
> –Rústico infiel, me engañas: aquí el oro
> Se esconde que me tienta—

Volcó la *tola* el blanco, halló el tesoro. . .
¡Tesoro de osamenta![30]

["What are you doing, foreigner? Stop! Halt!
Don't knock that idol down!
If you do, it will be your own fault
When the gods go to town.
Do you think there is gold on the inside?
Not true! Not true!
It's in the mountain tall and wide.
But this, so much greater than your greed can construe,
Like the bones of one I loved so much,
My heart seems turned to dust,
The vanity of an innocent crush
This was not some vulgar lust,
But the delight of life I could not rush."
. . .
"Stupid Indian! You trick me and tempt
Me to indulge in riches of gold or stones,
An attraction to which no one is exempt."
The Spaniard breaks the idol open and inside, true treasure:
my beloved's bones.]

Similar to *El último rei de los Muiscas* and *Enriquillo*, the conflict in Mera's poem comes down to competing notions of space and place. For the indigenous speaker, the *tola* contains priceless holy relics, but for the Spaniard, the sacredness of the place is secondary to the value of the gold he suspects is contained therein. While Mera's poem does not employ the language of economics to describe the conflict of competing values, Fernando Mires's articulation of the concept of "the economy of the absurd" is founded on a calculus of value that is at the heart of the matter expressed in the poem. While the Spanish foreigner's search for treasure in the form of gold was always able to find an expression of price—the going rate for gold on whatever markets were in place when the metal was found and placed into the economy—the worth of an undisturbed burial site where a loved one has been interred can never truly be valued. Thus, the desecration of that site to extract the gold (if it were even there), is never accounted for when the price of the metal as a commodity is determined. The later nineteenth century novel *Huincahual* by Alberto del Solar pursues a corollary to this argument about how land where gold may be found inevitably becomes a target for exploitation by "foreigners."

In del Solar's novel, a dialogue between Huincahual, a Mapuche chief, and his *cautiva* wife explains that the expropriation of land by conquistadors was not simply an act of settling or exploring empty space, but the forceful occupation of a place infused with Mapuche history for generations:

> Estas tierras eran nuestras; los señores de esos valles eran mis antepasados que en la edad antigua . . . las poseyeron. Tú no sabes, tal vez, mujer, que mis padres fueron dueños absolutos de esas regiones inmensas. . . . Mis abuelos aprendieron el caso de los suyos, y éstos de los anteriores. Y así he venido yo á saber por ellos el origen de mi raza y de mis derechos sagrados.[31]
>
> [These lands were ours; in ancient times, these valleys all belonged to my ancestors . . . they were the owners. Perhaps you don't know, woman, that my fathers were the sole owners of these immense regions. . . . My grandparents learned this from their grandparents and they from of old. And that is how I came to know about the origins of my people and our sacred rights here.]

In the novel, the eventual invasion by Spanish conquistadors was motivated by the same desire that drove the greedy foreigner from Mera's poem to break the sacred object: an insatiable lust for gold. But in *Huincahual*, the act of desecration is the annexation of ancestral Mapuche land to mine it on a far grander scale. In a cathartic flourish to emphasize the tragedy of Spanish gold lust as the primary cause for the injustice suffered by indigenous groups in Chile, del Solar has the Mapuche apparently uninterested in the metal:

> El milla [oro] á que daban importancia suma, abundaba en nuestras montañas y aun en la arena de nuestros ríos, sin que hasta entonces nosotros nos hubiéramos cuidado siquiera de su efímera existencia. ¿Para qué nos habría servido? Al ver llegar, por tanto, á aquellos extranjeros, ansiosos de tan ordinario metal, juzgamos que no había razón para hacerles resistencia y les dejamos hartarse á su sabor de milla y siempre milla. Pero cuando, poco después, observamos que, no contentos con nuestra hospitalidad y tolerancia, comenzaban á abusar de su fuerza y de los instrumentos de muerte que traían consigo, llevando la opresión y la afrenta al seno de nuestros hogares, ¡ah! entonces sentimos encenderse la sangre en nuestras venas, al mismo tiempo que el odio y la venganza empezaron á alentar en nuestros pechos. (del Solar, *Huincahual*, 47-48)
>
> [The *milla* [gold], which was so important to them and whose existence was not even that important to us, was quite common in our mountains and even in our rivers. What good would it have done for us? But to see these foreigners, so intent on such an ordinary metal, come and come again, we decided that there was no reason to resist them and we let them take all they wanted: gold, gold, and more gold. But, a little while later, when we saw that they were not satisfied with our hospitality and tolerance and they began to oppress us and take advantage of us in our very homes through deadly force of arms—oh!—then our blood began to boil and feelings of hatred and revenge began to grow in our hearts.]

These lines articulate the catastrophic implications of gold desire, a common enough element in *indigenista* literature, but here, the issue is contextualized in the terms of land, violence, and indigenous rights. Later in the twentieth century, *indigenista* texts, particularly from Bolivia and Chile, take up these themes as they relate to the development of industrialized mining, and even include explicitly environmentalist themes that deal with the toxic by-products of such industries. The implicit suggestion here is that the commercialization of nature—as minerals extracted from the earth—was to be resisted. And while a more contemporary emulation of non-industrial, non-commercialized societies does not find clear expression in del Solar's novel, it is apparent that the indigenous loss of land—in *Huincahual* because of economic interests concerning the mercantilization of gold—has always been a central issue in the genre whether for mining, agriculture, ranching, the petroleum industry, or any other interest that benefited (and still benefits) external agents. From an ecocritical perspective, the political ecology in *Huincahual* affirms the rights of lasting, unimpeded and unpolluted Mapuche access to their native lands over those of Spanish newcomers to exploit gold as a financial interest.

Almost every one of the novels analyzed in this chapter represents a variation on the ideology of *Huincahual* as it relates to the traditional lands of indigenous communities. These are appropriated by foreign entities and variously deforested, mono-cultured, and polluted or turned into roads, with small portions rented back to the original owners in exchange for unsustainable labor agreements, and eventually abandoned as desert when exhausted or permanently usurped with little or no compensation for the original owners. The subsequent fallout for indigenous communities took many forms in the actual circumstances of the conquest as it did in the literature. For example, *Tabaré* by Juan Zorilla de San Martín has the Uruguayan Charrúa identity articulated as closely linked to the indigenous community's relationship with nature. First, the land and water are described as places created specifically for the Charrúa:

> Para él está formada
> Esa encantada tierra
> . . .
> Para él sus grandes ríos
> Cantando se despeñan
> Los himnos inmortales de sus ondas.[32]
>
> [For him, this blessed land
> Had been created and formed
> . . .
> For him, these great rivers
> Bellow out hymns
> From the immortal waves of water.]

Later the Charrúa are compared to tigers, comparatively for their bravery and metaphorically as owners of the jungle:

No son tigres,
Aunque algo del ademán siniestro
Del dueño de las selvas
Se refleja en su fiera actitud.
(Zorilla de San Martín, *Tabaré*, 66)

[They are not tigers,
Although there is something sinister
About these lords of the jungle
And their fierce comportment.]

The Charrúa are also repeatedly referred to as children of the jungle and blood of the desert; this is their homeland, and the word used in the Spanish—*patria*—has connotations of fatherland and nation. Thus, any mistreatment of that land can be considered an invasion either in quasi-military terms or by construing Spanish invaders as thieves. From an ecocritical perspective, the logic of the argument in *Tabaré* and other *indigenista* texts runs along these same lines. The extermination of various tribes and the misuse, over-development, contamination, and conquest of their lands all mean the same thing. In *Tabaré* the idea is repeatedly reinforced through the image of a flower that has fallen from its stem into the water, a flower that eventually wilts, withers, and dies. The image is a poetic analogy of the conquest and extermination of the Charrúa; the connection between indigenous depopulation and the destruction of nature is inseparable.

In the last part of this chapter, I consider texts that articulate the problem of land for indigenous communities in terms of competing notions of place and space, beginning with Bolivian Alcides Arguedas's *Raza de bronce* (The bronze race) (1919) and then Ecuadoran Jorge Icaza's *Huasipungo* (1934), with *Huasipungo* perhaps the most canonical of all the *indigenista* texts. Then, in chapter 7, I focus on the ecological effects of the surrender of indigenous property to external elements, particularly in the context of the repurposing of these lands for extractive and large-scale agricultural industries.

INDIGENISTA ECOLOGY, STRAIGHT FROM THE CANON

I began the chapter with *Los caminos de Paxil* as an analytic frame from which to initiate an ecological reevaluation of the *indigenista* genre as a whole. In Arias's novel, the North American oil company's proposal to drill Mayan cornfields for oil involves the fundamental belief in "private property" as the primary principle governing access to and authority over the land to be drilled. From the company's perspective, the most efficient use of the resources at hand would come in the form of extracting the oil from below (rather than growing corn in the topsoil). The sentiment is shared by Halach Uinich Emerson, local representative of the oil company and the novel's fictional apologist for the logic of development. Even though the land in question was already being used by the indigenous community to grow maize, Emerson believes that to let farmland go undrilled when oil is underneath it would be to allow underdevelopment to reign. Emerson justifies his beliefs to the indigenous community as follows:

> Ustedes están desprotegidos, expuestos, son gente vulnerable, por eso piensan así. Y por eso son pobres. Necesitan nuestro valor fiero e impaciente para protegerlos. Y nuestro espíritu de empresa para desarrollarlos. Con el tiempo, ya les iré enseñando toda la magia de la libre empresa. Pero para entenderla, tienen que reconocer el sagrado principio de la propiedad privada.[33]
>
> [You are unprotected, exposed, vulnerable, that's why you think this way. And that's why you're poor. You need our fearless and impatient valor to protect you. And our enterprising spirit to develop you. With time, I will show you all the magic of free enterprise. But in order to understand it, you have to accept the sacred principle of private property.]

These criticisms come from a rejection of the principle of communal stewardship in favor of private ownership, a system more conducive to development.

The belief in the communal stewardship of property is dangerous for developmentalist thinking because the cultural basis for these beliefs rests on religious principles and not on models for investment and profit. As the indigenous community understands it,

> —La tierra no es de usted ni de nosotros. . . . La tierra es del Corazón del Cielo porque él la hizo y sólo él puede destruirla. Nosotros le agradecemos que nos permita trabajarla y que a cambio de nuestro sudor y nuestro esfuerzo nos conceda una buena cosecha de maíz. Y le pedimos perdón cada vez que por necesidad tenemos que destruir alguna de sus creaciones: cazar un venado, tumbar una ceiba. . . . Y uno no debe nunca cambiar ni el más pequeño granito de arena sin antes saber el bien y el mal que puede desprenderse de su acto. (Arias, *Caminos*, 65)
>
> [—The land is not yours nor ours. . . . The land belongs to the Heart of the Sky because he made it and only he can destroy it. We thank him that he lets us work it and in exchange for our labor, he gives us a good maize crop. We ask forgiveness every time it is necessary to destroy one of his creations: hunt a deer, cut down a ceiba tree. And no one should ever change even the smallest grain of sand without knowing what will come from such an act.]

The "sacred" principle of private property is heresy to the Mayan community's beliefs about the land. Arguments about development lack persuasive power for a community whose valuation of the land rests on the importance of maize harvests and the spiritual significance of ecological equilibrium, not the logic of profit and efficiency. From a practical, spiritual, and ecological perspective then, the development offered by Emerson and his company was to be rejected and resisted. In the following paragraphs, I demonstrate that this position, central to Arias's novel, has been a central element from the most canonical examples of the genre, the novels from the early and middle decades of the twentieth century.

Alcides Arguedas's *Raza de bronce* and Jorge Icaza's *Huasipungo* have maintained their status as canonical twentieth century texts in the *indigenista* tradition and provide evidence, from the position of canonical stalwarts, for my argument that the *indigenista* genre construes ecological concerns as a fundamental aspect of the protection of indigenous rights and the revaluation of native cultures and communities. In fact, in the earlier of the two, *Raza de bronce*, already assumes the classic parameters of the *indigenista* form, but also includes statements about the extinction of tree species, fisheries, avian life, and the sustainable exploitation of lacustrine ecosystems. Arguedas is frank in the description of the way in which the indigenous communities of Bolivia were categorically dispossessed of their land:

> So pretexto de poner en manos diligentes y emprendedoras la gleba en las suyas infecunda, arrancaron, con mendrugos o a balazos, la tierra de su poder, para distribuirla, como gaje de vileza, entre las mancebas y las paniaguados del mandón, cayendo así en su aridez de ahora, porque el brazo indígena, que por interés, codicia y sarcasmo, dieron en llamar inactivo los congresales de ese año triste de 1868, resultó más pobre, más ocioso, que el de los improvisados terratenientes, que sólo tuvieron la habilidad de encontrar en el indio un producto valioso de fácil explotación. . . .
>
> . . . Así, a fuerza de sangre y lágrimas, fueron disueltas, en tres años de lucha innoble, cosa de cien *comunidades* indígenas, que se repartieron entre un centenar de propietarios nuevos, habiendo no pocos que llegaron a acaparar más de veinte kilómetros seguidos de tierras de pan llevar. De este modo, más de trescientos mil indígenas resultaron desposeídos de sus tierras, y muchos emigraron para nunca más volver, y otros, vencidos por la miseria, acosados por la nostalgia indomable de la heredad, resignáronse a consentir el yugo mestizo y se hicieron colonos para llegar a ser, como en adelante serían, esclavos de esclavos.[34]

> [Under the pretense of employing more diligent and enterprising hands to plough the earth left fallow by indigenous farmers, these thugs usurped the land through force of arms in order to redistribute it as pay for dirty deeds done by the boss's lackeys and hangers-on. That's how the land eventually became arid and useless as it is today. To add insult to injury, those congresses from the sad year of 1868, made a show of saying that the indigenous communities had abandoned their lands (instead of recognizing the value of leaving them fallow for a time) and as a consequence of this greed and avarice, these same communities became poorer with little more to do than allow themselves to be exploited by newly minted "landowners" who found this arrangement a valuable way to make easy money. . . .

> . . . Thus, after three years of bloody fighting and tears, hundreds of indigenous communities were dissolved and became subject to countless new landowners among whom no small number were able to monopolize fertile tracts of land as many as twenty kilometers wide. Thus, more than three hundred thousand indigenous people were dispossessed of their lands. Many left, never to return, while others who had fallen into abject poverty and haunted by an overpowering sense of nostalgia for the lands that used to be theirs, resigned themselves to the yoke of the *mestizo* foremen who for their part were also quite poor. In this way, they became the peons of peasants and the slaves of slaves.]

I cite the passage at length because it encapsulates all of the abuses—corruption of legal proceedings, declaration of abusive laws, use of force, concentration of land holdings, exile, and entrapping servitude—that are variously criticized and denounced in the whole of the *indigenista* corpus. As the genre matures and its conventions begin to take shape, reprsesentations of victimization seem to consistently involve the ways in which traditional lands are taken from indigenous communities, but Arguedas has the classic elements already, in one of the earliest examples of the tradition. But as in the quoted lament about the desertification of the land after it passed from indigenous to non-indigenous husbandry, *Raza de bronce* also illustrates the way in which ecological ideas seem to be at the root of the injustice that is condemned in the typical *indigenista* text. In this case, there are references to the way in which, after the indigenous exile, the fish and trees of a lake and other resources throughout the country are poorly managed and will eventually become extinct:

> El mismo lago, siempre pródigo en dones, ahora se mostraba esquivo con sus riquezas de peces, aves y *totoras* explotadas sin medida ni control desde tiempo inmemorial, hasta el punto de agotarse día a día por falta de una rudimentaria legislación que resguardase el raro tesoro de su fauna y flora, únicas en el mundo. . . .
>
> . . . Vamos causando un daño irreparable a la riqueza misma del lago. Todo lo van explotando sin medida de él: su flora y su fauna. Ya la totora va despareciendo en la mayor parte de las orillas, porque se la siega incesantemente, año redondo, sin tomarse el trabajo de replantarla en las partes cosechadas. Los peces se van haciendo cada día más raros, porque también se les coge todo el año, sin respetar el período de la incubación. . . . En nuestras regiones montañesas han desaparecido las garzas, por codicia de los *aigrettes* para sombreros femeninos; . . . en las pampas arrimadas a la cordillera van desapareciendo las vicuñas y los avestruces con la cosecha de las nidadas que se hacen en todo tiempo. Aquí, en el lago, ya lo ven: quedan pocas aves y pocos peces, y dudo que en veinte años más se pueda hallar algunos, siquiera para muestra. . . . Y bastaban unas cuantas leyes y un poco de dinero en primas de protección para salvar del naufragio un caudal inagotable. (Arguedas, *Raza de bronce*, 151, 293–94)

> [The self-same lake, always rich in life, has recently been experiencing a scarcity of fish, birds, and bull-rushes—so long exploited without any measure of control—such that these plants and animals are now disappearing. And the government refuses to enact even the most rudimentary legislation to protect those unique treasures, fauna and flora that can only be found there. . . .
>
> . . . Here's the situation: either intentionally or because of ignorance . . . we are causing irreparable harm to the riches of the lake. . . . The bull-rushes are disappearing because they are cut down incessantly, all year long, and no one makes any effort to replant them. The fishes are becoming scarcer because they are being fished all year long, even during the egg-laying season. . . . In the mountains, the herons are on the decline because of the demand for their plumage for feathered hats now in fashion; . . . in the pampas alongside the Andean cordillera the llama-like vicuna and the ostriches are disappearing because their nests are raided all the time. Here at the lake it has become painfully obvious: so many fish and birds have disappeared that I doubt that in twenty years there will be any left, not even for exhibits in zoos or aquariums. . . . And all it takes is a little money and a few laws for conservation to protect what before seemed like an unending resource from utter collapse.]

Again a lengthy quotation from the narrative to document the litany of complaints against irresponsible ecological behavior and clear warnings about species extinction and poor harvesting practices. The references to legislation illustrate the statist political ecology that many authors believed was the solution to abuses of indigenous communities and to the ecological catastrophe that such victimization sometimes implies. These proto-environmentalist moments in the novel—concern about extinction, sustainable development, and improper use of resources—would be at home in any recent "green" mission statement or platform. But *Raza de bronce* contains such extensive ideas at a rather early moment in literary history, expressed through the conventions of an *indigenista* ideology that would inform the genre in the years to come.

Jorge Icaza's novel is more specific than *Raza de bronce* with regard to political ecology as it exemplifies the process by which indigenous groups are forcibly separated from parcels of property that had seemingly been legally granted them in exchange for labor or in swaps for other land. In *Huasipungo*, the narrative takes the form of a criticism of liberalism as an economic system where the constitution of Ecuador at that time allowed for foreign interests under the guise of economic development to take control of the Ecuadorian sierra and resulted in an alliance between these external agents and wealthy, domestic landowning classes to the detriment of indigenous communities. Although it has been suggested that Icaza's text was more a criticism of economic systems,[35] it is also apparent that political ecology represents an important ideological element: the unjust prohibition of access to traditional lands (small though they may be) so that foreign interests can profit.

The *huasipungo* from which the novel takes its title refers to the plot of land which the landowners would grant peasant indigenous workers in exchange for their labor. When the foreign interests represented in the novel propose the construction of a road to better access potential petroleum fields in the east, part of the proposal is the construction of temporary buildings to direct the construction on land granted to the indigenous community: "El socio no quiere dar un paso sin antes estar seguro de las mejoras indispensables: . . . un carretero para automóvil. La compra del bosque de Filocorrales y Guamaní. Limpiar de huasipungos las orillas del río para construir quintas cómodas para ellos" ["The investor will not give one cent until he is sure of the necessary improvements: a vehicular highway, the purchase of the Filocorrales and Guamaní forests. The demolition of the *huasipungos* sitting on the riverbanks in order to construct comfortable office and living space for the executives and other VIPs"].[36] Because the terms of the contract require the displacement of the *huasipungos* and their inhabitants, the result is an uprising, a violent challenge to foreign appropriations of local property.

The novel also articulates competing visions concerning the use to which the land will be put. While indigenous communities mainly want to preserve title to the tiny plots in exchange for their labor, even that inherently unjust recompense can be threatened when the actual land may be developed in more profitable ways: "Esa naturaleza es privilegiada. Se puede perfectamente abastecer a todos los ferrocarriles de la República" ["These lands are good. They can quite easily provide the space necessary for all the railway lines in the whole Republic"].[37] The construction of transportation infrastructure takes precedence over the *hacienda* as an economic model because of the profit it can imply; the rights of indigenous communities do not figure in such calculations nor do their conceptions of how the land *should* be used. *Huasipungo* provides a view from the 1930s of the beginnings of the massive transformation of Latin American landscapes by the auto industry. The use to which the road will be put in *Huasipungo* to provide access to oilfields means that the economic stakes are doubled: both the automobile and petroleum industries stand to lose if the *huasipungos* are allowed to stay. With its condemnation of officially sanctioned uses of indigenous land for the economic benefit of foreign interests, the text suggests that the dispute is more ecological than anything else and proposes a reformation of the way in which access to resources is negotiated. But given the irreconcilable differences between these competing notions of land, the authority for its utilization is inevitably going to come down to a contest decided by the violence of force and power. In *Huasipungo*, the conflict is an indigenous insurrection that concludes with the almost total liquidation of that community:

> Y al amanecer, entre las chozas deshechas, entre los escombros, entre los cadáveres tibios aún, surgieron como . . . espigas de cebada, que al dejarse acariciar por los vientos helados de los páramos de América, murmuraron con voz ululante de taladro:
> –¡Ñucanchic huasipungo! (Icaza, *Huasipungo*, 160)
>
> [And at sunrise, from among the ashes and ruin of the shacks and from among the still warm corpses . . . like the sound from among wind-blown ears of wheat in the bleak plateaus of the Americas came a staccato, ululating wail:
> –*Ñucanchic huasipungo*!]

The last sentiment—"*ñucanchic huasipungo*" (our huasipungo)—conveys the reformation proposed in the novel: rather than simply eliminating indigenous communities through force of arms, indigenous perspectives on something as important to them as their connection to the land must be respected and not ignored.

Huasipungo has achieved canonical status, most likely because it involves many of the most salient features of the *indigenista* novel: the unjust influence of foreign interests, the highly unequal exchange of indigenous labor for a tenuous claim to a place to build shelter, the eventual forcible removal of native workers to appropriate their land for other purposes. However, the subjugation of indigenous groups also spells trouble for the land itself, particularly when the dislocation of native communities leads to soil exhaustion as when traditional cultivation practices have been abandoned in *Raza de bronce* or to deforestation as when indigenous small-holders are evicted to make room for road-building in *Huasipungo*. The agenda of the *indigenista* novel is generally perceived as part of the effort to relieve injustices suffered by native communities. A central part of the effort to restore traditional lands to their original occupants has significant potential for a more favorable ecological outcome.

NOTES

1. Bartolomé de las Casas, *Brevísima relación de la destrucción de las Indias* (Madrid: Cátedra, 2001), 69.
2. Arturo Arias, *Los caminos de Paxil* (Guatemala City: Editorial cultura, 1990), 47.
3. David Stoll, *Rigoberta Menchú and the Story of All Poor Guatemalans* (Philadelphia: Westview, 2008); Arturo Arias, *The Rigoberta Menchú Controversy* (Minneapolis: University of Minnesota Press, 2001). There are dozens more articles on the controversy dealt with in Stoll and Arias's volumes, but much less on ecology in the Menchú testimony. See Mayumi Toyosato, "Grounding Self and Action: Land, Community, and Survival, in *I, Rigoberta Menchú*, *No Telephone to Heaven*, and *So Far from God*," *Hispanic Journal* 19, no. 2 (Fall 1998): 295–311; and Randy Malamud, "Poetic Animals and Animal Souls," *Society and Animals* 6, no. 3 (1998): 263–77.
4. Elizabeth Burgos-Debray, *Me llamo Rigoberta Menchú* (Habana: Casa de las Américas, 1983). When Burgos-Debray had transcribed Menchú's testimony and arranged her account for publication, she mentions unnamed readers who advise her to leave out the explanations of

cultural, spiritual, and ecological belief of the Quiche community, but that these are left intact "para hacernos conocer su cultura" ["so we can get to know their culture"], 21.

5. Ibid., 51, 134.
6. Ibid., 31.
7. Ibid., 85.
8. Robert L. Thayer Jr., *Gray World, Green Heart: Technology, Nature, and the Sustainable Landscape* (New York: John Wiley & Sons, 1994), xiv, 55.
9. Arias, *Los caminos*, 23, 93.
10. Cristobal Colón, *Textos y documentos completos*, ed. Consuelo Varela (Madrid: Alianza Editorial, 1992), 110.
11. Susan Paulson, "Land-claims: Racialized environmental struggles in Latin America," in *Ecology and Power: Stuggles over land and material resources in the past, present, and future*, ed. Alf Hornborg, Brett Clark, and Kenneth Hermele (New York: Routledge, 2012), 263.
12. Shawn William Miller, *An Environmental History of Latin America* (New York: Cambridge University Press, 2007), 54–55.
13. Ibid., 70.
14. Individual born in the Americas, but of Spanish (as opposed to indigenous or genetically mixed) heritage.
15. Antonio Cornejo-Polar, "*Indigenismo* and Heterogenous Literâtures: Their Double Sociocultural Statute," in *The Latin American Cultural Studies Reader*, ed. Ana del Sarto, Alicia Ríos, and Abril Trigo (Durham, NC: Duke University Press, 2004), 115.
16. Mabel Moraña, ed. "Indigenismo y globalización," in *Indigenismo hacia el fin del milenio: homenaje a Antonio Cornejo-Polar* (Pittsburgh: Instituto Internacional de Literatura Iberoamericana, 1998), 245.
17. Jean Franco, *An Introduction to Spanish American Literature* (New York: Cambridge University Press, 1994), 242.
18. las Casas, *Brevísima relación*, 77.
19. Ibid., 113.
20. Ibid., 115.
21. Ibid., 148.
22. Walter Shaw, "El neoindigenismo y *Hombres de maíz*," in *Cien años de magia: Ensayos críticos sobre la obra de Miguel Ángel* Asturias, ed. Oralia Preblei-Niemi (Guatemala: F & G Editores, 2006). Shaw makes reference to Concha Meléndez's *La novela indianista en hispanoamericana* to distinguish between nineteenth century *indianista* and twentieth century *indigenista* literature: "para muchos críticos la diferencia básica entre la *literatura indianista* e *indigenista* es el énfasis romántico e idealizado del primer período comparado con la presentación más realista de la opresión en el período posterior" ["for many critics, the main difference between *indianista* and *indigenista* literature is the emphasis in the first period upon idealized romanticism as compared to more realist representations of repression in the later period"], 98. While Shaw and Meléndez can be consulted for further distinctions, I believe that there are idealizing tendencies in twentieth century texts and nineteenth century works also certainly contain representations of oppression; thus, the distinction seems moot and I will prefer the term "*indigenista*" throughout this chapter.
23. Non-indigenous individuals kidnapped or taken captive by native tribes.
24. Jesús S. Roza, *El último rei de los muiscas* (Bogotá: Echeverría Hermanos, 1864), 8.
25. Ibid., 9.
26. Ibid., 234, 256.
27. Manuel de Jesús Galván, *Enriquillo* (Santo Domingo: Ediciones de Taller, 1993), 390.
28. Ibid., 392, 411.
29. Ibid., 331.
30. Juan León Mera, *La virgen del sol. Melodías indígenas* (Barcelona: Escudillers, 1887), 337–38.
31. Alberto del Solar, *Huincahual* (Paris: Garnier hermanos, 1911), 43–44.
32. Juan Zorilla de San Martín, *Tabaré* (Montevideo: Biblioteca de autores uruguayos, 1888), 7.
33. Arias, *Caminos*, 66.

34. Alcides Arguedas, *Raza de bronce* (La Paz: González y Medina, 1919), 124–25.

35. Teodosio Fernández, introduction to *Huasipungo* by Jorge Icaza (Madrid: Cátedra, 1997).

36. Jorge Icaza, *Huasipungo* (Madrid: Cátedra, 1997), 9.

37. Ibid., 9.

Chapter Seven

Indigenous Land: Place, then Space

As the twentieth century waxes in the decades that precede its midpoint, the conventions of the *indigenista* genre begin to coalesce around competing notions of space and place; such notions are central to the analysis of Bolivian Alcides Arguedas's *Raza de bronce* (The bronze race) (1919) and Ecuadoran Jorge Icaza's *Huasipungo* (1934) discussed at the end of chapter 6. As in these two canonical examples, the literary conflict invariably takes the form of powerful external elements appropriating formerly indigenous land. But such texts also mark the moment in Spanish American literary history when the meaning of Argentine Domingo Faustino Sarmiento's categories of civilization and barbarism began to slip toward total ambivalence. For indigenous peoples, the supposed benefits of civilization were for them only a process by which their achieved state became one of poverty and degradation. Thus, the terms "civilization" and "barbarity" each acquire contradictory connotations and are rendered nearly meaningless. From the decline of "civilization" as a meaningful term in the literature, *indigenista* texts derived justification to challenge the legitimacy of the rights of land-"owners" to control the lives of the indigenous people that inhabited traditional lands that had then been lost. In the first part of this chapter, I analyze the representation of various manifestations of resistance to such losses and in the second part, I consider a corresponding impingement of the land-use practices employed by new, foreign "owners," particularly when these individuals and the corporations that employ them repurpose the land for large-scale agro-enterprises or logging or the industrialized extraction of metals and oil.

In all of the preceding chapters, I have set aside some space to evaluate the relationship between the ecological facts on the ground as they existed in the literary moment when the novels, short stories, and poetry under consideration were written. The development of imported agro-industrial models, the guano rage, the growth of cities and their impact on air, water, and land, the rise in demand for rubber, the discovery of oil, and deforestation to clear the tracts for agriculture, fields for livestock, and paths for the construction of roads: these were the environmental impacts that took hold in Latin America since the Conquest. And as I have documented in previous chapters, many of these made an appearance in the literature of the region and gave rise to the literary expression of political ecology, conservationism, environmental justice, and other ecocritical postures before such ideas were called by that name. In the *indigenista* literature of the twentieth century, most of these ecological issues remain relevant with many of them exacerbated as Latin America underwent a process of industrialization that had widespread environmental fallout. To outline the ways in which the contamination of water, the development of the oil industry, or the ongoing pace of deforestation in the Amazon are relevant to the *indigenista* novels of the mid-to-late twentieth century would be to simply repeat the causes and effects of such practices as they were summarized in the first six chapters. However, I have not yet addressed the profound effects of the especially ruinous mining industry for the injury it has unleashed on indigenous communities widely recruited to labor in the mines or for the damage these mines have caused to the lands where they are located.

While the Spanish lust for gold was a principal motivator of the Conquest, the discovery of silver from a few particularly rich veins in Mexico, Peru, and especially Potosí in Bolivia made the Spanish empire exceedingly wealthy and enabled its expansion to the four corners of the world. Initially the miners employed a traditional technique to separate the silver from rock and other less valuable metals through the use of Incan technology: fire-based processes using a *guayra* (smelting stone). The extraction of silver using the *guayra* required so much fuel to heat the stone that extensive deforestation was common all around each mining site. According to Shawn William Miller, "the solution of the problem of fuel scarcity arrived in the form of mercury"[1] which was combined with pulverized silver ore, water, salt, and iron ore to form an amalgam from which the mercury and other elements were vaporized to leave masses of pure silver. If the toxic effects of mercury, which are now well-known, were not entirely clear at the time, they soon become so, as droves of miners began to die. And those who lived in and around the mines were similarly afflicted by other toxic by-products, exposed as they were "to a variety of heavy metals, mostly in their water. Mine shafts and tailings . . . leached zinc, lead, copper, and arsenic, all of which entered local streams and ground water, and eventually human tissues."[2] This was the norm for the mining industry since the seventeenth century; and while the use of mercury for the extraction of silver was widely abandoned in the nineteenth century, industrial mining is still a dangerous job, responsible for thousands of deaths in the last 200 years.[3]

Silver mines, oilfields, pasture for the ranching of livestock, fields for the production of agricultural exports, stands of trees for timber, and dammed waterways for the production of electricity were all invariably located in places that had formerly been the traditional lands of various indigenous groups. The catastrophic disappearance of most native cultures from the microbial assault during and after the Conquest followed by the subsequent demographic recovery left an ever larger remnant of landless, ethnically indigenous individuals whose "legal" title to territories that had belonged to their ancestors for generations became ever more tenuous. The only recourse was to go to work on those same lands with no claim to ownership and a vague hope that the usurious "salaries" would stave off starvation, even if for a time. On the part of those who managed and owned mines, derricks, ranches, and plantations, the only interest was for profit. As it was in the mines of Potosí—"silver was more important than Indian lives, so the deadly work continued"[4] —so it was in every developing industry short of labor. The recovery of land for native groups, advocacy for relief from the horrid working conditions in mines, fields, and oil installations, and the condemnation of an ever more industrialized modern economy are the core concerns of the *indigenista* texts that are the focus of this chapter. Because these issues are intricately bound up with ecological effects, the novels adopt the contours of what would presently be called an environmentalist ideology. But for the authors of the texts considered below, connections between the natural world and the *indigenista* literary form was simply part and parcel of the ethical advocacy.

THIS LAND WAS MY LAND: INDIGENOUS CLAIMS TO TRADITIONAL PLACE

After the Guatemalan Nobel Laureate Rigoberta Menchú, Peruvian José María Arguedas is likely the most familiar author in the genre and his first novel, *Yawar fiesta* (Blood festival) (1940), contains several examples of the way in which *indigenista* texts represent connections between land, space, place, ecology, and advocacy for the rights of indigenous communities. In one instance in Arguedas's novel, there is a sharp comparison between what is of value to the *mistis* (Quechua for overlord) and to the native peoples represented in the text. The *mistis* simply cannot perceive the status of place in natural phenomena that are highly valued by the indigenous community:

> En las cascadas, el agua blanca grita, pero los mistis no oyen. En las lomadas, en las pampas, en las cumbres, con el viento bajito, flores amarillas bailan, pero los mistis casi no ven. En el amanecer, sobre el cielo frío, tras del filo de las montañas, aparece el Sol; entonces, las tuyas y las torcazas cantan, sacu-

> diendo sus alitas; las ovejas y los potros corretean en el pasto mientras los mistis duermen, o miran, calculando, la carne de los novillos.[5]
>
> [In the waterfalls, the white water shouts, but the *mistis* don't hear. In the hills, on the plains, and on the peaks, yellow flowers dance in the small breeze, but the *mistis* almost can't see any of it. In the cold skies at dawn, behind the profile of the mountains, the Sun appears; the wood pigeons in the cedars sing, shaking their wings; the sheep and the horses prance in the pastures while the *mistis* sleep or only look at the potential for meat on the sides of the steer, calculating their possible worth.]

Figuring the economic benefit of cattle interferes with non-native individuals' ability to perceive the value of places precious to the indigenous community. I call attention to this moment in the novel because it illustrates the way in which feminist ecocriticism can be utilized as a critical idiom to examine past moments in literature where feminist leanings were not immediately apparent. One aspect of ecofeminism is to approach natural phenomena such that there is a reevaluation of "what counts"; in this case, the water, the flowers, the sun, the birds, and other animals are prized not for profit but for their capacity to inspire a certain way of thinking: awe at the majesty of nature. But such things do not figure in the economic calculations among the *mistis* and when the dynamics of economic development are in play, differences in value become paramount:

> Casi de repente solicitaron ganado en cantidad de la costa, especialmente de Lima; entonces los mistis empezaron a quitar a los indios sus chacras de trigo para sembrar alfalfa. . . . Los mistis que llevaban reses a la costa regresaban platudos. Y casi se desesperaron los principales; se quitaban a los indios para arrancarles sus terrenos. (Arguedas, *Yawar fiesta*, 21)
>
> [Almost at once, there was a huge increase in demand for livestock from the coast, especially from Lima; so the *mistis* began to take away the fields of wheat from the Indians in order to plant alfalfa. . . . The *mistis* who ran herds of beef cattle to the coast returned loaded with money. Then, the large-holders really began to go crazy with greed; they simply got rid of the Indians to more easily take away their land.]

Through force of arms, corruption, and dubious or outright illegal claim to the land, the indigenous community is removed from its traditional lands so they can be exploited in ways that will most benefit the livelihood of the *mistis*.

In *indigenista* literature, the role of tenuous title to property ownership, conflicts about land authority, and access to resources play a significant role in the suffering of indigenous peoples. As Alf Hornborg has documented, "the management of land and land-based resources can serve as strategies for social domination, from more obvious cases of local land tenure through the political demarcation of national and colonial boundaries to unequal global flows of natural resources."[6] In most cases, the *indigenista* literary critique targets "social domination" of indigenous communities in the two ways identified by Hornborg: where land tenure is denied to those with historical claims and where, as the true reason behind these evictions, resources are extracted for export on a global level. The domination continues through local economies of exchange where labor on the haciendas is exchanged for small tracts for farming and shelter, but as these are improved and become productive, foremen reassign title and annex now-fertile plots for the hacienda. Indigenous tenants have no recourse; they are left without harvest, salary, or even a tiny plot among the vast tracts that belonged to his or her ancestors in their entirety.

Classic formulations of social domination are found with some variations in *Yanakuna* (1952) by Bolivian Jesús Lara. The story details the suffering of the young woman Wayra after her father dies unexpectedly and she is sold as an indentured servant only to suffer unspeakable abuse. The detailing of harsh treatment spares almost no institution, including the farcical system of justice lampooned throughout the novel; the implicit suggestion is that landownership arrangements and the enforcement of title through military and police units, corrupt politicians, and the bribery of judges must be replaced. The argument posits an equivalency between theft of property, enslavement of labor, and corruption of officials and the categorical injustice suffered by indigenous peoples. The first part of the formula—theft of land—invokes a largely moral argument about landownership and management: "En otros tiempos eran los indios quienes poseían las tierras. Vinieron los wiraqochas [los blancos] y se las arrebataron. Entonces, de dueños y señores que eran, los indios se convirtieron en yanakuna, en esclavos de los usurpadores" ["In earlier times, the Indians owned the lands. But then came the *wiraqochas* [the white people] and they took it away. So from owners and lords, they became *yanakuna*: slaves of the usurpers"].[7] The argument here makes an explicit connection between control of the land and slavery; the Spanish conquistadors and later, *mestizo* (individuals of mixed Spanish/indigenous heritage) settlers were able to enslave the original inhabitants through the assumption (or theft) of landownership. The logic of strict equivalence between slavery and ownership of traditionally indigenous lands is made explicit in a court scene later in the novel:

> Todo era despepitarse contra la raza india, atribuyéndole taras y abyecciones. Esa raza había nacido para labrar la tierra y no podía zafarse del imperativo de su destino. Un tiempo había trabajado en beneficio del Inca; éste fué sustituído luego por el español, y ahora le tocaba el turno al que con su dinero compraba

> la tierra. El indio era parte constitutiva de la tierra y por tanto su persona pertenecía de modo irrestricto al propietario. (Lara, *Yanakuna*, 395)
>
> [It was a whole lot of going on and on about the defects and deficiencies of the indigenous race. They said that they were born to work the land and that they could not escape the fate of their destiny. Once, they labored for the Inca who was replaced by the Spaniard and now by whoever had enough money to buy the land. The Indian was an essential part of the land and now his person was irrevocably tied to whoever owned the land.]

Anthropological bases for such analysis by the lawyer in the novel are surely dubious and the relationship between geographic phenomena and indigenous inhabitants certainly seems more complex than these quaint ideas about an "essential" identity of Indians derived from the land; however, it is instructive to see the ecological basis for the most central moral argument of the *indigenista* genre (and in *Yanakuna*, it is laid out quite explicitly): the theft of their property explains many of the great evils suffered by indigenous inhabitants but also explains the ecological damage to their traditional lands.

Furthermore, Lara's novel advances two other ecological arguments with regard to the injustices suffered by indigenous peoples. One comes from the same moment in the text as the above-cited passage; that is, from the closing arguments of the lawyer who defends a band of insurrectionist Indians that had violently resisted continued abuse by the landowners and their representatives. In this case, the lawyer makes reference to pre-Columbian Incan society:

> El Inca sabía comprender y respetar la dignidad humana y bajo su gobierno se hallaban proscritos el hambre, la desocupación, la prostitución, etc. Los súbditos del imperio eran agricultores y el producto de la tierra era para ellos y no beneficiaba a ningún usurpador del esfuerzo ajeno. (Lara, *Yanakuna*, 398–99)
>
> [The Inca knew how to understand and respect human dignity and under his government, there were regulations against the imposition of hunger, the stealing of land, prostitution, etc. His vassals were farmers and what they cultivated was produced for each and every one of them, not for some foreign usurper.]

The prohibition against the dislocation of individuals from traditional parcels and the share of laborers in the fruit of their labor makes a curious combination between capitalist and socialist ideals, but both are conjoined by their specific reference to customs of land access and authority.

The second idea comes from a moment early in the narrative just as Wayra, the long-suffering protagonist, is sold to the first of many abusive "owners." She clearly does not want to go with him and prefers the freedom of life in the hills as a shepherdess:

> Porque esto de servir a los amos no tenía semejanza alguna con el cuidado de las ovejas en el cerro, por ejemplo. Allí era lícito jugar, reír, cantar, inclusive pelear. Allí a nadie le importaba que uno se moviese de un lado para otro, o que no se moviese. Allí se disponía del aire, del sol, del viento. Allí se era libre. (Lara, *Yanakuna*, 62)

> [Because this whole idea of serving the owners was nothing like taking care of the sheep up in the hills, for example. Up there it was perfectly fine to play, laugh, sing, even fight. Up there, no one cared where anyone else went or if someone didn't go anywhere at all. Up there, it was nothing but air, sun, wind, and total freedom.]

The argument here is that a life of employment and servitude is far inferior to being out in nature. While the *indigenista* text makes forceful condemnation of the quasi-slavery of native peoples one of its fundamental points of departure, these lines contain something still quite relevant to the environmentalist rhetoric of the present. The sheer joy of living, playing, laughing, and even fighting in the fresh air makes an argument for the preservation of places where such activities are still possible. In *Loving Nature*, Kay Milton makes a similar argument: "the process of valuing things in the world is inseparable from the emotions and feelings they induce in us; without these emotions and feelings there would be no value."[8] According to Milton's formulation, the emotions described by Wayra in her interactions with the hillside comprise an argument about the value of the land.

Lágrimas indias (Indian tears) (1920) by Bolivian Alfredo Guillén Pinto and *La raíz errante* (The moveable root) (1953) by Paraguayan Natalicio González make the case for the intimate connection between indigenous communities and their ancestral land in a slightly different way: the attachment between native peoples and their homeland is evoked in particularly spiritual terms. Guillén Pinto's novel has the conventional condemnation of the way in which lands were distributed under the corrupt *hacienda* system:

> Hernani tomó para si lo que le pareció mejor y lo que más grande, y, como el acto de mayor magnanimidad, como quien se desprende de la mitad de sus riquezas, cedió a los indios pequeñas parcelas, las más serializadas y pedregosas, que en adelante debían esterilizarse de tanto dar pan para las numerosas familias.[9]

> [Hernani took for himself whatever seemed bigger and better, and, as if it were an act of great kindness or as if he were giving away half his wealth, let the Indians have small parcels, the furthest away and rockiest, which would eventually become useless because they had to be used to feed so many families.]

Within the ideological confines of the *indigenista* text, the problem of access to resources, land, and livelihood were central concerns. The fact that indigenous individuals were always afforded the worst parcels (as explicitly referenced here) represents an aspect of environmental injustice that derives from the unfair distribution of land. In current environmentalist thinking, the focus emphasizes an unequal pollution of land; but *Lágrimas indias* demonstrates that environmental injustice has as much to do with a denial of access to resources as with indemnity for the toxic by-products of resource extraction that cause air, land, and water contamination.

However, the environmental injustice suffered by indigenous inhabitants also has to do with the idea that the possession of land may have had supernatural meanings. At one point, the narrative indicates that "el indio se desprendió de lo único que poseía: su nido y su relativa independencia personal" ["the Indian gave up the only thing he or she ever had: home and his or her relatively independent personal freedom"].[10] The novel makes an explicit connection between a place of one's own and personal independence, a notion invoked when the narrator explains why Indians reject traditional Catholic instruction; the spiritual beliefs of indigenous peoples had a lot more to do with nature: "Su verdadera religión es la del suelo que le da algo que mascar sin hacerle promesas quiméricas y su verdadero templo es el de la Santa Madre Tierra" ["Their true religion has to do with the earth which gives them something to eat without making ephemeral promises and their true temple was that of Holy Mother Earth"].[11] This moment articulates the immediate, physical necessities of indigenous groups which have no use for seemingly empty "religious" promises as compared to the tangible benefits of working the ground.

The author of *La raíz errante*, Juan Natalicio González, represents still another example of Spanish American literary figures that ascended to high political positions in their respective countries. González was president for several months in 1948 until he was deposed by his own *Colorado* party on December 30. The political ideology of both the author and his party was inspired by European fascism[12] and the long-enduring dictatorship of Alfredo Stroessner in Paraguay was supported by the *Colorados*. So to speak of González as an *indigenista* author that sought to protect the rights of a persecuted minority would be inaccurate. Nevertheless, as the author's political leanings were profoundly nationalistic and the culture of the nation of Paraguay was strongly influenced by the Guaraní people, it may be expected that González would include them in his work. In *La raíz errante*, the land derives its perceived value through indigenous connections to it in non-economic forms. First, there is the sentimental attachment to a place that has belonged to several generations of the same family:

> En el hombre del agro la tierra se humaniza, entra a integrar la propia personalidad, es actora anónima de las vidas humildes que se deslizan sobre ella. Por

> eso el valor sentimental que representa para su morador tradicional el más pobre solar no es cotizable en ninguna moneda del mundo.[13]
>
> [For the man that farms the land, it seems almost human and begins to develop a kind of unique personality like an anonymous actor in the lives of those humble people that tread upon it. That is why, for its inhabitants, even the humblest site cannot be valued in any national currency.]

Even if a prevailing real estate market could determine the value of property according to non-indigenous economic models and even if the Paraguayan Guarani were given a fair price (something that does not happen in *La raíz errante* nor in any other *indigenista* text), the land cannot be valued from a indigenous worldview because of the people's intimate connection to it. In fact, in González's novel, the spiritual connection to the land goes back to Guarani deities who protect it from over-exploitation: "el árbol de la yerba mate, o sea el ca'á cuenta con el amparo de la ca'á yary'í [abuela de la yerba], genio protector que divaga en su contorno y lo defiende de las explotaciones irracionales que lo exponen a la destrucción" ["the *yerba mate*-tea tree, that is, the *ca'á* requires the protection of the *ca'á yary'í* (grandmother of the *yerba mate*-tea), a divine protector that inhabits these lands and defends the tree from irrational exploitations that expose it to destruction"].[14] Here, indigenous religious belief invokes doctrines directly related to conservationism: the spirit of the *yerba mate* protects the plant and the plot on which it grows from over-exploitation, destruction, and other harmful practices. And despite (or perhaps because of[15]) Natalicio González's militantly nationalistic political leanings, the representation of nature protectionism accounts for a significant part of his novelization of Guarani spiritual notions of the land.

In the Guatemalan context, Miguel Ángel Asturias's *Hombres de maíz* (Corn men) (1949) also emphasizes ancestral and mythological elements; for the Mayans of Asturias's novel, the cultivation of corn is central to spiritual belief. Specifically, the land is anthropomorphized in Mayan mythology and the worst barbarity that one can do is plant corn to sell it; one of the Mayan elders explains it thus:

> El méiz cuesta el sacrificio de la tierra que también es humana; . . . los que se han entregado a sembrar méiz para hacer negocio, dejan la tierra vacía de huesos, porque son los huesos de los antepasados los que dan el alimento méiz, y entonces la tierra reclama huesos, y los más blanditos, los de los niños, se amontonan sobre ella y bajo sus costras negras, para alimentarla.[16]
>
> [The maize requires a kind of human sacrifice; . . . those who plant corn as a business venture leave it totally devoid of ancestral bones because these are what nurture the maize. Then, the land will begin to demand new bones, the

> bones of children, the sweetest, softest bones, which will pile up underneath the black soil to keep it productive.]

According to Mayan mythology, child sacrifice is required to make the land fertile if the corn is sold. As for most cultures, violation of these requirements is taboo; but such a grisly alternative would only be necessary if the cultivation of corn is turned into an industrial rather than a small-scale program. The implication in the text is that any conversion of the land would require religious compromise on the part of the Mayan community and must not be done. Asturias's novel expresses the idea that spiritual and not economic or developmentalist logic was often behind conceptions of the land among indigenous communities. In an article on the connection between spiritual belief and the environment in actual Mayan villages in Guatemala, anthropologist Anne Hallum notes "the common practice of asking permission of the Earth before it is cultivated" and "the belief that the Creator placed them and their ancestors on a particular piece of land."[17] The representation of Mayan views concerning traditional territories, parcels, and property as part of a system of religious belief in Asturias are true to their deep-rooted connections to the earth and the sense of absolute injustice when these connections are denied, especially when the denial of access is founded upon a desire for profit.

Entre la piedra y la cruz (Between a rock and the cross) (1948) by Asturias's countryman Mario Monteforte Toledo is similar to the representation of Guarani or Mayan ecological spirituality as in *La raíz errante* and *Hombres de maíz*. In Monteforte Toledo's text, a mythical connection to the land is emphasized, but the narrative also implies that the fate of indigenous communities is irrevocably tied to processes that may result in land exhaustion:

> Los indios también eran pedazos de la tierra. Al verse los pies, macizos, adheridos al camino o al surco tibio, no podía decir dónde terminaba el mundo y dónde comenzaba su propio cuerpo. . . . Uno trabajaba la tierra inútilmente, hasta cansarla y dejarla exprimida y seca como la arena, y nunca podía reunir ningún dinero.[18]

> [The Indians were also pieces of land. To see their stout feet upon the furrows and rows of the crops, one could not tell where the earth ended and where the human body began. . . . One could work the land in this way ceaselessly until it no longer was capable of producing anything because it had gone sterile and dry like sand and then the money also dried up.]

The quoted passage starts out as an almost picturesque gesture toward the physical description of a typically indigenous body: construed quite literally as a piece of the land: one with the soil, feet, and earth indistinguishable. The image is admiringly wrought, but the *indigenista* literary discourse imposes itself upon the narrative and challenges the pastoral aspect of the description of an Indian as "one with the land"; later in the same paragraph, the land is described as tired with its life squeezed out of it; desertification has become the central element and none can profit. If indigenous individuals are equated to the land, then they too will eventually be used up, wrung out, and die penniless. Such is the state of affairs that has the narrative explicitly propose a political ecology that calls for the state to step in, protect and educate indigenous farmers, and create a better Guatemala:

> Mientras el gobierno no se encare con el problema de la tierra, el país no progresará. El ejido debe protegerse, y los indios deben comprender que las tierras nacionales son en realidad del pueblo, de la gente que las trabaja. Si se fundaran por estos lados institutos de capacitación rural que enseñaran a trabajar a los indios, la producción se multiplicaría y todos viviríamos mejor. Entonces no habría mercado de hombres, ni odio, y Guatemala sería grande y rica. (Monteforte Toledo, *Piedra y cruz*, 213)

> [As long as the government doesn't care about the problem of the land, the country cannot progress. The common lands must be protected and the Indians must understand that federal lands truly belong to the people, to those that work them. If Centers for the Benefit of Rural Farming were to be established that could teach the Indians how to more efficiently work the land, production would increase ten-fold and everyone would be better off. Then, there would no longer exist a slave-like market for human labor, hatred would disappear, and Guatemala would become great.]

While the faith in a government program to rescue the Indian, promote responsible land use, and eliminate slave-like working conditions seems naïvely idealistic in retrospect, it does represent the central argument that was at the heart of the *indigenista* literary moment. As the above quotation attests, ecological elements such as care for the land, an emphasis on proper husbandry, and a political ecology that advocates for universal access to resources are the very central elements of the genre.

Economic difficulties facing indigenous communities in Chilean Lautaro Yankas's *El vado de la noche* (The void of night) (1955) explain another ecological element of the *indigenista* genre: the tragic abuse of their own land by an indigenous community because of intense poverty. The Mapuche in Yankas's novel are forced to completely clear the land, not for farming, but because there is nothing left to use for fuel:

> Tierra pobre, escoriada, arrasada, donde no hay otra nota verde que la de algún renuevo en la quebrada profunda a donde cuesta trabajo llegar. El indio no deja mata viva, porque necesita fuego en invierno y la leña se acabó hace años en las reducciones. En otro tiempo, la abundancia de bosques lo mareó; . . . Ya

> nada queda de eso; la leña hay que mendigarla en los fundos vecinos, o robarla si se niega la limosna.[19]
>
> [What wretched land, sore, demolished, a place where even the hint of green was hidden deep in ravines or crags that were almost impossible to access. Needy as he is for fire in the winter and living in a place where the wood had disappeared years ago, the Indian doesn't leave a single bush standing. In earlier times, the abundance of forests was overwhelming; . . . but now nothing of that remains; firewood can only be gotten by begging in neighboring lands or stealing it if that becomes necessary.]

The clear-cutting is a product of the necessity of the reduction of Mapuche communities to the lands that were left after the Conquest, Independence, and the subsequent landgrabs by wealthy *huincas* (invading foreigners). In the novel, things are exacerbated by the extreme poverty in which the Mapuche live; not only have the *huincas* taken the best land, but they also have the means to improve it:

> El huinca venido de lejos cogía tierras mejores, . . . con vegas donde las máquinas han hecho milagros contra la piedra laja y la greda maldita. . . . ¿Que la tierra no da la semilla? Qué hacerle. El indio no tiene cómo abonarla, la trabaja a medias porque no puede comprar la buena semilla; le falta todo. (Yankas, *Vado*, 46)
>
> [These outsiders always take the best lands, . . . with fertile plains miraculously tilled from among the stony, clay-filled land. . . . What? The land does not produce? What to do? The Indian does not have the necessary additives and fertilizers, he cannot make it produce because he can't buy the best seed; he lacks for everything.]

Access to farm machinery, fertilizer, and other products that facilitate a successful harvest are inaccessible to the Mapuche for lack of economic means. Yankas's novel criticizes the injustice perpetrated upon indigenous communities for the loss of their land, but also details the ecological consequences of the extreme marginalization of an entire population. When humans are left in absolutely desperate conditions, conservationist or sustainable strategies of land exploitation become secondary to the raw facts of survival. The implicit argument is that the economic exploitation by *huincas* is to be faulted as much for its injustice toward indigenous communities as for the ecological byproducts of their impoverishment.

El Amauta Atusparia (1930) by Peruvian Ernesto Reyna takes a different tack; the emphasis in Reyna's novel is not the tragedy of indigenous poverty, but the possibilities for a response from those communities to its causes. The novel recounts an indigenous uprising from 1885 in Peru in which the Amauta Indian Atusparia envisions a new South American indigenous society based upon the ideal that "la tierra es para todos" ["the land is for everyone"].[20] But the narrative is not limited to arguments about the injustice of the theft of Amauta land to justify the uprising; it also expresses an alternate political ecology that attempts to remedy these losses. The model was the pre-Columbian Incan empire:

> El Imperio de los Incas contaba con veinte millones de hombres, que vivían felices dedicados a la agricultura en su mayor parte. . . . Hombres sabios y buenos que dieron al Imperio la organización comunista, que hoy no pasa de una utopía. . . . Los Conquistadores destruyeron todo, hasta el espíritu mismo de la raza, encadenaron alma y cuerpo. La tortura de las minas, la esclavitud, la guerra, la peste y el hambre diezmaron al gran pueblo indio. (Reyna, *Amauta Atusparia*, 55)
>
> [The Incan empire had twenty million men that lived happily and were largely employed in agricultural pursuits. . . . Wise and good men who seemed organized according to what we would call a communist system, a system that from present accounts seems nothing less than a utopia. . . . The conquistadors destroyed all of this, from the concept of the essential spirit of race to the servitude of soul and body. The torture of the mines, slavery, war, the plague, and starvation massacred the great Indian peoples.]

The praise of Incan agriculture versus the imposition of mining and the subsequent famine among indigenous communities puts these concerns in explicitly ecological terms that have everything to do with who controls the land. In fact, the reference to communism in the passage cited above reminds readers of the noted Peruvian Marxist critic José Mariátegui whose *Siete ensayos de interpretación de la realidad peruana* (*Seven Interpretive Essays on Peruvian Reality*) (1928) contains a section entitled "El problema de la tierra" (The problem of the land) that makes the same argument concerning the land and indigenous societies as the fictions cited above.

Sumag Allpa (1940) by Ecuadoran G. Humberto Mata Ordóñez displays a more conventional *indigenista* denunciation of land theft similar to many of the other titles surveyed above, but like *El Amauta Atusparia*, it also presents a defense of violent resistance as a valid response to non-indigenous usurpations of land. In Mata Ordóñez's novel, a certain don León Hazera systematically robs small properties from many of the original indigenous inhabitants of his hacienda for which he is kidnapped by his victims and burned alive. The novel concludes with Hazera's death but only after the narrator proclaims a vehement defense of such tactics as appropriate for a race in defense of its lands:

> En guardia, potente Raza que estalla cuando quiere para apretujar a todos aquellos que intentan estrangular el sexo de su albedrío puesto sobre la cara Morena de la Tierra. "De la tierra Nuño! Ñuño Allpa! Huinachic Allpa!"

Sol!
Sol Rojo. . .
Sumag Allpa![21]

[On guard, strong Race that can burst forth at will to crush all those that try to strangle the virility of your will on the dark face of the earth. "The land, *Nuño*! *Ñuño Allpa*! *Huinachic Allpa*!"
The Sun!
The red sun. . .
Sumag Allpa!]

Mata Ordóñez's novel breaks with the typical schema of the victimized Indian who takes revenge but whose resistance is put down violently. Instead, the revenge is justified as a legitimate response, a violent defense of continued access to traditional lands, awakened when foreigners seek to usurp "Nuño Allpa," "Huinachic Allpa," "Sumag Allpa": the nurturing, literally "mother"-land.

OIL, MINING, WATER, TIMBER: INDUSTRIALIZATION AND *INDIGENISTA* LITERATURE

Recall my definition of "political ecology" from the introduction: "analysis (approaching advocacy) of the way in which certain communities engage with structures of political power for access to natural resources and interactions with the environment." When these negotiations are substantiated upon traditional claims, religious belief, or "illegal" violent defense (as in the novels from the last section) versus economic "legal" arguments adjudicated in a corrupt system, the latter will win out. The result is almost always landlessness for indigenous communities and a consequent descent into poverty, victimization, and hopelessness. That is, by force of arms at the Conquest and subsequently, through any number of societal factors arrayed against them including bribed judges and politicians, immoral *hacendados* (owners of haciendas), brutal foremen, class stratification, racism, and outright thievery, indigenous individuals are utterly powerless to fairly negotiate property transactions, establish legal title to traditional holdings, or even sharecrop according to terms other than those imposed by external entities. The novels in the previous section feature various representations of the way in which access to traditional native lands had been completely cut off. In this section, I will take up another important element of political ecology: the negotiation of access to natural resources within bodies of water, underneath mountain ranges, and from various other Latin American landscapes. As Hornborg observes,

> different kinds of environmental load displacement (through trade) will accompany specific kinds of capital accumulation. We thus need to use different measures of ecologically unequal exchange for different historical and geographical contexts. What they all have in common is a concern with the factor of production referred to as *land.*[22]

The "environmental load displacement" in the form of global markets for water, timber, oil, and precious and industrial metals results in the accumulation and subsequent environmental contamination of land from which such resources are extracted. The traditional understanding of the *indigenista* literary tradition emphasizes the injustice of the loss of land to any external force that accumulated it. My approach in the rest of this chapter will be to highlight those texts from the tradition where such losses also involve environmental degradation as a by-product of the industrialization of land to remove the resources that were in, on, and below it.

As in chapter 6, I frame the analysis here with an episode from Guatemalan Arturo Arias's *Los caminos de Paxil* (The ways of Paxil) (1990). Like in each of the novels from the first section of this chapter, Arias's novel sets up competing notions about access to and ownership of the same tract of land as central to the conflict between the Mayan community and the oil company. But in *Los caminos de Paxil*, the conflict hinges specifically on perceptions of value for what the land contains and what it can or should produce. From the perspective of the representatives of the North American corporation, the regions inhabited by the Mayan community had great potential value and certainly not as cornfields: "Tienen petróleo. Valen muchísimo más" ["They've got oil. They're worth much more"].[23] But from the point of view of the indigenous community, none of that was worth it: "A nosotros el petróleo no nos sirve. Ensucia la tierra, la arruina para plantar maíz. Y encima nos caen todos esos fuereños como buitres, buscando despojarnos" ["Oil has no use for us. It fouls the ground and ruins it for planting maize. And what's more, all these outsiders descend like vultures wanting to get rid of us"].[24] The notion of ecological damage—that the oil will foul the ground—is closely linked to the issue of rights to the land: when oil is discovered, *fuereños* will come and try to take both oil and land. As in Arias's novel and in several of the other *indigenista* texts that I consider in the following pages, the extraction of oil and metal, the damming of waterways and the clear-cutting of forests, and the conversion of land for large-scale agricultural production by foreign corporations is condemned as unjustly paramount to the rights and property claims of indigenous communities that reside on lands destined for these types of exploitation.

Oil

As I mentioned above, a central facet of the ecological ideology inherent in the *indigenismo* of *Los caminos de Paxil* has to do with the environmental contamination associated with large-scale extractive industries. And all of the same environmental effects associated with the oil industry that were central to the regional petro-novels from chapter 5 are no different from the *indigenista* examples in that subgenre: deforestation, air pollution, water contamination, and toxic spills. These ill effects are explicitly articulated in the novel by Arias, but in *Barrancabermeja: novela de proxenetas, rufianes, obreros y petroleros* (Barrancabermeja: Novel of pimps, hooligans, workers, and oilmen) (1934) by Colombian Rafael Jaramillo Arango, there is a significantly earlier expression of the ruinous impact of the petroleum industry upon indigenous peoples:

> Aquí podemos decir todos los lugares comunes de la patriotería, sin exagerar un comino: el capital absorbente, la soberanía entregada en manos mercenarias, la mancha indeleble del aceite. . . . Y qué? Nada. Nada para nosotros, para los indígenas que no sabemos explotar, ni siquiera guardar, ni siquiera vender lo que tenemos.[25]
>
> [Here, we can mention all the usual jingoistic moves without exaggeration: the notion of a sort of ever-sucking capital, the idea of national sovereignty handed over to mercenaries, the permanent stain of the oil. . . . And? Nothing. Nothing for us, for the indigenous peoples who have no idea how to exploit, or save or even sell what we have.]

The tendency of oil to make a mess is here put in the explicit terms of environmental injustice with indigenous communities as the victims: the oil will do its environmental damage and the native peoples whose lands get ruined see no benefit, only the toxic by-products of extraction. In other parts of the novel, the ecological aspects of such victimization are reiterated: what is left of their land is inexorably changed: "aquí no había árboles" ["there were no longer any trees here"]; the industry is seen as much more of a curse than a blessing: "todo, todo, todo lo que se mancha de este aceite queda maldita" ["everything, everything, everything stained by the oil becomes hopelessly contaminated"]; the very air becomes tragically polluted: "las calles, negras de petróleo crudo, despiden vahos calientes que hacen más insoportable la temperatura. El aire es irrespirable" ["hot vapors rise from the oil-blackened streets and make it unbearably hot. The air becomes unbreathable"].[26] Like in *Los caminos de Paxil*, the oil extraction industry is perceived as an utterly useless, actively corruptive influence on society, an irreparably contaminating source of injustice for indigenous communities.

In *Sobre la misma tierra* (The very same land) (1944) by Venezuelan Rómulo Gallegos the oil extraction industry is also central to the development of the narrative, but when Gallegos was writing in the 1940s the idea that oil drilling and transport could cause catastrophic environmental damage was still a new idea. Nevertheless, the representation of the exploitation of oil in Gallegos's novel was always qualified by a critical posture toward the industry, a posture that from an ecocritical approach anticipates several current environmentalist criticisms toward the oil industry. The first and most pressing concern was for the way in which land to be drilled has been appropriated by petro-companies: "una inmensa porción del subsuelo venezolano pasó al dominio del taladro extranjero" ["a huge portion of subterrestrial Venezuela was transferred to the control of foreign drilling enterprises"].[27] Secondary to these more political concerns is the damage that oil wreaks on the land itself; in the novel, there are spills, uncontrollable gushers, and contamination: "y brotó a chorros la providencial calamidad. Aventó válvulas, alzó negra columna gigantesca, inundó tierras, alimentó durante varios días lluvia pringosa esparcida por el viento y bajo la cual se ennegrecieron los campos y pereció ganado" ["and the gushing calamity seemed providential. Valves burst, a huge black column rose up, flooded the lands, and a greasy, black rain spread by the wind went on for days blackening the fields and killing livestock"]. The earth is covered with spilled oil that kills off the vegetation: "una depresión del terreno entre colinas circundantes de poca altura, en cuyo suelo, todo negro de petróleo, no crecía la hierba y se extendían aguazales iridiscentes" ["in an oil-covered depression in the land surrounded by short hills, there was no vegetation of any kind, only iridescent puddles"].[28] Gallegos's novel is set in the sprawling oil installations of the same Cabimas that set the backdrop for Venezuelan Ramón Díaz Sánchez's *Mene* (1936) discussed in chapter 5. Here, the narrative indicates that the place has been named after "un árbol de este país que abundaba mucho en esta región" ["a tree native to this country that used to be quite common in this region"].[29] The use of the past tense to describe how the tree used to flourish in that area describes the deforestation implied by the construction of the towers, systems of pipe, roads, buildings, and other infrastructure associated with an oil drilling enterprise. The text repeatedly evokes the way in which the land is slowly laid to waste, a narrative that, if appearing in a text more contemporary to our own times, would lead to environmentalist ideologies concerning conservationism, nature protectionism, the search for alternative energy sources, etc. But the *indigenista* context within which Gallegos was writing in the middle of the last century cannot have explicitly addressed such ideas. Rather, the ideology has a lot more to do with the ethics of environmental justice, specifically the rejection of the petroleum industry for the way in which it affects the culture and livelihood of the (in this case, Venezuelan Guajira) indigenous community.

One illustrative situation deals with a Guajira village that must expend a good deal of physical energy just to get clean water. The criticism is not that the oil is contaminating all the clean water, but that the wealth brought to the nation by the industry is not equally distributed:

> —¡Ah mundo! Esta pobre gente teniendo que caminar kilómetros para llevarse al rancho la chirigüita de agua sucia y el viento perdiéndose por toda la Guajira, sin la torre de un molino que lo haga trabajar para que el indio siquiera beba una agüita limpia. ¡Ah mundo! Y no muy lejos de aquí, en la zona petrolera, ese territorio ajeno enriqueciendo al musiú, sobre la misma tierra. (Gallegos, *Misma tierra*, 143)
>
> [—What a world! These poor people have to walk several kilometers just to return with a pail of dirty water. The steady winds blowing all over the Guajira region go to waste without a single windmill to power a pump so that the Indians can at least get clean water. —What a world! And not far from here, the oil installations sit on tracts of land that only enrich the *musiú* (foreigner), on this very same land.]

In general, the text laments the fact that so much territory is dedicated to the petroleum industry, areas that might be used as pasture for cattle to produce milk or meat or as farmland to cultivate other agricultural products or even for the establishment of limited wind-power endeavors. But this has not happened in Venezuela[30] and the negative appraisal in *Sobre la misma tierra* of political ecologies that so heavily favor a state-supported petroleum industry represents a still-valid criticism of the way in which the appropriation of land for oil drilling affects local communities.

Mining

Among the concerns of current environmentalists, few generate as much attention as the environmental damage, issues of human safety, and contamination from the extractive processes of industrialized mining. Mountaintop removal techniques, surface extraction, the disposal of slag and other wastes, the atmospheric and climate impacts of coal burning, and the unknown carcinogenic and other toxic effects of heavy and radioactive metals are all among the usual suspects when the ecological effects of mining are discussed. In the novels considered from this section, as in most *indigenista* literature, the primary concern is the human toll that an industry or economic system inflicts on individuals and families; as I document in the introduction to this chapter, mercury poisoning, silicosis, mine explosions and collapse, the contamination of land and water around mining sites, and the generally inhumane and unhealthful working conditions were the main culprits. These elements drove my analysis of *Mi camarada padre* (My father, my comrade) (1958) by Chilean Baltazar Castro from chapter 5, but a more extensive analysis of Spanish American mining novels comes here because in mining nations like Chile, and especially Bolivia, most miners are and were from indigenous communities.

In *En las tierras de Potosí* (In the lands of Potosí) (1911) by Bolivian Jaime Mendoza, the criticism of access to mining concessions develops along the same lines as several of the other *indigenista* texts with regard to the land in general: "La cuestión de la propiedad privada, tú lo sabes, está aún por resolverse. ¿De quién es la tierra? ¿De quién son, por ejemplo, las minas? De todos. Sólo por abuso unos cuantos apoderan de este patrimonio común" ["You know, the question of private property must still be resolved. Who does the land belong to? Or, for example, who is the owner of the mines? It belongs to us all. It was only through corruption that a few individuals seized what should be a national asset"]. Later, one of the characters criticizes the mining industry in general: "estas industrias a medias donde todo es incipiente y defectuoso, en que no se va sino a ganar pronto, a ganar de cualquier modo, a ganar aun con desprecio de la vida de los otros" ["these half-baked industries where everything is so amateurish and defective, where the only concern is to make money by any means necessary, to make money even if it means loss of life"].[31] As early as the second decade of the last century, the political ecology in Mendoza's novel calls for reform of the industry: a reevaluation of ideas about private property, a prohibition against the realization of profits at any cost: social, cultural, environmental. The narrative is focalized from the perspective of Martín Martínez, a middle class student from Sucre who decides to work in the mining town of Llallagua to earn some savings and continue his education. His actual experience could not have turned out more frustratingly hopeless as he discovers that the wild tales of quick wealth are vastly exaggerated. Nevertheless, Martínez's experiences provide the literary conceit from which to frame a narrative of mining: all of the industrial practices, employment, and characters from a typical mining town are developed in this way. However, because Martín is a non-indigenous character and the novel is more about mining per se than the indigenous experience of mining, the *indigenista* ideology does not mark the text as strongly as it does in *Socavones de angustia* (Shafts of anguish) (1947) by fellow Bolivian Fernando Ramírez Velarde. In the latter, both environmental damage and indigenous miner experience are explored, but with far more emphasis placed on the suffering of miners.

Ramírez Velarde's novel features a few brief, but powerfully cathartic moments where the narrative decries the environmental impacts of mining. In one, a miner longs to return to the unsullied agricultural region where he was born: "Allá la tierra es buena, porque apenas la acariciamos con el arado en la siembra; aquí se ha vuelto mala, porque la torturamos haciendo socavones con barrenos y pólvora" ["There, the land is still good because we hardly touch it, just some light plowing for planting; but here, the land is bad because we torture it with shafts and holes drilled by enormous augers then hollow it out with TNT"]. In another, the narrator describes the defoliation and creation of waste characteristic of industrial-scale mining: "El panorama de las minas 'grandes' como el de todos los centros mineros, es hostil. Cerros pelados y rojizos sin pizca de vegetación, rodean los campamentos, limitando el horizonte. Por aquí y por allá, enormes desmontes de caja amontonada en el transcurso de los años" ["The landscape around the big mines, like all mining projects, is hostile to the eye. Bald, reddish hills with no sign of vegetation ring the camps blotting out the horizon. Here and there, huge mounds of slag piled up over the years"].[32] The language in the description of the use of explosives, the total lack of vegetation, and the accumulation of unusable slag is a violent imagery of torture and hostility. The negative description of ecological contamination from the industry represents an implicit appeal for reform, but the true "message" of *Socavones de angustia*, what qualifies it as an *indigenista* text, is its advocacy for reform of the ecology of the mine itself as protection for the well-being of the indigenous miners of Bolivia.

In the novel, a particularly earnest and emotional description of the mine as an exceedingly dangerous place to work lays out the case for reform:

> ¡Mina! . . . ¡Entraña de la tierra dominada por el hombre y que se venga del hombre fatal y sañudamente! . . . Topes tan faltos de ventilación, que en ellos se vuelve amarilla la luz de las lámparas y se marchitan las vidas de los obreros por falta de oxígeno; . . . Partículas pequeñísimas de roca y átomos de polvo, que introduciéndose en el organismo, arteramente, contribuyen a esclerosar los pulmones . . . galerías negras y amenazantes como bocas de monstruos hambrientos: Socavones de Angustia! . . . Mina: ¡Vivero de la muerte! (Ramírez Velarde, *Socavones*, 137)
>
> [Mine! . . . Created by men from the entrails of the earth and that viciously takes vengeance on these same men! . . . Shafts with so little air that lantern lights turn yellow and human life begins to wither for lack of oxygen; . . . tiny particles of rock and atoms of dust get unknowingly inhaled and lead to lung disease . . . black, threatening caverns like the mouths of hungry monsters: Shafts of Anguish! . . . Mine: The Face of Death!]

The miner has multiple enemies to fear: unbreathable air, shaft collapse, pulverized shards of rock and metal that remain permanently lodged in the lungs. In fact, these fears are not mere flights of literary fancy because, currently, industrial mining counts for far more human casualties per unit of energy produced than any other form of energy.[33] But when the miners inevitably succumb in such conditions, their suffering does not end there. In the novel, new workers are forced to sign bills of health that include a diagnosis of pneumoconiosis, even when their lungs are perfectly clear:

"Yo, Juan Calle, declaro expresamente que ingreso a los trabajos de la Compañía, en el estado de salud indicado por el médico de la Empresa. Por lo tanto, no tengo derecho a reclamación alguna, por el progreso de la enfermedad que adolezco" ["I, Juan Calle, expressly declare that I begin my employment with the Company in the physical condition indicated on the medical report submitted by the Company doctor. Therefore, I may make no legal claim for the worsening of the chronic and progressive condition with which I already suffer"].[34] Companies intend to duplicitously protect themselves from civil suits arising from claims about worker respiratory disease by having employees falsely diagnosed with a condition they are most likely to contract as a result of work in the mines. The literary conventions of the *indigenista* text mean that Juan Calle, the miner who signs the quoted statement above, will eventually get lung disease, as indeed he does. Again, the unfortunate proximity between the experience of the fictional Juan Calle and real working conditions in Bolivian mines is corroborated by a footnote that reveals that the health waiver featured in the novel was "tomado textualmente de un proceso que cursa en los archives de la Oficina del Trajabo de Oruro" ["reproduced without alteration from actual proceedings archived in the Office of Labor in Oruro"].[35] Other Spanish American mining novels of the time seem to substantiate such claims; Mendoza's novel discusses the way in which the mined tin gets everywhere: "en la ropa de las personas, en su piel, en sus pulmones, en su estómago" ["in the clothes, the skin, lungs, and stomachs of the miners"];[36] in Peruvian Raúl Botelho Gosálvez's *Altiplano* (The highlands) (1945) about the Aymara from the central Andean highlands around Lake Titicaca, a chapter on mining cites the statistic that "el ochenta por ciento [de los mineros] caía con los pulmones convertidos en grumosos y amarillentos trapos de pus y sangre" ["eighty percent [of miners] were sickened with a disease that turned their lungs into yellow, lumpy bags of pus and blood"].[37] While the mining industry represents one of the most entrenched and ardently hostile antagonists to the environmentalist movement, the serious problems with mineshaft ecology in Bolivia had (and still has)[38] inordinately negative consequences for the indigenous Quechua and Aymara groups in that country. Specifically, the advocacy for these peoples in *Las tierras de Potosí, Socavones de angustia*, and *Altiplano* emphasizes the threats to their health and well-being from a decidedly ecological perspective.

Water

As I write, there are controversial plans for the construction or expansion of dams on every inhabited continent including the Illisu over the Tigris River in Turkey, the Gibe III on the Omo river in Ethiopia, the Xayaburi on the

Lower Mekong in Laos, a height extension of the Sardar Sarovar on the Narmada River in India, the expansion of the Olympic Dam in Australia, ongoing proposals to resurrect the Auburn project on the American River in California, and a five dam installation in the Chilean Patagonia region along the Baker and Pascua Rivers. Each of these projects is actively resisted by environmental and social justice groups who advocate for the communities that will be displaced by the massive reservoirs created by the waters behind the floodgates. Often these groups highlight the environmental impact that the construction of dams implies including deforestation, habitat loss, downstream water scarcity, ecosystem alteration, etc. In *Donde haya Dios* (Wherever God may be) (1958) by Argentine Alberto Rodríguez, many of these same issues are implicit for the indigenous Huarpe community living among the lagoons formed by the mouths of the rivers Mendoza, San Juan, Bermejo, and Desaguadero. Because of the over-exploitation of these rivers, the Huarpe are exiled after the lagoons had been drained, rendered unnavigable, and the surrounding land, which had previously been fertile and well irrigated, now barren. In the opening pages of the novel, Rodríguez cites a 1929 report from the Commission on Irrigation compiled by the Mendozan provincial government in Argentina:

> La anunciada construcción del nuevo dique sanjuanino, el acrecentamiento de los cultivos, y el consecuente aprovechamiento cuasi integral de las aguas del río San Juan, hará que las extensas lagunas de Guanacache, situadas en el límite nordeste de nuestra provincia, se vayan secando gradualmente.[39]
>
> [The forthcoming construction of the new dam in San Juan province, the growth in crop yields, and the consequent ability to more easily make use of the waters from San Juan River will cause the Guanacache lagoons in the northeastern part of our province to gradually go dry.]

Thus, the opening lines of the novel set up the environmental concern which is the focus of the narrative. In the rest of the text, this central issue is explored from the perspective of the cultural impacts on the Huarpe community and evoked in the terms of what would now be identified as environmental justice. On three separate occasions, dialogues among members of the communities living in and around these lagoons identify the desire by a wealthy, upriver class to dam the waters in order to irrigate vineyards, produce electricity, and create desirable vacation spots:

> Dice el maistro que no va venir maj'agua . . . quel gobierno ta haciendo un murallón pa que unos rico tengan más vino . . . que'stán plantando leguas y legua di'uva. . . .
>
> . . . Agora dice el masestro quel gobierno está haciendo un murallón pa que unos rico tengan luces. . . . ¡Quén sabe cómo prenderán las velas en l'agua. . . !

> . . . ¿Y cómo dice el mestro quel gobierno está . . . que'están haciendo un murallón, dice . . . pa que unos rico tengan dónde bañarse . . . y andar en bote . . . y pasar sus ocios. . . ? (Rodríguez, *Donde haya*, 36, 64, 105)
>
> [The engineer says no more water . . . that the government has made a huge wall so that the rich folks can have more wine . . . that they're planting endless acres of grapes. . . .
>
> . . . Now the engineer says they're making this huge wall so those rich folks can have lights at night. . . . Who knows how they can light candles with water? . . .
>
> . . . And how is it that now the engineer says that the government . . . is making a huge wall . . . so that those rich folks can have a place to swim . . . and go boating . . . and have fun. . . ?]

The regional, colloquial dialogue empowers the narrative with local perspectives on the human effects of dam construction, especially powerful as it originates from those whose homes and traditional lands will be inundated by the reservoirs held up by the new dams. In the novel, this perspective on the economic equation takes the form of a contrast between the real value of goods made possible by the dams versus the intangible value of a community.

The contrast between these values is made explicit in the dialogues contained in the narrative (in this case, in the words of the Huarpe Cristino) with expressions of what would now be called political ecology and ideas about environmental justice: "—¿Y quén ej'el gobierno pa hacernoj'eso? . . . —salió de su consternación el Cristino— ¿Cómo nos puede quitar l'agua pa echársela al vino? . . . ¿Qué nohotro somo meno que vino? . . . ¿Qué somo nohotro? . . . ¿aguapié acaso?" ["—And exactly who said that the government could do something like this? . . . —Cristino asked aloud seething with anger—How can they take the water away from us to make wine? . . . What? Are we worth less than the wine? . . . What do they think we are? . . . Do they think we're nothing but dregs?").[40] The demand to compare the values of a human community and its sacrifice for the production of an agricultural product to the construction of dams makes clear the injustice in such an arangement. This alternative conception—the political ecology that emerges from the novel—invokes such things as communal water rights and spiritual conceptions of nature: "–¿Y quén es el gobierno pa'hacerlos eso, ah? . . . El agua nu es dél . . . el agua es de todos . . . el agua ej . . . el agua es de Dios . . . ¡eso es . . . de Dios!" ["Exactly who said the government could do this to us, huh? . . . The water doesn't belong to them . . . it belongs to all of us . . . the water is . . . the water belongs to God . . . that's it . . . it belongs to God!"].[41] The rights and power of the wealthy and the government are challenged and the perspective of the indigenous community is given voice precisely because the conflict is ecological.

Donde haya Dios calls to mind Fernando Mires's ideas about the "economy of the absurd" as these dialogues demonstrate the impossibility of correctly valuing the production of wine, electricity, tourism, and other economic goods that can only be achieved through the subversion of an indigenous community's rights, their ancient connection to the land, and their spiritual conceptions of access to water. The latter come in cultural, moral, and social forms that simply cannot be calculated in economic terms. And when a distant government and its wealthy patrons attempt to undertake such projects nevertheless, resistance can be expected. At present, such resistance takes the forms of environmental watchdogs such as the Friends of the River Narmada coalition local to the struggle to contain dam construction along that river in India or the multinational focus of the International Rivers mission or, as demonstrated by Rodríguez's novel, it can take the form of a piece of writing in the *indigenista* literary tradition. The environmental groups mentioned above were created only after the maturation of notions about community organization, the creation of movements, activist political advocacy, or even the design of stunts to promote awareness or participation in outright physical resistance. I am reminded of the humorous public debut of the Earth First! activist organization that unfurled a tarp in the form of a large crack on the face of the Glen Canyon dam in homage to Edward Abby's *Monkey Wrench Gang* (1975). The characters in *Donde haya Dios* also mull the destruction of the dams that prevent the water from reaching their lands—"Vamos a romper ese dique . . . ¡y les vamos a mandar el río para ese lado!" ["We're going to knock that dam down . . . and we're going to let the river flow down here as it always had before"][42] —this is a sentiment that anticipates Abbey's novel by more than fifteen years, the founding of Earth First! by nearly twenty, and the current controversies surrounding dam construction on five continents by a half-century.

Timber

Another of the extractive industries that can be included among the targets of environmentalist activism is the practice of clear-cutting virgin forests, jungles, lowlands, and other nutrient-poor locales to briefly exploit a terminal harvest and then abandon the land after it has been exhausted. Among the negative consequences of these agricultural practices relevant to this chapter are soil compaction, erosion, and exhaustion, habitat degradation, reductions in biodiversity, and the loss of indigenous husbandry knowledge. Several of these elements are explicitly explored in *Mamita Yunai* (1941) by Costa Rican Carlos Luis Fallas, a novel in which the infamous United Fruit Company (known as "Mamita Yunai" in Central America with "Yunai" a corruption of "United" in colloquial Spanish), economically invades the region and completely disrupts economies, cultures, and ecosystems.[43] Valeria Grinberg

Pla and Werner Mackenbach consider Fallas's novel "el texto fundacional de la novela bananera" ["the foundational text of the banana novel"], a genre characterized by its "carácter antioligárquico, antiimperialista y anticapitalista, así como su protesta social contra las condiciones inhumanas de trabajo" ["anti-oligarchist, anti-imperialist, and anti-capitalist sentiments as well as for its social protest against inhuman working conditions"].[44] I propose the addition of environmentalist elements to Pla and Mackenbach's list given the way in which Fallas's novel emphasizes the severe ecological injustice perpetrated upon indigenous populations throughout Central America:

> Los gringos de la United . . . querían tierra y hombres bestias que la trabajaran. Y ya los pobres indios no pudieron contener el avance de la "nueva civilización." Llorando de impotencia vieron abatirse las montañas seculares, en donde por tantos siglos la Raza Heroica había cantado su canción de Libertad. Y ardieron sus palenques, se destruyeron sus sembradas y se revolcó la tierra en que dormían los huesos de sus bravos guerreros. (Buscaban esmeraldas fantásticas? No. Se iba a transformar el jugo de la tierra en bananos y en cacao que luego cambiarían por oro legítimo en los mercados extranjeros.)[45]

> [The gringos from United . . . wanted land and beast-like men to work it. And the poor Indians could no longer contain the advance of "our civilization." They were overcome with grief and impotent rage at the sight of the dismantling of ancient mountains where for many centuries the Heroic Race had sung a song of Liberty. Their plazas and fields were destroyed and the graves that contained the bones of their brave warriors were all dug up. (Were they seeking priceless emeralds? No. They were going to transform the essence of the earth into banana and cocoa trees which they would then indeed exchange for real money on foreign boards of trade.)]

The abuses, corruption, and political machinations of the real United Fruit Company are infamous, enumerated above as the institution of a system of pseudo-slavery, cultural insensitivity, and outright lust for profit despite the collateral damage.

These elements are central to Fallas's novel, but in several instances the abuses of United Fruit are articulated as ecological contamination, land exhaustion, and even climate change:

> Al poco tiempo la tierra se cansó de dar bananos y ya el cacao no significó nada para los yanquis. Entonces éstos levantaron sus rieles, destruyeron los puentes y, después de escupir con desprecio sobre la tierra exhausta, se marcharon triunfalmente hacia otras tierras de conquista. (Fallas, *Mamita Yunai*, 65–66)

> [In a short time, the land could no longer support the banana trees and the cocoa trees had ceased to be a good source of profits for those Yankees. So, they took apart their railways, knocked down the bridges, and, after spitting

> contemptuously on the exhausted land, went off in triumph to find other lands to conquer.]

Irresponsible land husbandry has long been a target of environmentalist activism; currently, in the Amazon, after cattle pasture creation and human settlement, clearing for commercial agriculture and infrastructure improvements (road and railway construction) needed to support it is the leading cause of deforestation.[46] The concern about any loss to the forests of the Amazon is central to current environmentalist priorities including many of those enumerated in *Mamita Yunai*. Clear-cutting and land exhaustion are particularly pressing in the novel, but in one of the most remarkably prescient moments in all of Spanish American literature, Fallas includes ideas about climate change:

> Son millones y millones de metros cúbicos de robles y cedros y laureles y de todas clases de maderas buenas que se pudren de abono p'al banano. . . . Pero, ¡qué l'importa la madera a los machos si no les cuesta nada! Hasta el clima nos van a cambiar botando las montañas.[47]

> [It's millions and millions of cubic meters of oaks and cedars and laurel trees and other kinds of good wood that has been knocked down and left to rot as fertilizer for the banana trees. . . . But that timber doesn't mean a thing to these tough guys because they haven't paid a penny for it! They are even going to change the climate on us by knocking down the trees on all these mountains.]

The passage expresses an idea similar to Fernando Mires's articulation of the "economy of the absurd" with the idea that wood from trees felled in a forest does not cost anything to a corporation granted the right to clear-cut a forest while the long-term effects of such deforestation can be devastating, culturally, socially, and economically. But when the narrative invokes the specter of climate change in the form of extended drought in a passage penned in the 1940s, it means the expression of ideas about climate and environmental degradation that anticipates the articulation of even the most technically obscure scientific statements of concern by at least a decade.[48]

José Tombé (1942) by Colombian Diego Castrillón Arboleda, like *El Amauta Atusparia* and *Sumag Allpa*, represents another of the *indigenista* novels that culminate in an armed and violent revolution by aggrieved Indians. The factors that spark the revolt take the familiar form of a gradual awareness of injustices perpetrated by landowners that have stolen land from the indigenous former inhabitants. However, in *José Tombé*, the narrative defines the conflict explicitly as a stark difference between native and "white" views on how land should be used. Fabio Gómez Cardona argues that the representation of these "white" beliefs occurs more in figurative or generic terms than through the development of specific characters:

> Como quiera que la trama central de la novela pretende representar el conflicto racial entre indios y blancos, es más que significativo que el blanco en tanto antagonista no pase de ser meramente aludido mediante referencias vagas. Se lo menciona como el objeto del odio secular de los indígenas, causante del despojo, es amo y propietario, pero no se concreta jamás en un personaje que intervenga activamente en las acciones y la trama.[49]
>
> [As the central plot in the novel attempts to represent a racial conflict between Indians and whites, it is rather significant that the white antagonists never really appear except for briefly or in vague references and allusions. They are mentioned as the object of near universal indigenous hatred, as the cause of the loss of traditional lands, as lord and landowner, but they are never concretely portrayed as fully developed individuals that actively participate in the action of the plot.]

Such universalizing techniques represent a strategy to portray the indigenous struggle as a communal rather than individual effort. But Castrillón Arboleda's novel makes an appearance here because part of the threat to the community involves environmental destruction. In a dialogue with Eloy Cuají, another member from his tribe, José Tombé frames the argument for revolt against the white landowners using, among other things, an argument critical of deforestation: "Hay vé, Eloy Cuají, lo mesmo que han hecho los blancos. Se han tragau toa la montaña, con los árboles y hojarasca y rastrojal y diay nosotros los dejamos" ["Look, Eloy Cuají, it's the same as what the white folks did. They knocked down everything on the mountain and left a wake of destruction in their path—fallen trees and leaves, bark, limbs, and branches—and we just let them do it"].[50] In this context, the image of the white landowners as devourers of the forest is a metaphor for the way in which foreign monopolies leave nothing for indigenous inhabitants. But while that may be a valid interpretation, the use of similar imagery later in the dialogue invokes the specific terms of deforestation:

> Antes me decía mamá, que podíamos sembrar donde queríamos, porque nacimos aquí y semos hijos de la tierra y siempre vivimos cultivándola y semos igual que los árboles . . . , y un día suben pu el quemau los blancos y derriban los árboles, y hacen potrero, y cogen las lomas pa dale la yerba a sus puercas bestias. (Castrillón Arboleda, *José Tombé*, 68)
>
> [My mother always used to tell me that we could plant wherever we liked because we were from here and were the very children of this earth, the ones to plant and reap here and that we were like the trees . . . , but one day these white folks came up here and slashed and burned and cut down and made enclosures for their horses and took over the hills so their beastly animals could graze.]

As in most of the other *indigenista* texts the implicit argument is framed in the terms of political ecology—the negotiation of access and authority over resources—but in this case the condemnation of the landowners' use of the land to create pasture for their livestock is articulated explicitly as a case of injustice and abuse. If members of the indigenous community are "just like the trees," then cutting down those trees to create pasture is nothing less than the murderous destruction of the native people themselves. The novel concludes, as do almost all the rest in this genre, with a destructive war won by non-native forces. But the reason for the belligerence is also an element that seems to underpin the ecological element of most of the *indigenista* texts that feature violent conflict. The narrative of *José Tombé* makes the ecological motivation explicit; that is, the fight for "la libertad para amar la tierra" ["the liberty to love the land"].[51]

While less well-known than several of the major *indigenista* authors, Guatemalan Carlos Wyld Ospina's short stories also deal with several of the ecological issues that surround the question of land authority and indigenous conceptions of nature. In particular the short story "Los lares apagados" (Disappearing lairs) (1958) explicitly articulates ideas critical of deforestation that have become central to present environmentalist concerns. The story narrates the misfortune of and vengeance by Bartolomé Yat after, in what is a classic convention of the genre, he is unjustly imprisoned, loses his land, and his wife takes up with the owner of the local rice mill. Earlier in the story, before his incarceration, Yat lives a simple agricultural life and expresses his close ties to the land: "He de amarla, . . . he de amarla. . . . He de cultivarla y no he de destrozar a sus criaturas" ["I must love it, . . . I must love it. . . . I must cultivate it and not destroy its children"].[52] The connection between responsible husbandry and love for land and animals takes on outright conservationist condemnation of anti-deforestation when Yat thinks about the clearing of trees in the forest near his hut to plant coffee:

> Un árbol . . . es una cosa santa, y por consiguiente, una cosa buena. Un palo de café es, sin duda, un vegetal también; pero no es propia de la tierra kecchí. En el bosque paternal, significa un intruso, usurpador de un sitio que no le pertenece; y por su existencia sacrifícase el derecho de los abuelos de la selva. Se viola la justicia porque el árbol no puede defenderse, como el animal, y vive únicamente amparado por la sabiduría del chabilgüinc [hombre bueno o santo]. El extranjero no comprende esto, pero el kecchí está obligado a entenderlo. (Wyld Ospina, *Lares apagados*, 50)

> [A tree . . . is a holy thing, and thus, a good thing. A coffee bush is a living thing also, of course; but it is not native to K'iche' lands. When we think about the original forest, it is an intruder, a usurper in a place where it does not belong; and what's more, the decision to plant coffee meant that my ancestors lost their right to the land. In the same way, the native trees cannot defend themselves like animals can, and they can only survive by the good will and wisdom of the *chabilgüinc* (a good or holy man). The foreigner cannot understand this, but we, the K'iche' people are obliged to understand this.]

There are several elements here that anticipate what has become standard in current ecological thinking: the sacred value of trees, a criticism of clear-cutting for agriculture, the right of indigenous peoples to control their resources, the inability of foreigners to understand special relationships with nature. These elements, communicated in the narrative as ideas of nature from an indigenous man, are inevitably a statement of political ecology that advocates for local control, articulates K'iche' conceptions of place, and more than anything, condemns deforestation in any form whether to transform trees into timber or remove them to plant other crops.

ECOLOGY AND NEO-*INDIGENISMO*

Part of the critical emphasis in Wilde Ospina's story analyzed in the previous paragraph was the fundamental inability of foreign individuals to understand indigenous perspectives. That inability was also a prominent element of *Los caminos de Paxil*, especially inasmuch as Arias's novel attempts to account for the indigenous perspective by configuring the narrator as a Mayan deity, a kind of minor lunar god and unseen witness to the events as they unfold. While typical elements of *indigenista* literature remain—labor, land use, property rights, and cultural values—this incorporation of mythology is one aspect of the development of what has come to be known as neo-*indigenismo*.[53] *Cauce sin río* (Dry riverbed) (1962) by Puerto Rican Enrique A. Laguerre is a good example of the way in which these traditional elements are combined with spiritual or supernatural elements as a way to communicate a certain ideology or political ecology. In Laguerre's novel there are repeated references by the first person narrator Victor to an awakening spiritual connection to Taino predecessors on the island: "ciertamente se ha despertado el cemí que siempre llevé dormido en mi conciencia" ["truly, the *cemí* [Taino god] that I always sort of carried with me in my subconscious was awakening"].[54] The reference to the "*cemí*," who for the Taino people were the gods of various natural, solar, and celestial phenomena, marks the connection between the narrator and the reformation of his thinking from a hyper-developmentalist to a nature-protectionist perspective.

The narrator is one of the partners in a prestigious and profitable construction firm contracted for the development of condominium complexes, advertised aggressively to foreign investors, and with plans to go multinational. Disillusioned by what seems like the empty life of a successful industrialist, Victor attempts to take his own life, fails at that, and returns to the family estate of Sanetién to recover where he embarks on a life of quiet reflection. Here is where several of the nature protectionist and anti-developmentalist ideas begin to emerge in the novel: "De seguir así la cosa, dentro de unos años los pocos recursos naturales del país estarían en manos de los inversionistas y especuladores de fuera. Los que nos aprovechábamos ya buscábamos disculpas para justificar los monopolios y la concentración de riquezas" ["If things go on like this, within a few years, the few natural resources in the country will be in the hands of outsider investors and speculators. Those of us here that took advantage were now apologizing for the way in which we tried to justify these monopolies and the subsequent concentration of riches"].[55] The narrator Victor yearns for a simpler life—"El exceso de afición por el regodeo material nos daña. Este inmoderado deseo de objetos y comodidades lleva al mundo hacia su propia destrucción. Bastaría con lo indispensable" ["The excessive desire for such perverse pleasures is harmful. The immoderate desire for luxury and comfort seems to lead the world to its own destruction. It would be enough just to have what one truly needs"]—despite the pressure from colleagues and his own identity as an industrialist: "Para ellos mi locura mayor consistía en querer resucitar un regionalismo de bueyes y carretas, luego de haber comprado la más moderna grúa del país" ["They think that the craziest thing about my ideas is that I want to resurrect the life of ploughs driven by oxen after having purchased the most modern crane in the whole commonwealth"].[56] These moments from the novel capture some of the most central elements of the current environmentalist project: the idea that natural resources are finite and should therefore be conserved, that overconsumption will eventually lead to a bad end, that a modernized, industrialized society may not be the highest human aspiration, and that certain pre-modern societies may have gotten it right with regard to the environment. All of this proceeds with an undercurrent of *indigenista* ideas translated to the ideological parameters of twentieth century Puerto Rican cultural identity and independence movements. The "*cemí*" that awakens in Victor's soul leads him to reject selling out his homeland: "No, no voy a sancionar con mi aquiescencia, en nombre de la libre empresa, la liquidación del país" ["I won't sanction by my acquiescence to the principles of free enterprise, the liquidation of my homeland"]. And his insistence "en recalcar nuestra responsabilidad de hijos de esta tierra" ["to emphasize the importance of our responsibility as children of this land"][57] thus unites ecological concerns about limited natural resources, anti-developmentalist ideas, Taino spirituality, and a discourse of Puerto Rican identity.

The publication of *El zorro de arriba y el zorro de abajo* (*The Fox from Up Above and the Fox from Down Below*) in 1971 (two years after the suicide of its author, José María Arguedas—also the writer of *Yawar Fiesta*, discussed above) represents something of an end point for the genre. While other novels in the tradition have been published since (*Los caminos de Paxil*, Rigoberta Menchú's testimonial novel), Arguedas's death saw the end of a writer who was among the foremost of the *indigenista* authors. However, his last novel also marks a highly interesting moment in the genre (or neogenre) for its explicit embrace of ecological elements as part of the suffering of native communities that *indigenista* texts sought to relieve. In *El zorro de arriba y el zorro de abajo*, the land can more justifiedly be associated with a cesspool infested by a crowd of fishermen, stench, and whores: a place where "degradation is taken to an extreme."[58] The indiscriminant fishing industry has caused the extinction of life in the Sea of Chimbote, once the largest fishery in the world, and the demise of the various fish species in the bay is a metaphor for Peru as a whole and for the individuals who inhabit the sea-coast—*cholos*,[59] displaced indigenous *serranos*,[60] and whores; all, like the sea, seem to have lost their life force; all have become victims of development. The *indigenista* text, as has been seen, represents injustice with the intention that these conditions should be relieved, reformed or abolished. Here Arguedas does the same in terms of ecological injustice; he represents the stream of *serranos* who descend from the mountains and must trade their largely indigenous ways for the filth and garbage that is life in Chimbote.

The toxic discourse of the narrative is particularly gothic with regard to the way in which the disposal of waste from fish processing plants dirties the sea, beaches, and air:

> La fetidez del mar desplazaba el olor denso del humo de las calderas en que millones de anchovetas se desarticulaban, se fundían, exhalaban ese olor como alimenticio, mientras hervían y sudaban aceite. El olor de los desperdicios, de la sangre, de las pequeñas entrañas pisoteadas en las bolicheras y lanzadas sobre el mar a manguerazos, y el olor del agua que borbotaba de las fábricas a la playa hacía brotar de la arena gusanos gelatinosos.[61]
>
> [The fetid seas dispersed the densely odiferous vapors from industrial pots where millions of anchovies were boiled and reduced to a kind of oily pulp that emitted a vaguely nutritive stench as they were boiled down. The odor of waste, blood, and trodden entrails on the decks of fishing boats were cast into the sea from hoses and the stinking water that leaked from the factories toward the sea in small rivers seemed to produce gelatinous worms in the sand.]

These images of entrails pumped through hoses back into the sea, worm-infested beaches, and the stench of blood, guts, and fish oil represent the hellish state of affairs in Chimbote. Later, the novel describes how smog and smoke from various fish processing entities has been transformed into what passes for the skies over the port:

> Tiene y no tiene luz, tiene y no tiene bordes, no se apaga jamás. Se levanta de esas galerías largas, de todo ese laberinto de torres, minerales, sudores y luz eléctrica, de las tripas más escondidas de tanta maquinaria; le cuesta levantarse, pero parece que nadie, ni las manos de los dioses que existen y no existen podría atajarlo. (Arguedas, *El zorro*, 114)

> [The sky seems both light and dark, finite and infinite. In any case, it never gets turned off. The light seems to come from enormous caverns, from a labyrinth of towers, minerals, sweat and electric sources, from the hidden innards of machines; these lights shine out and it seems that even the hands of the gods that may or may not exist are powerless to control it.]

Such passages are typical of many that appear throughout the novel to describe the contamination of the air, land, and seas of Chimbote, representations that decry the ecological injustice of the city. The poor live in polluted, trash-ridden neighborhoods while the rich owners, the ones more directly responsible for the ecological carnage, live away from their own filth; they erect walls and plant trees to separate themselves from the pollution: "Los ejecutivos viven muy lejos del humo y la pestilencia. . . . El resto, unas . . . digamos treinta barriadas, doscientos mil vive en la basura" ["The executives live far from the smoke and stench. . . . The rest, let's say two hundred thousand souls in perhaps thirty slums . . . , live in filth"].[62] The ecological injustice inherent in the living arrangements dictated by class takes on more of a geo-political sense later in the novel with representations of economic disparity through exploitation by multinational corporations of the resources of developing countries:

> Los norteamericanos . . . están embruteciéndose y en camino de la podredumbre. El dominio y el desprecio directo o meloso sobre las naciones de medio mundo los pudre y embrutece, porque en lugar de aprender de los viejos pueblos como éste, solo quieren fomentar rencillas y el caos en ellos y entre ellos con el propósito insensato e imposible de meterlos en un molde y bebérselos después como si fueran una botella de Coca-Cola. (Arguedas, *El zorro*, 222)

> [The Americans are fooling themselves because their dominion and saccharine-like deceit of half the world are forces for decay and brutality and poverty in U.S. culture itself. That is, instead of learning from ancient, native societies, there seems to be nothing more than the promotion of discord and chaos with the senseless aim of putting these societies in a certain mold in order to down them later like so many bottles of Coca-Cola.]

The idea of international demand for the products of developing nations has been seen previously in *indigenista* texts dealing with oil, guano, mining, and large-scale agricultural interests: global demand for something can have hidden but insidious local effects as in several of the novels considered above. Arguedas makes a similar argument in *El zorro de arriba y el zorro de abajo* with its representation of the Chimbote fishery: the rendering of Peruvian *anchoveta* into fishmeal leads to the ecological catastrophe so disturbingly described by Arguedas in the passages cited above. The actual catch between 1968 and 1970 was more than twelve million tons[63] and the demand for labor to process the bone into meal caused internal migrations with permanent cultural consequences. Part of the urgency of the issues discussed in this chapter—the loss of traditional lands, the exploitation of indigenous communities by extractive industries, the articulation of political ecologies more protective of such communities—arise from the lament about the lost societies of indigenous communities absorbed into or syncretized by foreign agents imposed from the very first adventures of the conquest to the neo-imperial economic intervention of North American corporations. In *El zorro de arriba y el zorro de abajo*, the internal migration from their traditional homes in the mountains and jungles of central and eastern Peru to the cesspool that is Chimbote only means another form of the exploitation of and loss of identity for indigenous communities as one of many by-products from the appearance of any number of extractive industries in Latin America.

I find Arguedas's novel something of a touchstone for the transition of ecology in the *indigenista* tradition from a somewhat implicit emphasis to what is far more explicit. Understood in this way, the active criticism of the environmental fallout from industrialized oil drilling in *Los caminos de Paxil* represents the culmination of a long history of ecological emphasis in the *indigenista* genre. As the list of novels enumerated above and in chapter 6 demonstrates, Arias's novel only represents the latest expression within the *indigenista* genre of alternative political ecologies, of gestures toward conservationism, and of concern about deforestation, the toxicity of mining, species extinction, and soil exhaustion. In the third part of the book, I find that explicit expressions of environmentalism such as are present in *Los caminos de Paxil* and in Chilean Luis Sepúlveda's *Un viejo que leía novelas de amor* (*The Old Man who Read Love Stories*) (1989) become ever more common. So much so, I argue that they and others like them make up a distinct current within the production of literature after the "Boom" of Spanish American literature in the 1960s and 1970s.

NOTES

1. Shawn William Miller, *An Environmental History of Latin America* (New York: Cambridge University Press, 2007), 88.
2. Ibid., 89–90.
3. Ibid., 91.
4. Ibid., 90.
5. José María Arguedas, *Yawar fiesta* (Austin: University of Texas Press, 1985), 16–17.
6. Alf Hornborg, "Accumulation: Land as a medium of domination," in *Ecology and Power: Struggles over land and material resources in the past, present, and future*, ed. Alf Hornborg, Brett Clark, and Kenneth Hermele (New York: Routledge, 2012), 13.
7. Jesús Lara, *Yanakuna: Novela quechua* (La Paz: Libreria y Editorial Juventud, 1958), 242.
8. Kay Milton, *Loving Nature* (New York: Routledge, 2002), 100.
9. Alfredo Guillén Pinto, *Lágrimas indias* (La Paz: Casa Editora Mundial, 1920), 112.
10. Ibid., 184.
11. Ibid., 192.

12. See Manuel Vargas Talavera, "Costado Fascista de la ANR," *ABC Color*, November 29, 2012.

13. Natalicio González, *La raíz errante* (Mexico City: Editorial Guarania, 1953), 119.

14. Ibid., 103.

15. As I indicate in the discussion of the ecological elements in work by Argentine *modernista* Leopoldo Lugones in chap. 3, the connection between early Latin American conservationism and fascist ideology was often closely linked. Natalicio González's writing began within the same *modernista* context as Lugones, but one that developed somewhat later in Paraguay than elsewhere in Spanish America.

16. Miguel Ángel Asturias, *Hombres de maíz* (Nanterre, France : ALLCA XX, 1992), 177.

17. Anne Hallum, "Ecotheology and Environmental Praxis in Guatemala," *Nova Religio: The Journal of Alternative and Emergent Religions* 7, no. 2 (2003): 63.

18. Mario Monteforte Toledo, *Entre la piedra y la cruz* (Guatemala City: Editorial "El Libro de Guatemala," 1948), 16.

19. Lautaro Yankas, *El vado de la noche: Novela de la raza* (Santiago de Chile: Zig-Zag, 1963), 45.

20. Ernesto Reyna, *El Amauta Atusparia* (Lima : Ediciones de Frente, 1932), 56.

21. G. Humberto Mata Ordóñez, *Sumag Allpa* (Cuenca, Ecuador: Biblioteca Cenit, 1940), 126.

22. Hornborg, "Accumulation," 18.

23. Arias, *Caminos*, 16.

24. Ibid., 17.

25. Rafael Jaramillo Arango, *Barrancabermeja: novela de proxenetas, rufianes, obreros y petroleros* (Bogotá: Editorial E.S.B., 1934), 37.

26. Ibid., 7, 14, 27.

27. Rómulo Gallegos, *Sobre la misma tierra* (Buenos Aires: Espasa Calpe, 1961), 79.

28. Ibid., 72, 117.

29. Ibid., 111.

30. Bureau of Western Hemisphere Affairs, "Background Note: Venezuela," U.S. Department of State, http://www.state.gov/r/pa/ei/bgn/35766.htm (accessed December 15, 2011). According to U.S. Department of State economic figures from 2009, the petroleum industry in Venezuela accounts for 95 percent of all exports and nearly 12 percent of GDP as compared to 4 percent GDP for agriculture.

31. Jaime Mendoza, *En las tierras de Potosí* (Sucre: Talleres de la imprenta universitaria, 1976), 30, 65.

32. Fernando Ramírez Velarde, *Socavones de angustia* (La Paz: Biblioteca paceña, 1953), 17, 181.

33. S. Hirschberg, G. Spiekerman, R. Dones, *Project GaBE: Comprehensive Assessment of Energy Systems: Severe Accidents in the Energy Sector* (Villigen, Switzerland: Paul Scherrer Institute, 1996). According to Hirschberg, Spiekerman, and Dones's research from 1996, coal has been responsible for 342 fatalities per terawatt-year (a time/energy unit that assumes it takes a year to produce a trillion watts of energy) as compared to nuclear power which has only suffered 8.

34. Ramírez Velarde, *Socavones*, 176.

35. Ibid.

36. Mendoza, *En las tierras de Potosí*, 32.

37. Raúl Botelho Gosálvez, *Altiplano* (Lima: Ediciones Nuevo Mundo, 1967), 145.

38. See Actualité Scientifique, *The Impact of Mining in Bolivia* (La Paz: Institut de recherche pour le développement, 2010).

39. Alberto Rodríguez, *Donde haya Dios* (Buenos Aires: Lautaro bolsillo, 1958), 11.

40. Ibid., 88.

41. Ibid., 64.

42. Ibid., 109.

43. Walter Rojas Pérez, *Costa Rica violada. El caso de* "Mamita Yunai" (San José, Costa Rica: Editorial Porvenir, 2006), 21–44. In this book length analysis of Fallas's novel, Rojas Pérez traces the ecological ineptitude of foreign corporate intervention to the Soto-Keith

contract of 1871, a land concession to the North American Minor Keith to construct a railway corridor to the Atlantic. Keith also received 800,000 acres for personal use, tax-free for a period of 99 years. And, "por mal planeamiento agrario … logra en tiempo record desgastar los micronutrientes de la superficie" ["because of poor agricultural planning . . ., manages to completely exhaust the micro-nutrients of the soil in record time"], 24. The Keith contract typifies subsequent foreign corporate intervention in Costa Rica.

44. Valeria Grinberg Pla and Werner Mackenbach, "*Banana novel revis(it)ed*: etnía, género y espacio en la novela bananera centroamericana. El caso de *Mamita Yunai*," *Iberoamericana*, 2nd series, 6, no. 23 (September 2006): 161, 163.

45. Carlos Luis Fallas, *Mamita Yunai* (Mexico City: Fondo de Cultura Popular, 1957), 65.

46. Rhett A. Butler, "Deforestation in the Amazon," Mongabay.com, http://www.mongabay.com/brazil.html#colonization (accessed December 13, 2011).

47. Fallas, *Mamita Yunai*, 119.

48. Spencer R. Weart, *The Discovery of Global Warming* (Cambridge, MA: Harvard University Press, 2003). This moment in Fallas's novel must be interpreted as a denunciation of human impacts on local weather (persistent drought or regular flooding) rather than global climate change (planetary warming). However, the association between weather change and the removal of trees by humans in the novel seems to look forward to the understanding of the greenhouse effect as caused by the proliferation of heat-trapping gases occasioned by human activities in the more recent notion of global warming. The expression of dismay about changes in climate or weather in *Mamita Yunai* pre-dates the amateur scientist G. S. Callendar's claims concerning a theory of greenhouse gases and warming by several years and it was not until the advent of the environmentalist movement in the 1970s that this idea became scientifically commonplace. See the full history in Weart's monograph.

49. Fabio Gómez Cardona, "Silencios y ocultamientos en la novela *José Tombé*," *Poligramas* 33 (January-June 2010): 193.

50. Diego Castrillón Arboleda, *José Tombé* (Bogota: Editorial Antena, 1942), 67.

51. Ibid., 76.

52. Carlos Wyld Ospina, *Los lares apagados* (Guatemala: Edit. Universitaria, 1958), 23.

53. See chap. 6, note 17.

54. Enrique Laguerre, *Cauce sin río* (Río Piedras, Puerto Rico: Editorial Cultural, 1991), 210.

55. Ibid., 128.

56. Ibid., 158, 169.

57. Ibid., 171, 128–29.

58. Priscilla Archibald, "Urban Transculturations," *Social Text* 25, no. 4 (Winter 2007): 97.

59. In Peru, the term "*cholo*" is used to identify any individual with any amount of indigenous ancestry.

60. Inhabitants of the sierra, often of *mestizo* or indigenous heritage.

61. José María Arguedas, *El zorro de arriba y el zorro de abajo* (Nanterre, France: ALLCA XX, 1990), 40.

62. Ibid., 115.

63. Lucero de Vivanco, "Modernidad y apocalipsis en *los zorros* de Arguedas," *Revista chilena de literatura* 78 (April 2011): 59.

Part Three

Literature, Environmentalism, Ecology

Chapter Eight

Nature after the "Boom": Ecology and Environmentalism in Late Twentieth Century Spanish American Fiction

As the major genres—the regional novel, *indigenismo*, *modernismo*—eventually matured and coalesced into new forms (or neo-forms), overlapped with other literary periods, or simply disappeared, more recent authors began to experiment with notions of genre, content, form, and narrative expression. These experimentations eventually culminated with the "Boom" of Spanish American literature: a still-controversial designation covering literature produced primarily by Colombian Gabriel García Márquez, Mexican Carlos Fuentes, Argentine Julio Cortázar, and Peruvian Mario Vargas Llosa, but also by Mexican Juan Rulfo, Paraguayan Augusto Roa Bastos, Uruguayan Juan Carlos Onetti, Cuban Alejo Carpentier, Chilean José Donoso, and others. The principal point of departure for grouping these authors and their texts as a "Boom" of Spanish American literature resides in the development of the "*nueva novela*" (new novel), fictional works that feature any of the following characteristics:

- la sustitución del principio causa-efecto por la narración fragmentaria y de la secuencia por la simultaneidad
- la desintegración del personaje . . . al estatus de un antihéroe o un ser marginal
- la coexistencia . . . del orden real con el sobrenatural
- la activación del lector

. . .

- la inclusión de meditaciones metaliterarias en el *corpus* narrativo;

- la interpretación del lenguaje no como reflejo, sino como refracción arbitraria de la realidad y, en consecuencia, el ejercicio de la libertad lingüística por medio de la vertiginosa experimentación formal
- la transgresión del regionalismo hacia el nivel más universal
- la preservación de la tendencia crítico-social . . . pero con más hincapié . . . en la eficacia crítica del humor
- la sucesiva incorporación . . . de las manifestaciones literarias o culturales anteriormente marginadas.[1]

- [the substitution of the cause-effect principle by fragmented narrative and chronological sequencing by simultaneous action
- the disintegration of character to the status of anti-hero or marginalized individual
- the coexistence of actual and supernatural events
- reader participation

. . .

- the inclusion of metafiction in the narrative corpus
- the idea of language not as a reflection, but as an arbitrary refraction of reality, and therefore, the practice of free linguistic association through wild experimentations in form
- the disappearance of regionalism replaced by more universal concepts
- the continued use of social criticism . . . but with more emphasis . . . on the efficacy of humor
- the evermore common inclusion . . . of literary and cultural phenomena which had previously been marginalized]

These elements, particularly the use of non-chronological narration, meta- and intertextuality, and the articulation of various perspectives, narrative focalizations, and points of view led to explorations of the limits of language, fiction, literary structure, and form. But the problem with the identification of the "Boom" as the moment of first use for these literary elements is that several earlier authors were already doing such things: Argentine Jorge Luis Borges, Colombian José Eustasio Rivera, the Spaniard Miguel de Cervantes. . . . And the problem with my list of authors is that there is always debate about whom should be included in the list—Borges (too early?), Ernesto Sábato (too little?),[2] Manuel Puig (too late?)—and with the pesky fact that several of the authors associated with the "Boom" outlived it and are living still.[3] I leave the discussion to others who have already rehearsed it well and better than I have;[4] my concern is with those novels from about the second half of the twentieth century, particularly those in the decidedly post-"Boom" period of the 1990s, that demonstrate a more explicit literary expression of environmentalist ideas. I have already discussed several of these elements in chapter 4 with *novelas de la selva*[5] by Vargas Llosa, Chilean Luis Sepúlveda, and others and in chapters 6 and 7 with the overlap of *indigenismo*, ecology, and recent texts.

In his noted essay titled for this "*nueva novela*" of Spanish American literature, Carlos Fuentes seems to want a definitive break from these more regional novels:

> "¡Se los tragó la selva!", dice la frase final de *La vorágine* de José Eustasio Rivera. La exclamación es algo más que la lápida de Arturo Cova y sus compañeros: podría ser el comentario a un largo siglo de novelas latinoamericanas: se los tragó la montaña, se los tragó la pampa, se los tragó la mina, se los tragó el río.[6]
>
> ["The jungle has swallowed them!" says the last line of *La vorágine* by José Eustasio Rivera. The exclamation is something more than just an inscription on the gravestone of Arturo Cova and his friends: it is what could also be said about an entire century of Latin American novels: they were swallowed by the mountain, by the plains, by the mines, or by the river.]

Rejection of older forms is borne out in Fuentes's texts such that even his most ecological novel—*Cristóbal Nonato* (1987)—remains true to the author's iconoclastic form; it is more concerned with experimental approaches to narrative and culture than with the environmental degradation that appears periodically represented in its pages. References to acid rain and the greenhouse effect, desertification of the Lancandón rainforest, and ecological activism all make brief appearances in the text and few like Fuentes can so effectively manage the language of pollution as in the following description of Mexico City's bad air:

> El aliento mortal de tres millones de motores vomitando sin límites bocanadas de veneno puro, halitosis negra, camiones y taxis y materialistas y particulares, todos contribuyendo su flátula a la extinción del árbol, el pulmón, la garganta, los ojos . . . .
>
> . . . Mierda machada.
>
> Gas carbónico.
>
> Polvo metálico.
>
> Y todo ello a 2300 metros de altura, aplastado bajo una capa de aire helado y rodeado de una cárcel de montañas circulares: la basura prisionera.[7]
>
> [The gases from three million engines ceaselessly vomiting thick clouds of pure poison, black halitosis, buses and taxis and trucks and flatbeds and cars, all contributing to the extinction of trees, lungs, throats, and eyes with their flatulence . . . .
>
> . . . Smashed pieces of shit.
>
> Carbon dioxide gas.
>
> Metallic dust.

> And all this at 2300 meters above sea level, stuck under a blanket of freezing air and imprisoned by a ring of mountains as if the air and dust and waste were locked up in some kind of meteorological jail.]

In these few lines, Fuentes assigns blame and identifies victims for the massive ecological catastrophe that is the environment of the capital city, the largest in Latin America by population. But this is just one among many concerns in the novel: the marshalling of mass culture references from Madonna to Michael Jackson to Smurfberry Crunch, the intertextuality and metafictional flourishes so familiar to the Spanish American "new novel," the interplay of structure, convention, form, and language. These are Fuentes's forte and he does not abandon them to introduce something like a pervasively environmentalist rhetoric; rather, the ecological element is one among many that are repurposed for the innovative, experimental literary style for which the Mexican author was known.

Earlier, Paraguayan Augusto Roa Bastos's *Hijo de hombre* (*Son of Man*) (1960) also tackles material that, like Fuentes's novel and its representations of the pollution in Mexico City, would seem to suggest an ecocritical posture, but the content of the novel does not bear this out. Roa Bastos's novel deals extensively with the 1930s war between Paraguay and Bolivia for control of the northern Chaco region, claimed by both countries. The claim was important for Bolivia, as it entailed maritime access to the ocean which could be provided by the Paraguay River, but another important factor in the conflict was speculation that the region could contain oil. As the war unfolded, Standard Oil and Shell supported Bolivia and Paraguay respectively. Thus, the origins of the conflict have much to do with petroleum extraction and political ecology, but these concerns are almost entirely absent from the narrative. There are brief mentions of the absurdity of fighting for "los títulos y acciones de las empresas del petróleo" ["titles and shares of stock in petroleum corporations"] and for "las tierras de los gringos" ["lands that the *gringos* want"];[8] but there is nothing in the way of developing these ideas from an ecological perspective. Roa Bastos, like Carlos Fuentes, seemed more concerned with the kinds of textual experimentation that were typical of the style of writers from the "Boom": narrative representation as a refraction rather than a reflection of reality (a theme particularly fruitful, developed as it was in the novel through the portrayal of the fog of war), the incorporation of non-translated Guarani language utterances directly into the text, and the use of non-chronological, simultaneous, and variously focalized narrative moments.

It was either too early or unrelated to their artistic purposes for novelists like Roa Bastos (and others of the "Boom") to explore ecological issues in the 1960s even if those issues seem urgent in retrospect. But by the late 1980s, environmentalism as a movement had begun to mature and it should come as no surprise that environmental themes would show up in the literature. By 1983, noted Argentine-born, Chilean labor activist, historian, and intellectual Luis Vitale, whose observations are cited several times in the preceding chapters, published his *Hacia una historia del ambiente en América Latina*. In that volume, he makes reference to the beginnings of an ecological movement in Latin America, including references to certain issues, which, as he writes, were points of departure, but have now become central to the environmentalist movement. These include critiques of consumerist societies, mono-cultured agriculture, nuclear power, and commercialized fishing while espousing the preservation of green space, the ecofeminist critique, and sustainable development.[9] In 1990, Fernando Mires makes reference to a Latin American "Discourse of Nature" in his monograph of the same name which I cite in the introduction and thereafter. In the fictional literature, perhaps it was the highly literary qualities that characterize the texts of authors associated with the "Boom" that allowed for only limited exploration of ecological considerations as it did with *Cristóbal Nonato* or Vargas Llosa's *El hablador* (*The Storyteller*) (1987). And even in some relatively more recent works that centrally feature the natural world, ecological aspects are only apparent in the briefest of glimpses and to the most careful, perhaps biased, of readers.

La danza inmóvil (The immoveable dance) (1983) by Peruvian Manuel Scorza and *Colibrí* (Hummingbird) (1984) by Cuban Severo Sarduy are two examples of the limits of literary environmentalism in the 1980s. The first, while set in the rainforests of eastern Perú, focuses almost exclusively on the morality of revolution while *Colibrí* seems a parody of the *novela de la selva* genre with its echo of the famous conclusion of Rivera's novel—"—otra vez ése. ¿No se lo había tragado la vorágine?" ["—this guy again. Didn't the jungle already swallow him up?"][10]—its lost-in-the-jungle protagonist of Japanese ancestry like in Vargas Llosa's *La casa verde* (*The Green House*) (1966), and its snake and mosquito attacks which seem an element of all these texts. In *Colibrí,* the representation of the jungle is most often effected with sexualized and even homo-erotic tones: the forest sounds of are characterized by "monos en celo" ["monkeys in heat"]; one severed half of a snake spews venom "como en una ejaculación" ["like an ejaculation"]; the ruins of an abandoned house are taken over by the jungle "con sus sapos fornicadores" ["and its fornicating toads"].[11] These literary features—parody and sexualized imagery—are at the heart of the representation of the jungle in Sarduy's novel and it simply does not explore the kind of environmentalist commitment found in the work of other authors like Luis Sepúlveda, Mexican Homero Aridjis, and Costa Rican Anacristina Rossi publishing just a few years later.

But by the 1990s and 2000s, the reality of ecological contamination, species endangerment, and environmental activism begin to serve as backdrop for more and more Spanish American works. Novels like Guatemalan Rodrigo Rey Rosa's *Lo que soñó Sebastián* (What Sebastian dreamed) (1994) and Mexican Leonardo Da Jandra's "Coastal Trilogy" have protagonists that maintain a certain level of ecological awareness, feature narrators sympathetic to the environmentalist cause, and include story elements that make reference to such things. However, these elements do not realize a profound effect on narrative forms. In Rosa's novel, a murder mystery unfolds with signs of guilt pointing to a man who chooses to live alone on a plot of land in the jungle and who spends his nights keeping poachers of endangered animals at bay. Ecological issues appear commonly in the narrative—the removal of endangered caiman and tapir skins from hunted animals, the attempts by the protagonist to "proteger a los felinos" ["protect the feline species"], an awareness that military operations "que por achicharrar dos subversivas queman no sé cuantas caballerías de la selva" ["in order to roast a couple of subversives burn down who knows how many acres of forest"][12] —but do not affect the literary conventions of the text.

Da Jandra's trilogy—*Huatulqueños* (1991), *Samahua* (1997), and *La almadraba* (The maze net), (2008)—features a narrative that unfolds in an ecological reserve—"El Parque Nacional Huatulco"—but the trilogy does not experiment with ecological elements as explicitly as Anacristina Rossi's novel *La loca de Gandoca* (*Mad about Gandoca*) (1992) (also set in an officially "protected" but often violated wildlife refuge; I analyze Rossi's novel more extensively in chapter 9). That is, there are references to "la lucha contra burócratas y desarrollistas ecocidas" ["the fight against bureaucrats and developmentalist ecocides"][13] and to deforestation: "lo que era cedro, caoba y demás maderas de valor y a nada más quedaban los secos o los comidos por comején" ["there were cedars and mahogany and other valuable woods but now there was nothing left of them but dried out and termite-infested trunks"].[14] The closest that any of the novels in the trilogy come to the unique articulation of narrative register to express an ecological idea is when some of the residents living near the reserve ask about some biologists sent by the government:

> —¿Son gringos? —preguntó Tiburcio atraído por la algarabía de los biólogos.
> —No, son los que andan viendo de proteger a los animales y a los árboles.
> —¿Y pagan bien? —dijo Tiburcio interesado.
> —A ellos los manda el gobierno para que no chinguemos lo que aún queda. (Da Jandra, *Huatalqueños*, 216)

> [—Are they *gringos*?—asked Tiburcio drawn into the conversation by the gibberish of the biologists.
> —No, they're the people that go round trying to protect the animals and trees.

—And do they pay well?—asked Tiburcio, suddenly interested.
—The government sends them to stop us from fucking up what's left of the jungle.]

The use of the expletive to describe environmental destruction represents one of several times in Spanish American literature when profanity is used as a narrative device to articulate the profanation of nature.[15] But the above-referenced moments are isolated in Da Jandra's trilogy and do not represent anything like an innovation with regard to the literature. However, there were several authors—Luis Sepúlveda and Arturo Arias (discussed in chapter 4 and in chapters 6 and 7, respectively) as well as Homero Aridjis, Nicaraguan Gioconda Belli, Argentine Mempo Giardinelli, Anacristina Rossi and fellow Costa Rican Fernando Contreras Castro—whose ecological emphasis contributes to the creation of new forms, styles, and conventions of literature.[16]

THE NARRATIVE OF ECOLOGY: LITERARY INNOVATIONS IN ENVIRONMENTAL NOVELS

What I propose for an approach to this more recent batch of environmentalist novels rests on the notion that ecology can function as a generating principle of literature in the course of the expression of its ideology. For example, in an avowedly environmentalist text such as Sepúlveda's *Un viejo que leía novelas de amor* (*The Old Man who Read Love Stories*) (1989) the ecological elements affect the content *but also* the literary conventions in his novel. These may include narrative order, narrator reliability, character archetypes, the sequence of conflict and resolution within the parameters of a story arc, etc. In the case of a jungle detective story like Sepúlveda's novel, the logic of environmentalism affects the sequence of narrative such that readerly expectations for some conventions of the genre must be expanded.

In the novel, Antonio José Bolívar Proaño, the old man referenced in the title, decides to abandon his home in the sierra and participate in a governmental scheme to settle the Amazon. Eventually, when the plans go wrong and the settlers begin to starve, they are rescued by the indigenous Shuar tribe and Proaño decides to abandon the settlement project altogether and takes up with the Shuar. As he grows older and because of a tragic mistake that results in the death of a close friend in the tribe, Proaño returns to the settlement, which by now has grown into a full-blown pueblo complete with mayor, occasional voting in national elections, and even the infrequent appearance of an itinerant dentist. As seen in chapter 4, there are several moments where the narrative offers explicit statements of conservationist ideology, condemnation of deforestation, concerns about the extinction of species, and the articulation of several other elements central to the environmentalist project of the late 1980s and early 1990s. The content of the story allows the narrator all manner of situations from which to express such ideas and the representation of Proaño and the Shuar as responsible stewards of the rainforest functions to communicate the ethics of the novel: go and be like Proaño. But I am more interested in the way that Sepúlveda's novel functions as a catalyst for the articulation of a poetics of literary environmentalism. For example, the novel's non-chronological narrative sequences, unsettled character motivation, and inversions in the causal relationship between conflict and resolution are literary innovations that proceed from the environmentalism of the text.

Proaño's investigative work to solve the problem of the ocelot killings allows for categorization of the novel within the genre of "*novela policíaca*," a mystery novel or *policier* genre particularly popular in Latin America, often repurposed by Southern Cone writers for their own literary purposes: Argentines Jorge Luis Borges, Ricardo Piglia, Manuel Puig, Juan José Saer, and Chilean Roberto Bolaño. The genre features tales of mystery that often take place in urban settings with the detective displaying an ability to "read" clues from the place where the unlawful activity took place and from other leads that might come in during the course of an investigation. The crime scene itself, including bodies, weapons, or traces of the killer, becomes a text that the successful detective must read in order to solve the case. Such stories are almost exclusively urban in their telling, but in *Un viejo que leía novelas de amor* the jungle locale as well as the concerns of environmentalism influence the conventions of sequence that remain essential to the detective genre: clue-revelation and interpretation followed by identification of criminals, pursuit, and eventually some form of justice. The traditionally urban form of the mystery novel requires that the detective be able to interpret the signs of criminal activity in order to identity and apprehend the criminal. But by removing the novel from an urban context, Sepúlveda facilitates the expression of environmentalist ideals through a kind of sequential inversion.

When a crime has been committed, the perpetrators must be brought to justice; but the location of the plot of a mystery novel in a tropical forest produces ironies of continuity for the way that the crime must be investigated and the transgressor punished. Usually, the crime scene and other clues are unambiguous with regard to the villainous activity. Robbing a bank or killing a rival, lover, or innocent bystander all seem to fit unproblematically under the definition of criminal behavior; what's left is to find the culprit and bring him or her to the proper authorities. However, in Sepúlveda's novel, the crimes are not violations of any civil law, but transgressions against the natural world. From Proaño's perspective, the pursuit of the killer is not a just pursuit: dispatching the ocelot will not result in the righting of a wrong. Rather, the misdeed has been committed by an offender who has already paid the ultimate price; the *gringo* who killed the female's cubs and mate has already met his fate. In most murder mysteries, the body of the victim is analyzed to uncover the clues that will lead to the apprehension of the criminal, but in Sepúlveda's text, it is the body of the killer that must be examined to determine the crime.

Most mystery stories either turn out well for the "good" guys or end in tragedy. Usually, the detective gets the "bad" guys or (in the *noir* tradition), the criminals may get off on a technicality or escape. The conclusion to Sepúlveda's novel, however, features neither of these scenarios. Rather, the animal's death is portrayed as just one more in a tragic succession of ecological effects initiated by the *gringo*. In the end, the novel, despite the presence of many of the most characteristic features of the genre, departs from the conventional sequencing of the *novela policíaca* because the criminal suffers his just deserts quite early on and his victim is killed much later by an unwilling Proaño. When he is finally successful in tracking the animal and ending its life, the old man stands next to the great dead cat and feels a sense of profound sorrow rather than exultation at the completion of his task. The regret that he experiences is for the environmental destruction that claimed so many human and nonhuman lives with the implied ethical position that human invasions of the natural world ought to be discontinued for the great stress that such activity places upon the jungle.

The distinctive literary elements of Sepúlveda's novel arise as uniquely causal from the ethical advocacy: the representation of an ecological problem forces a narrative sequence that does not follow the typical conventions of the mystery novel. It should be remembered from the introductory paragraphs in this chapter that a non-chronological revelation of story elements, the development of plot through a series of non-consecutive episodes, unmarked leaps in fictional time and moment, and other literary devices with regard to sequence of narrative are common elements in the "new novel" tradition and in the "Boom" of Spanish American literature. Novels like Carlos Fuentes's *La muerte de Artemio Cruz* (*The Death of Artemio Cruz*) (1962), Julio Cortázar's *Rayuela* (*Hopscotch*) (1963), most of the novels by Mario Vargas Llosa, and particularly the richly complicated unfolding of the saga of the Buendía family in Gabriel García Márquez's *Cien años de soledad* (*One Hundred Years of Solitude*) (1967) are the most prominent of many examples in the region where sequential narrative had been abandoned in favor of a non-linear order. Yet, as I argue at the beginning of this chapter, these authors were concerned with a strictly literary evolution of narrative, what Fuentes described as a "ruptura de la insularidad tradicional de nuestra novela" ["a break from the traditional insularity of our novels"].[17] Yet in the case of Sepúlveda (although clearly an inheritor of the traditions of the "Boom" with regard to narrative sequence), the lack of a traditionally developed order of events in his novel is required more by the demands of the inclusion of environmentalist elements than by preoccupations with solely literary innovations.

ECOLOGICAL CHARACTER IN NOVELS BY LUIS SEPÚLVEDA

In *Un viejo que leía novelas de amor*, the human protagonist is a strange old man who knows the jungle well; in Sepúlveda's *Historia de una gaviota y del gato que le enseñó a volar* (*The Story of the Seagull and the Cat Who Taught Her to Fly*) (1996), it is a seagull orphaned because of industrial pollution; and in *Mundo del fin del mundo* (World at world's end) (1989), the narrator Ismaél is a political exile. The novels rescue the perspectives of disenfranchised outsider, victim, and independent political resistor from the globalized politics of mainstream environmentalist movements and multinational corporations. The novels' incorporation of exclusion, victimization, and dissent as figurative sites for the articulation of narrative voice contributes both to a environmentally political agenda but also to the development of unique forms for ecological representation.

Often, the alarming data recorded by controlled studies of environmental science are central to the environmentalist textual idiom. Rigorously determined statistics about global warming, carefully calculated reports about pollutant parts per million in the air and water, and formal censuses of threatened species populations represent scientific documents that can produce ethical conclusions about the stewardship of nature. Yet as "objective" as the methodology of environmental science may seem, it does not tell the whole story. Because factually scientific reportage does not employ the language of suffering, outrage, and trauma, it does not allow for empathy with victims of environmental destruction and may even exclude their voice from the ecological debate. Literature however, and Sepúlveda's work in particular, can focalize the experience of victimization as a consequence of the contamination of the natural world. Through a sort of return to the didacticism of Medieval literatures, Sepúlveda's environmentalist novels feature protagonists—the eccentric José Antonio Proaño from *Un viejo que leía novelas de amor*, the orphaned seagull Afortunada from *Historia de una gaviota y del gato que le enseñó a volar*, the exile Ismaél from *Mundo del fin del mundo*—cast as victims of harmful environmental practices in order to condemn the structures of power responsible for such policies. Their victimization allows for a fictional perspective from which the non-scientific side of environmentalism can effectively be told: the suffering of the outsider.

In Sepúlveda's novels, the protagonists' physical isolation from modernized society functions to provide privileged narrative positions from which to condemn beliefs and actions that have resulted in environmental destruction. The community at El Idilio considers Proaño an eccentric: an old man returned from nature and the indigenous Shuar community. He lives on the outskirts of town, observes the conservationist precepts of Shuar-living, and keeps to himself. Likewise, the narration of a story from the perspective of animals is a common device for children's literature and that of the cats and seagulls in *Historia de una gaviota y del gato que le enseñó a volar* affords a unique view on the contamination of nature. Animals are immediate inhabitants of nature and suffer its contamination in direct ways; given a voice, the animals personify the victimization of nature as a consequence of pollution. Finally, Ismaél, narrator of *Mundo del fin del mundo,* is a journalist and political exile from Chile. Years later, his endeavor to cover the story of whaling in the *Tierra del fuego* necessitate a return. Sepúlveda himself had escaped the oppression of the Pinochet regime by going abroad and he incorporates the necessarily critical posture of exile to the articulation of environmentalist elements in the novel.

The Eccentric

Sepúlveda's Proaño is a simple man with few possessions; he lives in a shack on the outskirts of town and makes meals of the prawns he can find on the banks of the river or of the animals he hunts in the jungle. His time among the villagers of El Idilio was spent in gathering from the forest only as much as was required for food or for trade to acquire his beloved romance novels. While he is never thought of as a part of the community at El Idilio, he is respected by the townspeople and plays a role in their affairs nonetheless. As gadfly and foil to the corpulent, bumbling mayor, the village turns to Proaño when the trouble with the ocelots becomes a real problem. Earlier, Proaño's time with the Shuar has alerted him to the fragile ecological balance of the natural world and effected a positive change on his body and mind. But there too, Proaño was never fully integrated: "En definitiva, era como uno de ellos, pero no era uno de ellos" ["Clearly, he was like one of them, but he was not one of them"].[18] In both instances, Proaño is outsider to the two communities, and as such, enjoys a distinct position from which to evaluate the attitudes and actions displayed within each one. As the narrative unfolds, the Shuar are praised as conservationist stewards of the natural world, while the specter of development, industrialization, and its by-products leads to Proaño's memorable curse upon the pimping of the Amazon.[19]

Wendell Berry has articulated ideas about holistic ecology similar to Proaño's convictions about simplified living. In an essay called a "Movement to Teach the Economy What It is Doing," Berry affirms that the "movement should content itself to be poor. We need to find cheap solutions, solutions within the reach of everybody, and the availability of a lot of money prevents the discovery of cheap solutions."[20] Proaño's existence is perfectly suited to Berry's vision: he is not a rich man; in fact, to most his standard of living would be considered close to poverty. While he is with the Shuar, life may seem like a utopian vision, but it revolves around a very specific manner of making a living that requires the perpetuation of the source of livelihood: animals to hunt, flora to gather. Proaño eventually returns to El Idilio and the realities of development and modernization but now with a highly critical perspective toward such ecologically harmful economic projects. The fictional articulation of the old man's outsider position reinforces the idea that limits to consumption are a necessary step to restoring the health of the planet, a metaphor for Berry's demand for "cheap solutions, solutions within the reach of everybody." Proaño is poor, yet offered as a model of ecological success: someone who does not just survive in the forest but flourishes. But in order to credibly suggest such radical positions as the severe limitation of consumption, resistance to development, and the undesirability of modernization, the novel must evoke narrative empathy with the voice of an eccentric and outsider. And while it may be unrealistic to assume that humanity can simply return to hunting and gathering, Lawrence Buell notes that "environmentalism of any sort cannot hope to achieve even modest reforms unless *some* take extreme positions advocating genuinely alternative paths."[21] In order to make the alternatives represented by Proaño's lifestyle more palatable, the novel features a protagonist so eccentric that it creates sufficient distance between the reader and that kind of character. The distance allows for emulation of the old man's ideals without necessarily implying that one must abandon hearth and home for the rainforest in order to achieve at least modest ecological reforms to one's behavior. An analogous critical distance is achieved in Sepúlveda's other novels through the use of similarly atypical perspectives: an animal-centered narrative and one from the marginalized status of political exile and non-mainstream journalist.

Animals

On the cover of the Tusquet's edition of *Historia de una gaviota y del gato que le enseñó a volar*, the subheading indicates that it is "una novela para jóvenes de 8 a 88 años" ["a novel for young people from 8 to 88 years of age"]. But like most books for kids, this story is not intended only for them. While the speaking animal is a common convention in children's literature, its use in Sepúlveda's novel serves to criticize (grown-up) human activities: the release of petroleum from oil tankers that slicks the seagull Kengah's wings and ultimately causes her death. By extending the power of speech and communication to animals, the text quite literally gives nature a voice and presents a vision of environmental damage from that unique perspective. When injured animals respond to the contamination of the environment in a very human way—their sorrow comes in the form of language as they eulogize a fellow creature's death—it offers a representation of nature as victim. Rather than an exploration of issues of species extinction as they pertain to conservation of forests in *Un viejo que leía novelas de amor*, *Historia de una gaviota y del gato que le enseñó a volar* deals with the consequences of industrialization upon the inhabitants of the city itself. The narrative goes beyond personification to the creation of empathy with nature in the representation of animals as sentient individuals.

The fictionalization of animal individuality articulates concern for nature from a more immediate perspective. In one part of the story for example, the pollution of the ocean by oil barges is described from that of a flying seagull, native inhabitant of the sea's coasts:

> Muchas veces desde la altura vio cómo grandes barcos petroleros aprovechaban los días de niebla costera para alejarse mar adentro a lavar sus tanques. Arrojaban al mar miles de litros de una sustancia espesa y pestilente que era arrastrada por las olas.[22]
>
> [Many times she saw from above how enormous oil tankers would take advantage of thick coastal fog and put to sea in order to clean their tanks. They would leave thousands of liters of a thick and pestilent substance slowly dispersed by waves.]

The view from seagulls is privileged because of their proximity to the ecological destruction: it is their habitat that is directly contaminated by the oil. The focalization of the narrative through the experience of seagulls and other animals allows for a more urgent perception of pollution. Animals are the direct victims of toxic industrial practices and it is the novel's ethical gesture to use their perspective to foretell our own eventual victimization as a result of the contamination of nature. The articulation of animal individuality in the narrative gives credence to the representation of pollution's effects from a highly subjective position. As the poisonous effects of oil tanker discharges are communicated within the narrative as the cause of the seagull Kengah's death, the language does not merely reflect an objective opposition to such practices, it expresses outrage:

> Con las alas inmovilizadas, pegadas al cuerpo, las gaviotas eran presas fáciles para los grandes peces, o morían lentamente, asfixiadas por el petróleo que, metiéndose entre las plumas, les tapaba todos los poros. . . .
>
> . . . La mancha negra. La peste negra. Mientras esperaba el fatal desenlace, Kengah maldijo a los humanos. (Sepúlveda, *Gaviota*, 28)
>
> [With wings immobilized and stuck to their bodies, seagulls were easy pickings for big fish, or they would slowly die, asphyxiated by the oil which would get in between their feathers and block all their pores. . . .

> . . . The black slick. The black plague. While she waited to die, Kengah cursed the humans.]

The narrative is emphatic in its condemnation of the contamination: the oil slick is called "la peste negra" ["the black plague"] or "la maldición de los mares" ["the curse of the seas"].[23] These descriptors are strongly nuanced and demonstrate how environmentalist rhetoric can influence the tenor of language within a text. In this case, the denunciation of what has happened to the water resorts to the language of curses and plagues. The terms of such description operate according to a logic of victimization that does not rely upon the statistical precision of science or the supposed objectivity of journalism, but upon the condemnatory or sorrowful registers of empathy. This and the way in which the rest of the novel features the traditionally feminine task of nurturing the orphaned Kengah invoke what Karen Warren describes as ecologically feminist content that gives "expression to a variety of ethical attitudes and behaviors overlooked or underplayed in mainstream Western ethics."[24]

In the novel, Kengah manages to escape the oil slick and fly to the apartment of the cat Zorbas in nearby Hamburg, but she dies soon after landing on the balcony. From the perspective of the poisoned bird, the pollution of the natural world can be represented in the terms of struggle, pain, suffering, and eventual death. By imagining nature as victim of human industrial practices, the text creates a sense of identification between readers and the natural world and reveals the ethical nature of Sepúlveda's environmentalist writing: an ecological ideology articulated within the framework of victimization always seems more intuitively compelling:

> –Los humanos son, por desgracia, imprevisibles. Muchas veces con las mejores intenciones causan los peores daños—sentenció Colonello. . . .
>
> . . . —¿Y qué decir del daño que hacen intencionadamente? Pensad en la pobre gaviota que murió por culpa de la maldita manía de envenenar el mar con su basura—agregó Secretario. (Sepúlveda, *Gaviota*, 85–86)

> [–Humans, unfortunately, are unpredictable. Often, with the best intentions they cause the worst damage—declared Colonello. . . .
>
> . . . —And what about the damage that they intentionally cause? Think of the poor seagull who died because of the damned obsession with poisoning the ocean with their garbage—added Secretario.]

When these cats fault humans for "la maldita manía de envenenar el mar con su basura," their status as victims of human negligence gives credence to the implicit claim that is at the root of the text: industrial activities that harm animals must be reformed. Writing the animal perspective in *Historia de una gaviota y del gato que le enseñó a volar* is not just a literary device nor a rehash of Quiroga's animal stories (considered in chapter 4), but the perspective from which a new kind of story can be told: the story of the environment as a victim of humanity.

The Exile

The final novel for this section is *Mundo del fin del mundo* where, rather than the passive victimization that characterized the description of the oil-poisoned seagull, Sepúlveda has the story describe a journalist-in-exile who escapes political and corporate censorship to cover acts of resistance against the environmental destruction of the planet. The reader is treated to a kind of environmentalist action novel where the villains are unfeeling whale hunters and the heroes affiliated with ecological activism. The narrator Ismaél, a freelance journalist and political exile from Chile, describes the intrigues of whale hunting versus species protection. The reference to *Moby Dick* is made obvious in the novel through the choice of the narrator's name but also in several moments where *Mundo del fin de mundo* adopts plot points directly from Melville's story including, most spectacularly, a sentiently vengeful whale. But Sepúlveda's work diverges from the classic North American novel by having the narrative focus on the protection and not hunting of the great whale. Ismaél's work centers on the activities of Greenpeace whose resistance to ecologically destructive activity is often mercilessly quashed by the perpetrators of such activities. Greenpeace's non-violent engagement philosophy often leaves its members vulnerable to violence; a circumstance represented in the novel when a correspondent for the news agency founded by the narrator becomes the victim of a hit-and-run attack for photographing an illegally registered Japanese whaling vessel, the *Nishin Maru*. Ismaél is dispatched to get the real story and as the narrative unfolds, the *Nishin Maru* (like Ahab's *Pequod* by Moby Dick) is itself attacked and disabled by the very whales it was hunting.

The novel adopts the perspective of political exile and bears a great deal of similarity to Sepúlveda's own autobiographical tale of exile in *Patagonia Express*. In *Mundo del fin del mundo*, the fear of physical violence enforces the narrator Ismaél's exile: "Siempre quise regresar a Chile. Tuve ganas, pero a la hora de la determinación pesó más el miedo" ["I always wanted to return to Chile. I wanted to, but in the end, fear always won out"].[25] Sepúlveda's experience of exile must have deeply colored the narrative in *Mundo del fin del mundo* as it is revealed in the novel that Ismaél, like the autobiographical Sepúlveda of *Patagonia Express*, was "uno de los tantos que conocieron la cárcel y huyeron del horror para reunir fuerzas en la tierra de nadie del exilio" ["one among many who knew what it was like to be imprisoned and fled from the horror in the no man's land of exile in order to recover"].[26] In the novel, the figure of political exile is an effective trope with which to tell the story of environmentalism. The exile has the advantage of perspective: free from the censorship of a native land controlled by a totalitarian regime, he or she is privy to information that his or her compatriots at home are not. Ismaél's decision to found an alternative news agency is based upon his desire to remain independent from "la prensa 'seria', interesada en los temas que afectan al medio ambiente solamente cuando éstos adquieren visos de escándalo" ["the 'serious' press, interested in environmental issues only when they reached the level of scandal"].[27] It is from the outsider position that Sepúlveda constructs a sympathetic representation of the efforts of Greenpeace and a harsh criticism of many forms of ecological destruction. Ismaél as a journalist in exile recovers a voice in the ecological debate, a voice denied him (and Sepúlveda) by the regime from which they are exiled.

It is fitting that Sepúlveda should choose to feature outsiders as the protagonists of his environmentalist novels. Although governments and political parties claim to support environmentally friendly policies like sustainable development and open space preservation, they do not support the kind of sweeping changes that many environmentalists believe are necessary for the long-term health of the planet. According to those on the progressive side of the environmentalist debate, NGO's like Greenpeace, activists associated with somewhat more radical movements like Earth First! and the Earth Liberation Front, and even ordinary but ecologically responsible individuals who have rejected the consumerism of modernized society are the ones that practice the most authentic and effective kind of environmentalism. It is with sympathy for these radical beliefs that Sepúlveda narrates his environmentalist novels. For progressive environmentalists, he wouldn't get the story right if he told it from the perspective of a government official, a corporate manager, or a political candidate because such individuals are not engaged in "real" environmentalism. Sepúlveda's novels are almost didactic in their advocacy of radical environmental reform and while didactic texts are often criticized when they are too blunt in the articulation of a political or ethical agenda, Sepúlveda envisioned his kind of writing as an antidote to the texts of the "Boom," described by Andrew Graham-Yooll as "large, impenetrable tomes plagued with obstacles."[28] In an interview with Graham-Yooll, Sepúlveda somewhat scandalously affirms the *modus operandi* of a new "circle" of writers: "I write adventure stories with the basic purpose of hooking readers and bring them back to reading. That is the function of the new writing. Latin American literature was becoming very boring before we came along."[29] It is not coincidental that the use of literature to convince readers of the importance of ecological activity was for Sepúlveda also a kind of literary reform. It may be too much to insist that Sepúlveda was all about aesthetic innovation when the author himself criticizes authors from the "Boom" generation who "forgot to tell stories while they were trying to show how clever they were."[30] And although he adopts non-chronological narration and invokes unique perspectives as did many of the writers in the tradition of the "new novel," Sepúlveda's work can also be characterized for its attention to the conventions of narrative storytelling in the service of an unabashed commitment to the ethics of environmentalism. The addictive, page-turning qualities of Sepúlveda's texts make for fun reading, and as such, highly effective communicators of environmentalist ideas.

NOTES

1. Elzbieta Sklodowska, *La parodia en la nueva novela hispanoamericana (1960–1985)* (Philadelphia: John Benjamins, 1991), xii–xiii.
2. Sábato only published three novels: *El túnel* (*The Tunnel*) (1948), *Sobre heroes y tumbas* (*On Heroes and Tombs*) (1961), and *Abbadón el exterminador* (*Angel of Darkness*) (1974).
3. When I began this project, Carlos Fuentes (1928–2012) was still living and Gabriel García Márquez, and Mario Vargas Llosa are with us still; Vargas Llosa won the Nobel Prize for literature in 2010.
4. Carlos Fuentes, *La nueva novela hispanoamericana* (Mexico City: Cuadernos Joaquín Mortiz, 1969); José Donoso, Historia personal del "Boom" (Barcelona: Seix Barral, 1983); Raymond L. Williams, *The Twentieth Century Spanish American Novel* (Austin: University of Texas Press, 2003); Jean Franco, *An Introduction to Spanish American Literature* (New York: Cambridge University Press, 1994). As evidence of the intellectual debate about labels for this period of literary history, Donoso includes the designation "Boom" in his title while Franco seems to scrupulously if suspiciously avoid the term entirely.
5. See Introduction, note 43.
6. Fuentes, *Nueva novela*, 9.
7. Carlos Fuentes, *Cristóbal Nonato* (Mexico City: Fondo de Cultura Económica, 1987), 91.
8. Augusto Roa Bastos, *Hijo de hombre* (Madrid: Espasa Calpe, 1993), 252, 368.
9. Luis Vitale, *Hacia una historia del ambiente en América Latina* (Mexico City: Editorial Nueva Imagen, 1983), 105–14.
10. Severo Sarduy, *Colibrí* (Barcelona: Argos Vergara, 1984), 176.
11. Ibid., 143, 152, 153.
12. Rodrigo Rey Rosa, *Lo que sonó Sebastián* (Barcelona: Seix Barral, 1994), 66, 84.
13. Leonardo Da Jandra, *La almadraba* (Mexico City: Planeta, 2008), 64.
14. Leonardo Da Jandra, *Huatalqueños* (Mexico City: Editorial Joaquín Mortiz, 1991), 198.
15. Other instances of this sort include when Luis Sepúlveda's José Antonio Bolívar Proaño at the close of *Un viejo que leía novelas de amor* (Barcelona: Tusquets, 1993) curses those that "emputecían la virginidad de su amazonía" ["pimped the virginity of his Amazonia"], 137; and frequently throughout Carlos Fuentes's *Cristóbal Nonato*.
16. In Laura Barbas-Rhoden, *Ecological Imaginations in Latin American Fiction* (Gainesville: University Press of Florida, 2011), the analysis includes several titles that I do not pursue here; these include seal hunters in *Un piano en Bahía Desolación* (1994) and in *Fuegia* (1991) by Argentines Libertad Demitrópulos and Eduardo Belgrano Rawson; a revival of the rubber extraction novel in *Fordilandia* (1997) by Argentine Eduardo Sguiglia; and the effects of modernization on local populations in *Calypso* (1996) by Costa Rican Tatiana Lobo. Despite the fact that they contain relevant environmental themes, the novels do not feature the unique literary forms tied to their ecological explorations that I use as a point of departure for the inclusion of texts in this chapter and in chapters 9 and 10.
17. Fuentes, *Nueva novela*, 97.
18. Sepúleda, *Un viejo*, 37.
19. Ibid., 137.
20. Wendell Berry, "In Distrust of Movements," *Orion* (Summer 1999): 18.

21. Lawrence Buell, *Writing for an Endangered World* (Cambridge, MA: Harvard University Press, 2001), 7.

22. Luis Sepúlveda, *Historia de una gaviota y del gato que le enseñó a volar* (Barcelona: Tusquets, 1996), 29.

23. Ibid., 27.

24. Karen J. Warren, *Ecofeminist Philosophy: A Western Perspective on What It Is and Why It Matters* (Lanham, MD: Rowman & Littlefield Publishers Inc., 2000), 103.

25. Luis Sepúlveda, *Mundo del fin del mundo* (Barcelona: Tusquets, 1994), 13.

26. Luis Sepúlveda, *Patagonia Express* (Barcelona: Tusquets, 1995), 87.

27. Sepúlveda, *Mundo*, 48.

28. Andrew Graham-Yooll, "Light at the End of a Tunnel: The New Writers of Latin America," *Antioch Review* 52, no. 4 (Fall 1994): 566.

29. Ibid.

30. Ibid., 570.

Chapter Nine

Eco-Satire: Green Humor, Contaminated Imagery, and Environmental Language in Recent Spanish American Fiction

In my introductory comments to the section on Chilean Luis Sepúlveda from chapter 8, I suggested that the analysis of Sepúlveda's work could be taken as a blueprint for the approach to the environmentalist novels to be studied here. The ironic and humorous tone of *Un viejo que leía novelas de amor* (*The Old Man who Read Love Stories*) (1989) can be found in Argentine Mempo Giardinelli's *Imposible equilibrio* (*An Impossible Balance*) (1995), a comical, mock epic about fugitive hippopotami kidnappers. The toxicity of industrial practices that are at the root of the conflicts in Sepúlveda's novels comprises a new kind of material for representation. The morbid appeal of death and destruction allows for the elaboration of novel images: the representation of the terrible beauty of oil-slicked oceans in Sepúlveda's *Historia de una gaviota y del gato que le enseñó a volar* (*The Story of the Seagull and the Cat who Taught Her to Fly*) (1996), resembles the language with which Mexican Homero Aridjis describes the stupendously abundant toxicity of Mexico City, AD 2027 and with which Costa Rican Fernando Contreras Castro evokes the mountains of trash at Río Azul dump outside San Juan, Costa Rica. Finally, the inclusion of previously marginalized ideas provides a whole spectrum of new possibilities for literature: in Sepúlveda's novels, the articulation of environmentalist principles from the perspective of outsider; in Costa Rican Anacristina Rossi's *La loca de Gandoca* (*Mad about Gandoca*) (1992), the inclusion of diverse narrative forms comes as impassioned love songs or withering parodies of the euphemisms from formal legal language that provide political cover for government officials whose policies prove disastrous for the environment. The novels considered in this chapter, like those written by Sepúlveda, exploit the possibilities of environmentalist elements for the production of new literary conventions, unique metaphors, and captivating stories.

HIPPOPOTAMIAN HUMOR: ECOLOGICAL RESISTANCE IN MEMPO GIARDINELLI'S *IMPOSIBLE EQUILIBRIO*

In Mempo Giardinelli's *Imposible equilibrio*, the comic power of farce presents an opportunity to challenge the established values of the ultra-serious rationality that is at the root of over-development and environmentally debilitating modernization. The story takes place in the Argentine Chaco, an arid, desolate land, represented in the text as a post-industrial region, abandoned after the promises of development did not materialize:

> Vastas selvas taladas insensatamente, en veinte años se habían convertido en desiertos—de millones de hectáreas—mientras que las zonas más húmedas y fértiles eran ahora inmensurables espejos de aguas podridas, lagunas y bañados, esteros y charcos.[1]
>
> [Vast jungles cut down senselessly over a period of twenty years had become deserts—millions of acres—while the wetter, more fertile regions had become immeasurable patches of fetid water, lagoons and swamps, inlets and pools.]

The consequences of these permanent landscape alterations—the legacy of a failed program of industrialization—are onerous: the stench of wood rot and the scarcity of potable water. To confront these difficulties, the regional government adopts a solution as ridiculous as the water is polluted: they will import hippopotami from an African nation to eat the branches, bracken, and other rotting debris that are impeding river navigation and the availability of potable water, "por la sencilla razón de que los hipopótamos comen toneladas de toda esa mierda" ["for the simple reason that hippopotami eat tons of all that shit"].[2]

Hippo-mania sweeps Argentina. Popular bars, restaurants, and sports teams change their names and/or logos to include hippopotamian imagery. A massive public poll is taken to solicit suggestions for names, provincial and national curricula are amended to include hippopotamic zoology; suddenly, everyone knows everything about the enormous African lake-dwelling mammal that will soon arrive in the backwaters of the Chaco. But from among a group of patrons of a local bar, there emerges a small contingent who opposes the idea: the serious problems of ecological destruction will not easily be resolved through the simple introduction of hippo-herbivores, large though they may be. Three of them decide to take action: ex-guerrillas, Victorio Largomarsino and Pura Solanas, and an exiled gringo, Frank Woodward, Pura's lover. An idealistic, jeans-wearing co-ed, Clelia Riganti, not involved in planning the operation, unexpectedly joins the group as they kidnap the hippos during the animals' ceremonial arrival in the Chaco and flee with the authorities in pursuit.

In the novel, the justification for rejecting the government's hippo scheme is implicit: the complex web of ecological relations so disastrously disrupted by the failed imposition of modernization upon the Chaco cannot simply be repaired by the introduction of a few hippopotami. The government's proposal is ludicrous; stealing the hippos, however, is even more absurd and illustrates the novel's derisive attitude toward Argentine history and politics. It is hard to keep a straight face as I write about abducting hippopotami, yet the use of humor is an effective strategy precisely because it makes people laugh: humor makes critical observations more palatable. Giardinelli's satirical narrative condemns an unsound ecological policy and examines the ethical limits for the use of violence by former guerrillas living in a democratic society, serious issues both. The palliative register of humorous satire however allows the ethical positions in the novel—the need for better stewardship of the Chaco, the offense of violence—to come across through less contentious, more artful literary imagery.

The humor with which Giardinelli resurrects the guerilla figure, for example, creates the necessary figurative distance to finally come to terms with these skeletons in Argentina's closet. Although the violence utilized by the Jorge Rafael Videla regime of 1976-81 to put down the threat posed by the armed guerrillas was far more brutal, widespread, and enduring,[3] the guerrilla insurgency was itself a violent movement. Political prisoners were kidnapped and executed or put into "people's prisons," often holes dug into the foundations of private homes. Torture was commonplace for these prisoners as well as food-, water-, and light-deprivation. The guerrillas took key sites such as military bases, headquarters of important corporations, and television and radio stations and they assassinated some high ranking political and military figures. But the shadow of a failed guerrilla past is reconciled in the novel by having Largomarsino and Solanas reject the old methods. And what creates a stark contrast between the fictional abduction of hippopotami and capturing police or military outposts during Argentina's Dirty War is the absolute absurdity of the hippo situation. There was nothing funny about tortured and disappeared political dissidents, but having the commando operation invoke a comic rather than tragic effect allows Giardinelli the freedom to represent a guerilla operation stripped of its ideological baggage. In this way, he incorporates the ethical advocacy of non-violence to the representation of an operation motivated by ecological concerns: the hippo-napping is more of a prank than an assault.

THE SUN'LL GO OUT, TOMORROW: ECOLOGICAL APOCALYPSE IN NOVELS BY HOMERO ARIDJIS

According to Mejica[4] mythology, the next fifty-two year cycle of time structure is scheduled to draw to a close in the year 2027, a year that Homero Aridjis adopts as the fictional setting in *La leyenda de los soles* (The legend of the suns) (1993) and *¿En quién piensas cuando haces el amor?* (Who do you think about when you make love?) (1996). These novels share a fictional world where Mexico City is renamed Ciudad Moctezuma and suffers from the social, cultural, and ecological problems it faces today, but Aridjis has them exacerbated, intensified, and more widespread. His vision of a perverse, corrupted, filthy megalopolis on the eve of a new cycle of time sets the tone for an exploration of the connections among political power, moral decay, and ecology. The novels are apocalyptic in that they conclude with the transformation of time from the Age of the Fifth Sun to the Age of the Sixth following the destruction of Ciudad Moctezuma in the magnitude of an earthquake so intense that the whole city collapses upon itself. But the End depicted in the novels is unlike Western conceptions of the apocalypse that predict the abolition of time and history. Rather, the circular structure of Mejica conceptions of cyclical time allows for the emergence of a new age from the destroyed material of the old: the conclusion of the epoch of the Fifth Sun signals the beginning of the Sixth. From the perspective of environmentalism, a literary representation of the destruction of the Old from which arises the construction of the New offers a paradoxical warning against the ecologically destructive practices of the present: at the same time that it condemns such excesses, it almost gleefully caricatures them in a mix of yearning for and dread of the future. Critic Miguel López-Lozano comments that "the metropolitan nightmare depicted in Aridjis's novels allow[s] for the discussion of specific limitations of Mexican modernity" and the representation of the future in the novels is "an ecological nightmare [that] stands as a resounding criticism of the results of modern industrialization."[5] These are valid critical responses to the issues raised in Aridjis's novels; the dystopian posture in the novels functions to highlight failures of current political ecologies of modernization and industrialization. However, in the following analysis, I also consider how a formal textual element such as the formulation of unique imagery arises from the process of the representation of ecological apocalypse and simultaneously articulates an alternative to the reform-oriented emphasis of contemporary environmentalism.

The imagery of environmental apocalypse in Aridjis's novels features a poetics of language heavily influenced by the kind of extreme ecological contamination that characterizes Ciudad Moctezuma, circa 2027. The narrative of pollution, moral depravity, violence, and corruption with which the two novels tell the story of the ecological and social ills of the future comprises what Lawrence Buell would call toxic discourse; that is, the novels' unflinching depiction of the irredeemable urban structures of Ciudad Moctezuma regularly employ "totalizing images of a world without refuge from toxic penetration."[6] Every element of the city—the sky over it, the air that envelops it, its lack of vegetation, the toxicity of its waste, the victimization of its inhabitants—are central to the representation of what the world might be like in thirty years. The sky and air of Ciudad Moctezuma are perhaps the most commonly occurring points of reference to communicate the deep sense of decay and toxicity that has overtaken the city. Throughout the novels, the sun, the sky, and the air are described in several different, but equally toxic ways: "No hay aire" ["There's no air"]; "no había ni huella de azul en el cielo, había manchas moradas, ciertamente, y una especie de rojo profundo" ["there wasn't even the trace of blue in the sky, there were purple patches, certainly, and a kind of deep red"];[7] "el cielo, semejante a la cara de una actriz que se ha hecho cirugía plástica, parecía que iba a romperse en mil pedazos si se sonreía" ["the sky, something like the face of an actress that has had too much plastic surgery, seemed like it was going to break into a thousand pieces if there was ever a smile"];[8] "[el cielo] parecía una naranja podrida y dolían los ojos con sólo mirarlo," . . . "parecía huevo podrido" ["[the sky] seemed like a rotten orange and it pained the eyes to look at it," . . . "it seemed like a rotten egg"].[9] The complete penetration of pollution and contamination to all parts of the sky, air, and the sun itself, indicates the total reach that ecological toxicity has achieved in Ciudad Moctezuma. There is no refuge from toxic penetration in the world described. It is not just that anyone can be exposed to the toxicity, but that it is everyone's most fundamental experience of the material world: "[el cielo era] una humillación pública" ["(the sky was) a public humiliation"].[10]

The numerous images of the contamination of air, sun, and sky are merely the most obvious examples that illustrate the lack of refuge from toxic imagery in the texts; other moments of toxicity are similarly widespread. Bodies of water are as contaminated as the skies: "no lejos un río sólido, como serpiente pútrida, fluía sin reflejos, sin alegría y sin leyenda" ["not far from there, a solid river flowed like a rotting serpent, without reflections, without joy, and without legend"]. The heat is inescapable and unending: "la oscuridad aún no se transformaba en incipiente azul, una frescura tímida aún no aliviaba del calor que había azotado a Ciudad Moctezuma durante la última década" ["the darkness had not yet become an incipient blue, a timid cool still could not relieve Ciudad Moctezuma of the heat that had lashed it for the last decade"].[11] Residents cannot escape breathing and exhaling the stench of the city: "su[s] boca[s] exhalaba[n] *smog* como si la contaminación le[s] saliera de las entrañas" ["their mouths exhaled smog as if it came from their very guts"].[12] There was no escape from the cacophony of noise that echoed continuously and ubiquitously: "el zumbido de los motores era interceptado por las alarmas de los coches que aullaban sin que nadie les hiciese caso, y por las sirenas de las ambulancias atrapadas en el tránsito" ["the drone of engines was drowned out by car alarms that went off without anyone paying attention to them and by the sirens of ambulances caught in traffic"].[13] Every physical experience of the city is corrupted by contamination of some kind: visual, aural, tactile, olfactory, gustatory.

With respect to literary representation, the extent to which the physical world is revealed as thoroughly corrupted by pollution is reflected in the language with which such phenomena are described. New images must be innovated in order to adequately portray a new reality, one that is beginning to be experienced in the more and more pervasively contaminated environment of the present but which in the novels occurs as the thoroughly befouled reality of 2027. As he extrapolates present environmental toxicity to extreme levels in the future, Aridjis appropriates well-worn tropes to describe water or sky and with the addition of toxic adjectives devises new ways to describe them. Comparisons of the sun in the sky to an orange or an egg, which have become trite in conventional literature, are modified to reflect an ecologically dystopic future and result in an innovative and disturbing language to describe the natural world. The toxicity of the language with which Aridjis describes the contaminated sun of 2027 is a parody of the innocent literature of an unpolluted world. No longer does a gentle river at dawn inspire a grandiose language of beauty and sublimity, only disgust and repulsion; credible metaphors are no longer beautiful, only images like rotting eggs or putrid snakes serve. In Aridjis, the romantic tendency to nostalgically personify nature is parodied as grotesque: the sky looks like the smile of an actress who has had too much plastic surgery.

Another innovation of language employed in the narrative is an instance of a different type of toxic discourse identified by Buell: the gothic description of the consequences of pollution where by "gothic" Buell means a graphic and morbid description of ecological victimization. The representation of agony and suffering are common in gothic literature and graphic portrayals of suffering as the result of environmental toxicity are widespread in Aridjis's novels:

> Los esqueletos de los árboles muertos del Paseo de la Malinche se fueron precisando a medida que se acercaba al Ángel de la Independencia, ángel de la libertad perdido en el neblumo.
>
> Los edificios de ambos lados de la calle estaban borrosos. Un niño, borracho de *smog*, tosía y lloraba parado en una esquina, sin saber qué dirección tomar. (Aridjis, *Leyenda*, 132)

> [The skeletons of the dead trees on Paseo de la Malinche slowly disappeared as you got closer to the Ángel de la Independencia, angel of liberty, lost in the mist.

> The buildings on both sides of the street were fading. A child, drunk on smog, coughed and cried as he stood on the corner, not knowing where to go.]

Typical of gothic description, a skeleton is utilized as a figure in the narrative, but in this case it functions as an image for dead trees juxtaposed with the figure of the suffering child, equally victimized by a toxic environment. The agony of the child as innocent and defenseless victim is juxtaposed in its own right with the image of arboreal skeletalization; the novelty here is that the suffering and victimization, which in most gothic work is endured by human victims, in Aridjis's text is also directed toward nature.

But even gothic toxic discourse fails to capture the total corruption of nature because it has not the capacity to represent the loss of what had come before. I refer to a language of representation that cannot account for absence, that cannot recreate the lack of nature, and that cannot portray the full ramifications of the memory of what has been lost:

> Por esa avenida venía un río, ¿cómo pintar ahora su ausencia, su cuerpo entubado, su carga de aguas negras?, ¿cómo pintar la desesperación de un río, el grito silencioso de la Naturaleza en agonía? (Aridjis, *Leyenda*, 164)

> [Somewhere by this avenue, a river once flowed, how to paint its absence now? Its tubed course and flow of raw sewage? How to paint the despair of a river, the silent scream of nature in agony?]

The dilemma here expressed by Juan de Góngora, a painter by profession, is the difficulty to represent that which has ceased to exist. A fundamental aspect of painting resides in the fact that the referent, the object being represented, is reproduced in some form. Even in the most abstract or expressionist of work where the relationship between painting and object might not always be clear, the relationship itself is not interrupted: the model exists as object of representation. But in the case of a portrayal of something that has been lost, the representative act fails because it can attempt to portray what had gone before, but it cannot capture the sense of loss implied by the fact that what is being represented can now only function as memory, as pure description: a signifier signifying nothing. The present ecological crisis resides precisely in this: the memory of what had existed before is now all that remains of what has been lost. When a river becomes encapsulated and buried in concrete, it loses almost all the characteristics of a river. Painting it becomes an exercise in futility: water flowing through concrete has nothing to do with what the river had been before and the loss is definitive. It is not that literary imagery lacks representative capacity, but that such representations fail to adequately convey what has been lost in the process of absolute change implicit in the reality of ecological contamination.

The discursive failure of language to adequately represent memory and loss illustrates a fundamental aspect of environmentalist thought. First, capitalist economics demonstrate a lack of symbolic language with which to correctly represent the value of an uncontaminated world. The sense of loss in the face of the destruction of nature is not taken into account nor is the economic value of human desperation in the face of a vanishing natural world. Here again is Fernando Mires's "economy of the absurd": capitalist economics does not and cannot calculate the true and total cost of the destruction of nature. It cannot account for the despair of lost nature nor can it figure the negative consequences that such a loss will imply for social, cultural, and spiritual values. The figure of the painter utterly unable to translate irretrievable loss into the idiom of painted image corresponds to the persistent failure to include the irrecoverable loss of nature as part of the economic calculus of development, modernization, and industrialization.

Of all the most recent authors surveyed in this chapter, Homero Aridjis is among the most politically committed. He founded the *Grupo de los cien*, a group of one hundred intellectuals, poets, and authors whose affiliation as a group is based upon their concern about environmental problems; the group's efforts include lobbying and petitioning for environmentalist causes in Mexico, Latin America, and the rest of the world. The ethical position is clear in his novels: turn away from the current course or suffer the nightmarish ecological future that they prophesy. Still, I cannot ignore the literary possibilities of the ecological elements in his novels: the frightening imagery of an apocalyptic future and the ironic beauty of the language of pollution. The innovation of unique images for environmental catastrophe is also an element of Fernando Contreras Castro's novel *Única mirando al mar* (Única looking at the sea) (1993). And like Aridjis, who is known as much for his activism as for his novels and poetry, Anacristina Rossi has produced a novel—*La loca de Gandoca*—in which the narrative has a decidedly non-fictional purpose: the preservation of the Manzanillo-Gandoca Wildlife Refuge in her native country. But also similar to the literature produced by Aridjis, Contreras Castro and Rossi have written novels that demonstrate the literary possibilities of the incorporation of ecological rhetoric, specifically the way in which the fictional space of these novels are elastic enough to continually allow for the innovation of imagery as well as to absorb disparate and heterogeneous ecological ideas.

NOT IN MY BACKYARD: FILTHY IMAGERY IN FERNANDO CONTRERAS CASTRO'S *ÚNICA MIRANDO AL MAR*

Fernando Contreras Castro's use of unique images to communicate ecological ideas in *Única mirando al mar* is perhaps even more urgent than in

Aridjis's novels because of the realities of waste disposal in supposedly "eco-friendly" Costa Rica, and in the rest of Latin America, an issue that appears repeatedly in the global press.[14] And, like in the novels by Aridjis, Contreras Castro invents some startlingly unique images to express the extent of the environmental contamination of the Río Azul dump in the interior of Costa Rica,[15] images that are especially startling given the pristine reputation maintained by the nation for its ecological responsibility, eco-tourism, and progressive environmental philosophy. As Jerry Hoeg observes, Contreras Castro's novel "does not focus on the landscape of the countryside, but rather on that of the city, and specifically on that part of the city that is most emblematic of the negative effects of the consumer society, Río Azul. The landscape of Río Azul is the 'after' landscape of global production and consumption."[16] The images of surfable waves off pristine beaches, impeccably preserved colonial architecture, or kayaking among the mangroves are not to be found here. Rather, Contreras's novel invites readers to consider what these manifestations of a globalized economy based on the fiscal input of tourism implies in terms of the disposal of outputs: the waste generated by Costa Rican society as a whole added to that of the externally focused tourism industry. Again, Hoeg observes that in "*Única* there is a call for an alternative to the industrial development and runaway consumerism that produces the environmental devastation symbolized by Río Azul."[17] Consumerism just as easily applies to the domestic urban development of San José as it does to that of the ever-growing eco-tourism industry, *the* Costa Rican export commodity, consumed domestically and thus responsible for a share of the waste that ends up in Río Azul. Contreras forces a reconsideration of the seemingly pristine reputation for Costa Rican eco-tourism: the exhilarating zip-lining above an unsullied rainforest may only be possible by removing the waste generated by these same tourists to a place they do not visit: the Río Azul landfill.

From the perspective of literary analysis, the images that must be innovated in the text to represent so much waste include a comparison between the operators of trash-moving bulldozers and the eternally pointless boulder rolling from the mythical story of Sisyphus: "de adelante hacia atrás, todo el día, como herederos del castigo de Sísifo sin haber ofendido a los dioses con ninguna astucia particular" ["forward and back, all day, like inheritors of the curse of Sisyphus but without having offended the gods for any trickery whatsoever"]. The dump is invaded by carrion insects—"su cortejo de moscas y zopilotes" ["a not-so-regal court of flies and buzzards"]—and its effluvium is a source of water contamination: "los ríos Damas y Tiribí condenados a beberse los caldos que se filtraban constantemente; pero sólo una parte de ellos, porque el resto iba a dar a los mantos acuíferos profundos, inyectándose de manera intravenosa en el cuerpo de la tierra" ["the Damas and Tiribí rivers were condemned to drink the foul soup that flowed into them regularly; but they shared this soup with underground aquifers that were infused as if from an intravenous needle that penetrated the flesh of the ground"]. The land that is fouled by the garbage is constantly compared to a body in ill health and to the stench of illness, an odor "que despedía la indigestión eterna de la tierra atragantada de basura" ["that reeked from the earth which suffered a choking indigestion from all the garbage"].[18] The notion of ill health is reinforced through these references to corporeal malady and tragically, the patient is terminal: the contamination of the trash leads to death. The land under and around the dump slowly succumbs, surrounded by soon-to-be bereaved local residents that observe its ineffable demise: "veinte años de estar soportándolo, viéndolo crecer y viéndolo morir en una agonía infinita de cadáver palpitante y enfiebrado" ["twenty years of dealing with this, watching the landfill grow and then, after a seemingly infinite agony, seeing it die, turned a feverish and quivering corpse"].[19] All of these images—references to mythological curses, carnivorous insects, intravenous drug use, ulcerated flesh, choking stench, and death—are similar to what Aridjis employed to imagine what the contamination of the near-future of Mexico City might look like. But what is particularly troubling about Contreras Castro's narrative is that the referent is an actual place and the ugly imagery, as disturbing as it is, pales in comparison to the real thing: readers and tourists do not see, smell, wade through, or taste the garbage like the actual Costa Rican *buzos*[20] do. From a literary perspective, the novel employs gothic toxic discourse, but in such a way that it cuts in two ways: the putrescence of the garbage dump requires a certain intensity of imagery in order to capture its foulness at the same time as the gargantuan proportions of the waste disposal problem in Costa Rica invites the creation of unique images of decay to communicate the feculence, impurity, and refuse of the dump. The strength of such images—hypodermic injections of putrid liquid, the exquisite agony of decomposing corpses, the construction of a mountainous topography of trash upon which humans suffer to eke out a living like modern-day Sisyfi—resides in their ability to communicate the urgency and the ecological catastrophe of garbage collection and disposal in Costa Rica hidden behind that country's reputation as an ecological paradise. And from the perspective of literary analysis, the foul reality of dumps and landfills creates the fictional space for a kind of symbiotic literary interaction that employs the realities of threatened nature to innovate unique images whose exceptional qualities likewise communicate the ethical concerns of ecology and environmentalism.

SAVE MY BACKYARD: CONSERVATIONIST ADVOCACY IN *LA LOCA DE GANDOCA* BY ANACRISTINA ROSSI

Like the all-too-real issue of garbage removal in Costa Rica, *La loca de Gandoca* by Anacristina Rossi had a specific and real-world Costa Rican issue as the focus of the ethical position in the narrative: the protection of the Gandoca-Manzanillo National Wildlife Refuge. Her novel represents a fictional manifestation of the tendency of environmentalist rhetoric to adopt, adapt, and co-opt various modes of expression. This could also characterize the narrative tendencies of Spanish American literature since before the "Boom." The short stories of Jorge Luis Borges, for instance, routinely incorporated straight narrative, the recompilation of other texts (some real, some fictitious according to the purposes of Borges's text), the invention of fictional encyclopedias that make reference to fictional worlds, the citation of actual works of literature but in ironic academic contexts, the invention of scholars, the recharacterization of actual historical figures, etc.[21] Many texts by Borges's fellow Argentine Manuel Puig were also articulated through narratives that took the forms of various means of communication. In perhaps the most well-known of Puig's works, *El beso de la mujer araña* (*The Kiss of the Spider Woman*) (1976), the novel includes the incorporation of dialogue uninterrupted by a narrator's framing conventions, police reports, the ekphrastic incorporation of film, transcripts from police surveillance, etc. The incorporation of ecological elements in Rossi's novel employs similar devices, a technique that provides the narrative space for the creation of heterogeneous narrative registers with a range that in *La loca de Gandoca* allows for varied expressions of environmentalism as it is practiced. Ecological language can be as diverse as legal briefs, the enumeration of statistics, protest songs, prayer, fiction and nonfiction texts, harmless pranks, ecotage; several of these are present in the novel. The most traditional are the statistics that the narrative marshals at several points and juxtaposes with sardonic commentary to provide some concrete contexts to the argument for preservation of the refuge:

> El informe era terrible. Agregaba que además, solo para el área de amortiguamiento de ese parque, había cuatrocientas solicitudes de proyectos turísticos. . . . Además, todos los parques tenían problemas de fondos, problemas con la propiedad privada y con la contaminación. Del Parque Nacional Manuel Antonio sacaban diariamente unas veinte toneladas de basura y ese parque recibía millones de metros cúbicos de aguas fecales. "Los extranjeros van y se meten a nadar y lo ven muy bonita," dijo Mariana. "Y no saben que están nadando en mierda."[22]

> [The report was horrifying. It indicated that just in the park's buffer zone, there were at least four hundred formal applications for the development of tourist

> projects. . . . Plus, all national wilderness reserves were experiencing budgetary problems, disputes over private property claims, and pollution. Every day, twenty tons of garbage were removed from the Manuel Antonio National Park alone and it absorbed millions of cubic meters in volume of fecal sewage. "The tourists come to swim and say how nice everything is," said Mariana, "but they don't know they are swimming in shit."]

There are also several careful articulations of legal arguments:

> En efecto, Mariana prueba que el ministerio está violando los artículos 6, 23, 24, 45, 50 y 89 de la Constitución Política de nuestro país, el artículo 18 de la Ley de Vida Silvestre, el artículo 1 del Decreto de Creación del Refugio, los artículos 3, 4, 9 y 10 del Decreto de Reglamento del Refugio, la Ley Forestal, La Convención Centroamericana de Ambiente y Desarrollo, [etc.]. (Rossi, *La loca*, 92-93)

> [In effect, Mariana has documented how the ministry is violating articles 6, 23, 24, 45, 50, and 89 of the political constitution of our nation, article 18 of the Wildlife Protection Law, article 1 of the Bill to Create the Reserve, articles 3, 4, 9, and article 10 of the bill that created the bylaws for the maintenance of the reserve, as well as the Forest Laws, the Central American Convention for Environment and Development, [etc].]

The laying out of these more deductive arguments based on statistics and compelling statements of legality and illegality with reference to statutes, international treaties, and various conventions and congresses with binding and non-binding findings concerning the refuge undergirds the serious nature of what is at stake. However, statistics and appeals to law are not always as persuasive as other forms of discourse and the novel plays with the idea that different registers of language can also affect the resolve of readers (or viewers or listeners) to come to different conclusions. In the quotation above about the quantity of sewage removed from the Manuel Antonio National Park, alarming statistics are cited, but it is perhaps more effective to point out that "Los extranjeros lo ven muy bonita [pero] no saben que están nadando en mierda. . . ." The transgressive use of profanity which would have no place in the formal language of reports, declarations, and communiqués or in the dry factuality of statistics finds an entirely appropriate outlet in a fictional text. The juxtaposition of volume statistics about sewage with the comical image of tourists swimming in their own feces not only puts two forms of narrative register in conversation with each other but also provides for immediate, visceral contexts for what the statistics actually mean.

The novel does not only change register with regard to language types, but in several instances, it reverts to a language of emotional persuasion and the invocation of spiritual aspects to communicate ideas about nature. There are several examples of a kind of romantic anthropomorphization of nature:

references to over-development as a sounding of "las campanadas de destrucción de estas latitudes" ["bells that toll the destruction of this region"].[23] Ecological advocacy has often taken the form of arguments that find their point of departure in support of principles of sacred value and spiritual motivation; Rossi's novel articulates such principles on several occasions, including a fortuitous meeting with:

> un bicho verde, greñudo y gordito . . . el "dueño de monte," entidad legendaria de cuentos y consejas. —Lo primero que te voy a pedir, Daniela, es que aclarés a los lectores que yo no soy una metáfora ni un recurso de estilo. Que esto no es realismo mágico. Mi presencia es verdad. (Rossi, *La loca*, 100)
>
> [a kind of green creature, fat and disheveled . . . the "hobgoblin of the jungle," a legendary figure from fairy tales and story books.—First, I must ask you, Daniela, to make it clear to your readers that I am not a metaphor or stylistic flourish. This is not magical realism. My existence and presence here in the jungle is absolutely real.]

The meta-fictional gesture by the "dueño de monte" to discourage readers from thinking that the appearance of supernatural actors in a battle over the preservation of the natural world represents another merely fictional gesture adapted by the novel is perhaps the posture that a work of literature is most well-suited to articulate: that is, the use of narrative conventions under the cover of fiction to playfully communicate a "message."

In another moment of meta-fiction, the narrative employs a reference to the absurdities of the real-world battle by Anacristina Rossi to enforce the protection of the Gandoca reserve.[24] The creation of the reserve was resisted at the governmental level through the arbitrary shifting of jurisdiction from various ministries of forestry and natural resources from the regional to the national level. In the novel, the bureaucratic morass is condemned by having the narrator exclaim, "ocho oficinas distintas en un año, los refugios han bailado por ocho oficinas y siguen bailando. Si esto fuera una novela y no la pura verdad, dirían que el autor abusa del *Deus ex machina*" ["eight different offices in one year, the reserves have danced in and out of eight offices and they're still dancing. If this were a novel and not the literal truth, critics would say that the author is abusing the literary device of *Deus ex machina*"].[25] The meta-fictional gesture is all the more poignant since writing the novel was an expression of direct environmental advocacy. According to Rossi scholar Sofia Kearns, the text:

> narrates the events leading to a real-life environmental and legal case, through which residents of that area tried to save the refuge from imminent commercialization and eventual destruction. The novel was written hastily in just four months after every legal channel to defend and save the refuge had been tried.

> The purpose of the novel was to blow the whistle on political corruption that threatened the destruction of the refuge.[26]

The tragedy is that the events narrativized in *La loca de Gandoca* had a catastrophic equivalency to what was actually happening in the Gandoca-Manzanillo Wildlife Refuge in the province of Limón in Costa Rica. In fact, the last-resort use of a quixotic novel dashed off in a few months was part of a larger strategy of political advocacy to avert the corrupt development in the actual refuge.

The use of the meta-fiction cited above was not to be construed as an abuse of artistic license; the generally "fictional" status of novels does not apply here. And if potential readers suspect an authorial "trick" or some contrived plot device, such doubts are dispelled with the claim to "la pura verdad." The borders between fiction and reality are utterly dissolved in the narrative, yet novelistic conventions are observed. It was as if all non-fictional arguments had lost their efficacy in the reality of Gandoca-Manzanillo refuge politics and the activists' last resort was to tell the story in the form of literature, a strategy that was to have no small measure of success. Kearns observes that the novel:

> instantly fueled an unprecedented national dialogue on the topic of environmental protection vs. development, . . . the impact . . . was immediately tangible in official circles as well as at the grass-roots level: first of all, while the legal case was still pending, the Costa Rican government ordered all large-scale construction projects within the refuge to be stopped. (Kearns, Introduction, *La loca*, 7–8)

In *La loca de Gandoca,* Rossi marshals several other techniques often associated with the "new novel"[27] of the "Boom" in addition to these meta-fictional flourishes to evoke persuasively charged images of nature, communicate ecological ideas, and reinforce its ideological positions. As enumerated by Elzbieta Sklodowska (whom I quoted in chapter 8), these include the use of intertextuality, meta-fiction, parody, irony, experiments with language, and ever-changing narrative registers. I already referenced the meta-fictional statement by the "dueño de monte" and the use of humor for social criticism in the form of the condemnation of development as nothing more than a bunch of tourists swimming in their own shit. But the narrative also contains an intertextual moment where a line comparing love to a burning coconut tree—"arden los cocoteros" ["the palm trees are burning"][28]—is quoted from a creole love song of Martinique, but its context is reset as ecological commentary by recontextualizing the line with a reference to trees as the trash of development: "Arderán los cocoteros como basura" ["The palm trees will burn like garbage"].[29]

Parody and irony are also used unsparingly in the scathing critical register that Rossi evokes throughout. In the fictional articulation of an amicus brief for a hearing on the enforcement of conservation in the Gandoca reserve, the narrative repeatedly employs a sarcastic phrase that addresses a constitutional tribunal as "excelso Tribunal Constitucional" ["esteemed Constitutional Tribunal"];[30] the honorific "excelso" is employed so many times that its use becomes a kind of ironic disdain for the ineptitude of such judicial entities when it comes to conservation. Finally, the novel parodies the use of words like "green" and "eco-" as a prefix or suffix for anything wanting to sound environmentally responsible (whether it is or not) by having ministers praise "un Burdine's Verde, un JCPenney Verde . . . un Mcdonald's Ecologico, una discoteca ecológica . . . al ser Verde y estar lo Verde de moda captaremos miles de turistas de otros sitios que vendrán a pasar una Navidad Verde" ["a green Burdines, a green J. C. Penney . . . an ecological McDonald's or an ecological dicotheque, . . . by claiming to be "Green" and with everything "Green" now so fashionable, we will lure thousands of tourists away from other places so they can come here to have a Green Christmas"].[31] The parody here brings in the terms "green" and "eco-" for absolute contempt for the way they can be used by government ministers as a cynical attempt at a kind of political cover for irresponsible development.[32] Once such terms are used to denote the opposite of their conventional meaning, anything that claims an "eco-" label is exposed as a sham if the policies the label disguises are actually just the opposite of environmentally friendly.

The ominous imagery detailed in this chapter focused as it is in the novels by Aridjis, Contreras Castro, and Rossi on garbage, sewage, and other refuse is particularly disturbing, especially from the perspective of readers sympathetic to environmentalism. It is all the more troubling when the objects of representation—the Río Azul landfill and Gandoca-Manzanillo Refuge in Costa Rica—are all too true-to-life. But when the focus of the narrative turns to what our ecological future might hold, the pessimism of Homero Aridjis and his vision for the end of the age of the Fifth Sun are truly chilling. In chapter 10, I consider novels by Nicaraguan Gioconda Belli and Chilean Fernando Raga that also envision what our ecological future may hold. Both novelists imagine the world divided North and South, but while Belli has things divided along the traditional axes of First and Third worlds (with all the economic, social, and ecological inequalities that such schema imply) in *Waslala* (1996), Raga's Gaia trilogy represents something of a contrarian view in recent Spanish American environmental fiction with its division of the world into two spheres of influence: totalitarian, ultra-conservationist Gaian societies versus the technologically advanced global South. While their novels are somewhat more optimistic than Aridjis's work, they provide a glimpse of an imagined future from the perspective of Spanish American authors that engage the environmental possibilities (for better or for worse) of current trends in globalization, technological innovation, the valuation of resources, and the search for solutions to ecological dilemmas.

NOTES

1. Mempo Giardinelli, *Imposible equilibrio* (Mexico City: Joaquín Mortiz, 1995), 15.
2. Ibid., 16.
3. Graciela Fernández Meijide, *Las cifras de la guerra sucia* (Buenos Aires: Asamblea Permanente por los Derechos Humanos, 1988).
4. Kay Almere Read, *Time and Sacrifice in the Aztec Cosmos* (Bloomington: Indiana University Press, 1998). I follow Read in the use of the term Mejica instead of Aztec for the sixteenth century inhabitants of Tenochtitlan whose mythology is referenced throughout this section. See her "Note on Terminology," xxiii.
5. Miguel López-Lozano, *Utopian Dreams, Apocalyptic Nightmares: Globalization in Recent Mexican and Chicano Narrative* (West Lafayette, IN: Purdue University Press, 2008), 185, 187-88.
6. Lawrence Buell, *Writing for an Endangered World* (Cambridge, MA: Harvard University Press, 2001), 38.
7. Homero Aridjis, *La leyenda de los soles* (Mexico City: Fondo de Cultura, 1993), 122, 163.
8. Homero Aridjis, *¿En quién piensas cuando haces el amor?* (Mexico City: Alfaguara, 1995), 14.
9. Aridjis, *Leyenda*, 84, 155.
10. Aridjis, *¿En quién*? 11.
11. Ibid., 39, 169.
12. Aridjis, *Leyenda*, 84.
13. Aridjis, *¿En quién*? 28.
14. See the front page of the "Money" section of the December 30, 2008 edition of *USA Today*; the article by Jeff St. John in the September 27, 2006 edition of the *Fresno Bee*; and the February 25, 2002 editions of the *San Jose Mercury News* (Julie Sevrens), the *Seattle Times* (Kyung M. Song), and the *Bergen Record* (Brian Bergstein).
15. Ana Patricia Rodríguez, *Dividing the Isthmus: Central American Transnational Histories, Literatures, & Cultures* (Austin: University of Texas Press, 2009): 208–14. The cited pages include research about how the dump has become an infamous blot on Costa Rica's reputation for eco-friendly policies.
16. Jerry Hoeg, "The Landscape of the Consumer Society: Fernando Contreras Castro's *Única mirando al mar*," in *Reading and Writing the Latin American Landscape*, ed. Beatriz Rivera-Barnes and Jerry Hoeg (New York: Palgrave Macmillan, 2009), 180.
17. Ibid., 183.
18. Fernando Contreras Castro, *Única mirando al mar* (San Jose, Costa Rica: Ediciones Farben, 1994), 12, 43, 49, 23.
19. Ibid., 83.
20. The term comes from the Spanish for scuba-diving: *bucear*. *Buzos* dive through the trash to find whatever may be salable or salvageable for food, fuel, or other domestic needs.
21. Examples can be found in "El jardín de senderos que bifurcan" (The Garden of Forking Paths), "Tlon, Uqbar, Obus Tertius," "Pierre Menard, autor del Quijote" (Pierre Menard, Author of the Quijote) among many other of Borges's stories found in the anthologies *Ficciones* (*Fictions*) and *El aleph*.
22. Anacristina Rossi, *La loca de Gandoca* (San Jose, Costa Rica: Educa, 1996), 103.
23. Ibid., 38.
24. Anacristina Rossi, "Author's note," *ISLE* 15, no. 1 (Winter 2008): 195–215. In her note on the re-printing of a selection of the authorized English-language version of the novel in *ISLE*, Rossi addresses "the links between the protagonist of the novel, Daniela, and my own

personal history" (195). The elements of advocacy and legal efforts to protect the refuge represented in the novel were all too real.

25. Rossi, *La loca*, 134.

26. Sofia Kearns, Introduction to *La loca de Gandoca/The Madwoman of Gandoca*, by Anacristina Rossi, trans. Terry J. Martin (Lewiston, NY: Edwin Mellen, 2006), 1.

27. See chap. 8, note 1.

28. Rossi, *La loca*, 86.

29. Ibid.

30. Ibid., 93–97.

31. Ibid., 120–21.

32. Jennifer Basye Sander and Peter Sander, with Anne Basye, *Green Christmas: How to Have a Joyous, Eco-Friendly Holiday Season* (Avon, MA: Adams Media, 2008). The parodic gesture seems more acute given that several websites advertise this "Green Christmas Getaway" to Costa Rica, some of which are documented in the Sander book.

Chapter Ten

Paradise Trashed: Utopian and Dystopian Ecological Scenarios in Gioconda Belli's *Waslala* and Fernando Raga's Gaia Trilogy

Imagine living exactly like Chilean Luis Sepúlveda's Antonio José Bolívar Proaño from *Un viejo que leía novelas de amor* (*The Old Man who Read Love Stories*) (1989). The appeal of simplified, ecologically responsible living is undeniable, but the jungle life would be asking too much of a population grown up on modernization. Instead, the positive contribution of Sepúlveda's novel lies in its celebration of limited, moderate consumption, ecological wisdom, and the value of nonhuman life. Similarly, if a movement to undertake commando operations like the one described in Argentine Mempo Giardinelli's *Imposible equilibrio* (*An Impossible Balance*) (1995) could gain mass appeal, societies would sink into violent anarchy, a scenario unlikely to benefit the environment. Yet in Giardinelli's novel, the criticism of a self-serving market economy and its allies in government is not invalid and can encourage readers to attempt to alter such policies. Also, the reading audience for Arturo Arias's *Los caminos de Paxil* (The ways of Paxil) (1990) would most likely not be associated with an indigenous community and would not consider the violent elimination of corporate executives a viable solution to the ecologically damaging industrial practices of Big Business. However, most can readily identify with the feeling that one's community could be targeted by corporate interests and might consider a variety of options for resistance. Something similar happens with the representation of future utopias and dystopias in *Waslala* (1996) by Nicaraguan Gioconda Belli and in the Gaia trilogy by Chilean Fernando Raga: the failure of utopian schemes throughout history and the pessimistic vision of coming-catastrophe in the novels does not generate confidence in finding workable solutions for actual ecological problems. Yet the very possibility of idealized future societies or the fear inspired by apocalyptic visions can function to inspire belief: belief in the possibility of alternatives to and in the necessity of turning from the economic, social, cultural, and ecological priorities of development and modernization.

WRETCHED REFUSE: UTOPIAN IDEALS IN GIOCONDA BELLI'S *WASALA*

Gioconda Belli's *Waslala* is about a destiny of deterioration suffered by the cultural, social, and ecological conditions of the Third World.[1] In the future of *Waslala*, robots perform domestic chores, humans have orbited Mars, solar energy powers domestic appliances, even household furniture is equipped with computer consoles to control temperature, angle of recline, and orientation. But things are not all rosy: the text envisions a world dominated by technology and ecological exploitation extrapolated by a factor of one hundred years. In the novel, a tale of drug trafficking, suicide, and the quest for a seemingly unreachable utopia in the fictional Central American country of Faguas is the stage upon which an ecological critique is articulated. The quest to locate Waslala is the quest for ecotopia, but the idea that utopian principles are incompatible with the economic demands of a technological society is a central argument of the text: an ecotopia may be possible as an alternative to the logic of development; however, it must be pursued on the scale of individual comportment and small community.

Waslala contains elements of violent struggle like in Arias's *Los caminos de Paxil*, but while things end badly for the Mayan community in that novel, Belli's text concludes with a kind of metafictional argument for the existence of utopia within literature. The story is set in a time of ecological decline for the industrialized world as underdeveloped countries like Faguas are fetishized on the basis of their ability to produce clean air and forced to preserve their forests so that oxygen is available for fully deforested Northern nations. Meanwhile, two brothers operate a family-owned drug ring known as the Espada cartel; the family uses a garbage processing industry that imports trash from the First World to Faguas as a cover for their network of narcotics distribution. (As was seen in chapter 9, waste disposal looms large in recent Spanish American environmental fictions.) But when a local crew employed at the landfill find a discarded receptacle of cesium-137 and are exposed to a lethal dose of the radioactive material, they are given less than a week to live. As their last act, they strap on some dynamite, invade the Espada compound, and blow themselves up. The end of the Espada brothers' control over the country serves as a frame for a more personal secondary story in the novel: a couple, Melisandra and Raphael who seek the mythical land of Waslala, are finally able to travel freely and they eventually find it.

The novel deals with many elements central to the current environmentalist debate including the role of violence as a legitimate form of resistance, conceptions of space and place, and local versus global legitimacies of environmental activism.[2] Lest it be believed that the novel advocates morally justified violence as the only avenue to good communal living, it should be noted that the story does not end with the suicide-bombing liberation of Faguas, but provides a non-violent model for the future by offering an account of Melisandra's eventual encounter with Waslala. As the possibility of utopia requires a suspension of disbelief, the science fiction in the novel invites acceptance of the fictional parameters required to imagine a place like Waslala where "el amor, la cooperación y el bien común serían los pilares para erigir una felicidad . . . jamás conocid[a]" ["love, cooperation, and the common good would be the pillars of a happiness . . . never before known"].[3] These parameters include the almost blindingly naïve idealism that a community of humans could coexist on such principles, but by the same token, the very idea of Waslala is evoked in magical and spiritual terms throughout the novel: in one scene, Raphael asks Florcita, a waitress in a bar in Las Luces, the best way to Waslala, and she responds: "tenés que tener fé" ["you have to have faith"].[4] The novel has fantastical elements—parrots who read fortunes, the legend of the ghosts of Wililí, a stretch of the river that reveals an individual's thoughts in images on the water—that serve to prepare readers—through the use of a kind of meta-fictional flourish—to approach an idea like Waslala not for its plausibility as a real alternative but as fictional inspiration for spiritually based ecological ideals.

The element of the spiritual introduced in the quest for the utopian Waslala would be ridiculed in a First World locale where the values of logical rationality reign supreme, but in a country like Faguas, it fits right in:

> Mientras más primitivo el entorno, más fácil creer en mitos, lugares mágicos, felices, intocados. Waslala podía existir en Faguas. El pensamiento científico, la técnica, la educación, habían marcado el fin de lo mitológico. El progreso amenazaba la imaginación. (Belli, *Wasalla*, 165)
>
> [The more primitive the environment, the easier it is to believe in myths and magic places. Waslala could exist in Faguas. Scientific thought, technology, and education have marked an end to mythology. Progress threatened the imagination.]

The ideal of Waslala offers the mythological alternative of magic and imagination as alternatives to the rationality of modernization and development. The scenario represented by Waslala requires a measure of spiritual belief and thus marries ecological ideals to notions of faith. For a professional ecologist, this may represent something utterly distinct from the scientific objectivity derived from statistics or repeatable experiments; but spiritual approaches to the environment are not all that unusual in ecological thinking. For example, William Cronon observes that the spiritual tenor of Henry David Thoreau's nature writing "took the physical mountain on which he stood and transmuted it into an icon of the sublime: a symbol of God's presence on earth."[5] Similarly, *Waslala* describes a kind of return to innocence, an Eden free from ecological sin, corruption, and injustice:

> Hay que crear el núcleo original, descontaminarlo a través de varias generaciones hasta que sólo lo conformen hombres y mujeres que nunca hayan conocido la ambición, el poder, la avaricia, la violencia, el mal. Se trata de construir la primera célula, la partícula, el primer organismo vivo. (Belli, *Waslala*, 60)
>
> [It is necessary to create the original nucleus, decontaminate it through various generations until it is only made up of men and women who have never known ambition, power, avarice, violence, or evil. It's nothing less than the construction of the first cell, the atom, the first living organism.]

In *Waslala* the path to ecological purity is made possible by a unique, almost magical element described as a "ranura en el tiempo-espacio" ["a rift in space/time"][6] : clearly fictional but credible in the novel for the extensive use of other similarly fantastic devices. The location was revealed to the community in a dream come to one of the poets about an area of great natural beauty in the North of the country, "[un] lugar privilegiado por la naturaleza" ["a place favored by nature"].[7]

Now to pause and consider the utility of the idea of an ideal ecological community for the purposes of the environmentalist cause, there seems to be little practical use for visions of an Edenic place, hidden by a rift in space-time. While the requirements for uncovering a kind of terrestrial paradise can be satisfied in a fictional text, in the real world, a population of seven billion and the infrequent or nonexistent status of the fantastic makes such a vision impractical. Except for the small community that could fit within its confines and find and pass through the time-space rift, the very nature of a place like Waslala demands that the rest of the world be excluded. The resource requirements, moral turpitude, and sheer quantity of humans automatically disqualify Waslala as a catholic model for humanity. And even on a small scale, its inadequacies are apparent. In the novel, the reality of these concerns combined with the inexplicable inability of couples to reproduce within the rift in space-time leads to a decrease in numbers for the Waslala cell and to its threatened disappearance. In an attempt to compensate for the community's decline, the poet founders decide to approach their experiment from a different angle, an angle that does not necessitate the opening of Waslala to the masses, but that also does not compromise the ideal of societal reformation proposed in the act of its founding. The plan was to allow rumors of Waslala filter throughout Faguas in the hope that the very idea of such a place would improve the world outside:

> Waslala ya no era solamente el vacilante experimento que habíamos construído. Era una leyenda, un punto de referencia, una esperanza. Aun antes de que se comprobara su eficacia, se había convertido en un paradigma. Cumplía la función de un sueño capaz de movilizar los deseos y las aspiraciones de quienes ansiaban un destino colectivo más acorde con las mejores potencialidades humanas. Comprendimos entonces que la fantasía había adquirido tanto valor como la realidad. (Belli, *Wasala*, 316–17)
>
> [Waslala was no longer the tentative experiment that we had constructed. It was a legend, a point of reference, a beacon of hope. Even before its efficacy could be proven, it had become a paradigm. It carried out the function of a dream, capable of mobilizing the desires and aspirations of those who desired a collective destiny more in line with the highest human potential. We understood that the fantasy had acquired as much validity as reality.]

The founders believed that the abstract ideal of Waslala—the spiritual belief in the importance of the common good and in the possibility of an ideal future society—could function to inspire behavior outside of the place itself. In her analysis of the novel, Marisa Pereyra notes that Waslala functions along the lines of "the idea of utopia as impulse, as a motivating force and not as unique plan or geographic place."[8] This invites readers to consider the contents of the novel in the same way that the poet/founders intend that residents of Faguas should think of Waslala: "a dream, capable of mobilizing desires and aspirations." The failure of utopian projects across the ages indicates that readers could "never physically go" to a place like Waslala, but Belli's act of environmental imagination in writing a place like Waslala at least offers a vision of an ideal, ecologically responsible community. It offers an account of how utopian literature can function in the ecological debate and demonstrates the close relationship between ethical advocacy and the literary qualities of environmentalist fiction.

Within the novel, philosophical and literary utopianism exist side by side: philosophical observations shed some light on the metafictional relationship between writing and ethical advocacy:

> Although the utopian writer may do nothing to improve society, he may still deem it worthwhile to preserve the concept of the ideal. . . . To insist on the distinction between the acceptable and the ideal can have a chastening influence on those who govern as well as on those who happily go along.[9]

The desire to "preserve the concept of the ideal" in *Waslala* operates as metafiction because it both represents the explicit purpose of the founders in allowing the concept of Waslala to circulate throughout Faguas and functions as the ethical idiom specifically aimed at getting readers to form a certain belief:

> The long series of utopian texts enlarge the world by suggesting new character types and new social milieus in which each type could emerge. They also enlarge the world by their claim that the societies of the world ignore, repress, distort, or destroy human potentialities that have not yet been fulfilled. [10]

When the founders decide to allow the concept of utopia to take the place of Waslala in the novel, it functions as metafiction because that is exactly what happens to readers of environmentalist fictions. The ideals of Waslala are a potentiality for the fictional country of Faguas in the same way that the ideals of *Waslala* are a potentiality for actual countries outside the realm of fictional worlds. The founders' proposal—"un modelo de sociedad totalmente nuevo y revolucionario, basado en una ética que repudiaba el poder, la dominación y concedía a cada individuo la responsabilidad de la comunidad" ["a totally new and revolutionary model for society based upon an ethics that repudiated power and domination and conceded to every individual the responsibility of the community"][11]—is based upon the same ethical position espoused by the novel. It functions as an extratextual criticism of the actual state of globalized political exploitation of the Third World by the First. I do not mean to suggest that these novels always offer real alternatives that can or should be copied; rather, I want to suggest that they offer scenarios that rely upon the articulation of alternative values and it is these values that can be adopted as ecological principles in our own world.

CRITICAL LITERARY ENVIRONMENTALISM: THE GAIA TRILOGY BY FERNANDO RAGA

One of the elements that I explore in my analysis of *Waslala* in the article from *Ecozon@* referenced in note 2 from this chapter was the idea of forced conservationism in the novel as a form of anti-developmentalist neo-imperialism:

> Conservationism [in Waslala] is a form of neo-imperial intervention in that the government and residents of Faguas must comply with First World mandates as the only way to secure electricity and other goods. Concern for the preservation of nature had never been a factor in the economic development of the First World, but now a limited oxygen supply demands extensive conservation and economic disparity dictates the retardation of development for the Third World. [12]

The critique in *Waslala* has to do with the fundamental injustice of forced conservationism upon undeveloped nations so that citizens from more wealthy and powerful regions could go on living as they ever had without the fear that depleted domestic resources would result in a global catastrophe. Thus, like Costa Rican Anacristina Rossi's *La loca de Gandoca* (*Mad about Gandoca*) (1992), *Waslala* represents a process of maturation in the development of the Spanish American environmentalist novel, a maturation that includes critical perspectives regarding environmentalism and explorations of their possible implications in the fictional space of a novel. Mexican Homero Aridjis's novels seem to suggest that not taking ecological problems in the present seriously enough will result in something like the nightmare of Ciudad Moctezuma, circa 2027. Belli's novel implies that the imposition of ecological mandates from without is not only a fundamental injustice upon those who are the targets of its imposition, but it also represents an injustice in the form of unequal power relations and underlying hypocrisy: those who impose foreign political ecologies upon local populations to "save the planet" often do nothing to reform their own societies' behavior. Continuing the more critical tradition with regard to ecological principles in Spanish American literature, Chilean Fernando Raga's Gaia trilogy—*Los hijos de Gaia* (The children of Gaia) (2005), *Revolución en Gaia* (Revolution in Gaia) (2010), and a third entry yet to be released—develops a fully skeptical perspective with regard to many of the current prevailing beliefs of environmentalism. The novels seem to take their inspiration from Danish political scientist Bjorn Lomborg's *The Skeptical Environmentalist* (1998) as they narrativize many of the arguments from the Dane's book including the expression of significant doubts about the urgency of issues like deforestation, overpopulation, the disappearance of endangered species, the scarcity of natural resources, and the severity and danger of global warming.

Still, various characteristics of the Spanish American literary tradition also seem to appear as a legacy in Raga's novels: an articulation of the double-speak of totalitarian regimes from Guatemalan Miguel Ángel Asturias's *El señor Presidente* (*The President*), (1946) and Augusto Roa Bastos's *Yo, el supremo* (*I, The Supreme*) (1974), the cosmopolitanism of Jorge Luis Borges's short stories, even several elements of the *indigenista* genre. But Raga's work unfolds as a conventional political thriller and the ecological discourse does not result in new forms of literary expression as I have argued is the case for *Waslala* and the novels discussed in chapters 8 and 9. Nevertheless, Raga's skeptical posture toward many seemingly unassailable tenets of environmentalism is an element of the literature not yet seen in the ecological fictions of the region; I include a consideration of his work here for the uniqueness of the novels' critical perspectives in the context of Spanish American environmental fiction. *Revolución en Gaia*, the second entry in the trilogy, offers the backstory to what happens in *Los hijos de Gaia*, two centuries in the future. The novels detail the rise and practice of an extreme environmentalist philosophy called "gaianism" that takes hold in Europe, in the former United States, and in several other countries aligned with them after a prolonged, debilitating economic recession in the early decades of the twenty-first century. Eventually, in *Los hijos de Gaia*, the geo-political alliances are set. One is a society in what is called the Green Belt, controlled by the Gaian party, a totalitarian regime opposed to progress, to the unnatural extension of life, and to other activities or behaviors that could in any way threaten the natural world. In the geographical South, an alliance of Chile, Argentina, Uruguay, South Africa, and New Zealand as well as several non-aligned countries, have embraced technological exploration and research and have thus surpassed the formerly highly developed Northern countries in terms of life expectancy, per capita income, medical advancement, and several other categories that would normally be thought of as progress. The novels unfold as tales of espionage, political intrigue, and the way in which these elements affect a cast of characters from the three regions (the "Green Belt," the allied Southern countries, and the unaligned nations) as each seek dominance in a new world order.

It would be faithless to give away the ending of a spy novel in these pages (and nearly impossible with *Los hijos de Gaia* and *Revolución en Gaia* given the numerous narrative threads that Raga weaves into each story). Instead, I prefer to examine the novels' ideological positions (which are somewhat less numerous) and analyze the way in which they advocate a restrained environmentalism that charts a middle ground between what in the novels is termed the "Age of Depradation" and the extreme asceticism of the Gaian party. In the analysis of Mexican Rafael Bernal's mosquito conspiracy novel *Su nombre era muerte* (His name was death) (1947) from chapter 4, I cited historian Donald Worster's criticism of a heinous ecological morality that yearned for pre-modern societies whose growth was controlled "by practicing a brutal infanticide or by suffering warfare, hunger, and disease that restrained . . . populations."[13] But this morality is what Raga has as the basis of the new Gaian societies of the "Green Belt": medical interventions that "unnaturally" prolong life are prohibited; couples are limited to two children; the use of wood has been outlawed as contributing to global warming; killing predatory species like wolves, even in self-defense, is punishable by death; religion has been banned as overly anthropocentric. Many other similar regulations are enumerated as the fictional parameters of Gaian society in the novels and the caricature of possible consequences from a totalitarian imposition of the tenets of current environmentalism represents the novels' principal critique.

The characterization of the Gaian philosophy begins with an admirable sentiment of self-control, critical of those who:

> [sigue] comportando como si solo existiera cada uno de ellos, buscando exclusivamente su placer y bienestar, y destruyendo el planeta, atropellando los derechos no sólo de los demás, sino de las generaciones futuras, y de tantos otros seres que no pueden defenderse por sí solos, y que son criaturas de Gaia.[14]

> [continue to behave as if they were the only ones on Earth, looking out for number one and destroying the planet, crushing not only the rights of all others but also those of future generations; that is, so many other individuals that cannot defend themselves, the children of Gaia.]

The suggestion is that Gaian philosophy would not seek only its own well-being, but protect the planet and not trample the rights of others. In part, the novel represents it as such, but also examines what the philosophy might look like taken to a totalitarian extreme. The fear before the Gaia party comes to power in *Revolución en Gaia* is that "el ecologismo era en realidad una corriente política potencialmente tan fuerte como lo fue el comunismo, y que bastaba un liderazgo carismático y un cierto grado de organización para que controlaran el mundo" ["environmentalism had the potential to be a political model as strong as communism had been and that all it lacked was a charismatic leader and a certain level of organization before the devotees of that movement could take over the world"].[15] The Gaian party does indeed gain absolute control in *Los hijos de Gaia* and begins to construct a kind of environmentalist totalitarian society with "green" regulations for all aspects of human life in the areas they control. Some of the more chilling elements of the agenda are summarized in a lengthy moment where a Gaian philosopher defends that way of thinking: because of the anthropocentric prejudice of religious belief and the resultant emphasis upon humanity as the center of the universe, "se pensó en desalentar las prácticas de estas religiones antiguas. Como estaban profundamente arraigadas en lo cultural, fue necesario actuar con mayor energía prohibiendo tales cultos" ["it was thought a good idea to discourage the practice of these ancient religions. Inasmuch as they were so culturally entrenched, it was necessary to decisively prohibit these kinds of religious ceremonies"].[16] Because of the incessant tendency among humans to consume, dispose of, and acquire ever more technologically advanced objects and products, the Gaian party undertook two actions: the limitation of citizens to "tecnologías que son consideradas necesarias y razonables para la vida cotidiana" ["only those technologies considered essential for daily life"] and the rejection of "la idea del industrialismo, con sus grandes producciones en masa y su forma de vida impersonal" ["the idea of industrialization with its mass production and impersonal form of life"].[17] All of these prohibitions have the ring of ecological responsibility, but their imposition on the citizenry was only possible through the creation of an authoritarian regime construed as "protected" democracy where any non-Gaian idea was strictly verboten.

The extreme prejudice with which criticism of the Gaian regime is put down functions as an allegory for the perception that, presently, environmentalism as a discourse seems beyond the pale even though there may be legitimate criticisms that can be raised to challenge the received wisdom of ecological ideas. The critical posture in Raga's novels comes straight from Lomborg's skepticism and his infamous challenge to the validity of environmentalist ideas in many contexts. The connection is made explicit in *Los hijos de Gaia* where the Dane is cited by name; a cache of prohibited literature is found to include "un interesante libro de principios del siglo XXI de un tal Lomborg" ["an interesting book from the early 21st century by someone named Lomborg"].[18] As Raga's novels rehearse several of the arguments from that book including the idea that environmental problems are not as serious as portrayed by environmentalists and that the creation of wealth in impoverished countries tends to protect the environment more than the external imposition of conservationist political ecologies, the narrative ideology begins to emerge. Clearly, the call for radical changes in lifestyle and governance to protect the environment is challenged in *Los hijos de Gaia* and *Revolución en Gaia* through the invocation of Lomborg and the caricature of radical environmentalist ideology as a slippery slope to fascism. The extreme representation of a frightening future if something like a Gaian party were to get control of major geo-political regions and population centers suggests that the novels condemn the environmentalist project wholescale, but Raga does allow for several of the valid aims of environmentalism, including some radical voices: "[los Gaianos] han tenido mucho que ver con quc hoy hayamos resuelto gran parte de los problemas ambientales serios. Han obligado al resto de nosotros a pensar en serio en el medio ambiente" ["[the Gaians] have clearly contributed to the rectification of several serious environmental problems. They have forced the rest of us to think seriously about the environment"].[19] Raga's first book—*Los hijos de Gaia*—advocates something of a middle way that will maintain a certain degree of protection for the natural world without the excesses of a radical Gaian totalitarianism nor the isolation of a "Green Belt." In the fictional future of the novel, *las Comunidades Unidas* (formerly the United States of America) employ a system of debilitating taxation to discourage anything more than small-scale agrarian production, a system that inevitably leads to fiscal insolvency, even for farmers with very small parcels. One of these farmers affirms his distaste for the conspicuous consumption of the societies of the Southern Alliance, but expresses hope for something other than the absolute rule of the Gaian party that had so extensively curtailed freedom and dissent:

> Pensé que en las Comunidades esto se lleva demasiado lejos. Sentí que el costo que había que pagar en términos de las libertades de las personas era desproporcionado. Y comencé a pensar que tenía que haber algún otro sistema que permitiera proteger adecuadamente el medio ambiente y a la vez permitir una mejor vida a la gente. (Raga, *Los hijos*, 189)
>
> [It occurred to me that the Communities were taking this too far. I felt that extreme sacrifice of personal liberty was too high a price. And I began to think about the idea that there could be some other system that would both adequately protect the environment and allow for a better life for all people.]

This other system, what I take to be the ecological "message" of the trilogy especially in *Los hijos de Gaia*, is the optimism in future solutions and technological advances that will be able to relieve some of the most acute causes of an ecological crisis that the books suggest will occur sometime in the first half of the twenty-first century.

Among these advances are the use of cloning to preserve endangered species, the laboratory production of meat products which will render the livestock industry obsolete, the advent of the electromagnetic motor for automobiles, the genetic manipulation of trees to produce timber in designated sectors of forests at a rate that outpaces demand, the advent of low temperature nuclear fusion for energy, and the use of light capture technology for illumination of cities, towns, homes, and buildings. The faith in the future development of these "silver bullets" is considered somewhat dubious among most leading environmentalist thinkers of the present. For example, Wendell Berry and Peter Sauer condemn this type of thinking and express the importance of locating the causes of environmental degradation in the way that individuals and societies think about nature.[20] The role of advanced technology as an element of nature protectionism may have its place, but faith in the promise of such advancements is naïve according to thinkers like Berry and Sauer. While I am personally more sympathetic to the holistic environmentalism of the latter two, it is not my aim here to resolve these fundamental tensions within competing elements of the environmentalist project. Rather, Raga's trilogy comes near the end of the titles considered in these pages because I believe it demonstrates how literary environmentalism has come of age in the novels of the region. The advent of competing, critical narrative currents and the recent, nearly simultaneous appearance of several unique textual articulations of various ecological tendencies demonstrate an expansion in the contours of genre and a consequent coalescence around the idea of Spanish American ecological literature as a valid category of texts.

With this as the point of departure from which I began the analysis in the Introduction, I have come full circle; the array of novels includes authors from the Southern Cone to Central America and Mexico and explores ecological flashpoints from the remotest seas of the South Atlantic to the deepest interior of Amazonian rainforests to the Valley of Mexico and across the globe. The environmentalist tendencies in recent Spanish American fiction represent a necessary expansion of the notion of the characteristics and concerns of post-"Boom" literature. Raga's trilogy simply represents the very latest in the expansion, even if (or especially if) the narrative is less experimental than the texts examined in part 3 thus far. From the variations in narrative order in *Un viejo que leía novelas de amor* to the mythically poetic apocalyptic novels of Homero Aridjis and from the humor of *Imposible equilibrio* to the conventional forms and straight-forward exposition of Raga's novels, the ecology of Spanish American fiction is now expressed in the vocabulary of full-blown environmentalist language. The literature is not limited largely to political ecology like the texts of the nineteenth century and the novels of *indigenismo*; there are now no longer isolated expressions of ecological ideas like in the regional literature, but a fully developed sense of nature and nature protectionism in a growing collection of fictional works that express and address environmentalist advocacy. And in many cases, the ecological elements may represent an innovation in the literary forms themselves; while not all recent environmentalist novels in the tradition achieve as much, I have argued that, at the level of form, convention, language, and formal narrative, most of the novels discussed in chapters 8, 9, and 10 exemplify the way in which the ideological expression of environmentalism can impact the formal aspects of literary texts.

BACK TO THE FUTURE: ANTECEDENTS TO RAGA'S TRILOGY IN J. A. OSORIO'S *BARRANQUILLA 2132*

I conclude with an anachronism: the all-but-forgotten novel *Barranquilla 2132* by Colombian J. A. Osorio Lizarazo from 1932 makes for an intriguing precedent to Raga's text. The story is that of Juan Francisco Rogers, a man who has discovered a chemical process to leave human bodies in a state of indefinite inanimate suspension. He injects himself with the potion and has himself buried in a wall of concrete to be discovered two centuries later by an explosion that uncovers his body.[21] When he revives, Rogers finds himself in a world that has advanced technologically, but also culturally, and in ways that make him somewhat uneasy. Like in Raga's Gaia trilogy, there is a strong condemnation of the use of fossil fuels: "—Ah! Cree usted, acaso, que necesitamos un combustible tan peligroso?" ["Oh, you don't believe that we need to use such a dangerous fuel, do you?"]; seemingly inexhaustible atomic power in the form of "un procedimiento químico que descompone el aire en sus moléculas elementales" ["a chemical reaction capable of splitting air molecules into their atomic parts"]; and the problem of overpopulation cast in overtly ecological terms: "¿Pero no comprende usted lo que implica la crianza de un hijo y la cantidad de energías que requiere, energías que pueden invertirse en más provechosas finalidades . . .?" ["Don't you comprehend what is implied by taking on the energy costs required for raising children? This is energy that would be better invested in other pursuits"].[22] The last element, a seemingly legitimate question about responsible stewardship of resources, is similar to the fascism of Raga's trilogy for the draconian enforcement of laws meant to control the population:

> La natalidad está limitada. . . . Cada ciudad debe sostener una clínica para prevenir el exceso de natalidad. Además, sólo pueden tener hijos los individuos perfectamente conformados . . . si ni ella ni él reúnen las condiciones exigidas por esta disposición, el germen será anulado oficialmente. (Osorio Lizarazo, *Barranquilla*, 96)
>
> [The birth-rate has been limited. . . . Every city must maintain a clinic to control excessive birth among its residents. Plus, only ideally suited individuals are permitted to have children . . . if some defect is discovered . . . the pregnancy will be terminated by official decree.]

The euphemism for the imposition of abortion is just as sinister in Osorio Lizarazo's novel as the empty ecological platitudes that excuse authoritarian extremes in Raga's trilogy; in both cases, the resolution of societal distress through totalitarian solutions represents a decidedly unwelcome future. In Raga's trilogy, the critical posture toward a radical reduction in energy consumption rests on an enduring belief in the alternative of "silver bullet" solutions; at the close of *Los hijos de Gaia*, this comes in the form of discoveries about low-temperature nuclear fusion as key to stopping the advance of Gaian political thought.

Osorio Lizarazo's novel is perhaps a little more circumspect when it comes to the relationship between ecological social problems and the hope for progress in the future as an antidote:

> No era la desaparición de las bombillas y el procedimiento para transmitir la energía eléctrica por las ondas radiantes en lugar de los antiguos alambres metálicos. . . . Ni era tampoco el aprovechamiento del movimiento eterno de las olas marinas para aplicaciones industriales. . . . Nada de esto era lo que esperaba encontrar en su regreso a la vida: era la depuración de los sentimientos nobles del hombre, el perfeccionamiento de las sociedades, los grandes progresos espirituales, la evolución ascendente del arte, la purificación de los más elevados conceptos metafísicos, la existencia de una comprensión recíproca entre los hombres y de los más exaltados sentimientos de amor y de caridad. (Osorio Lizarazo, *Barranquilla*, 109–10)
>
> [It wasn't the obsolescence of light bulbs nor the technology for transmitting energy through radio waves instead of by metal wires. . . . Nor was it the use of

> energy from ocean tides for industrial applications. . . . He did not expect to find any of this when he returned to life: he expected to find the purification of noble human emotions, the perfection of societies, great spiritual progress, an ever-more-transcendent evolution of art, the refinement of the highest metaphysical concepts, the existence of a sense of mutual trust and understanding among human beings, and an exalted sense of the importance of love and charity.]

The critical posture in *Revolución en Gaia* and *Los hijos de Gaia* has a similar ideology as its basis: the problem with the fictional Gaianism in Raga's trilogy is the same as the problems found in the future of *Barranquilla 2132*: both imagine societies devoid of spiritual progress and cultures in which a basic sense of love and charity has disappeared; both predict the evolution of cleaner energies but at the price of an insidious debasement of the human sense of nobility, fraternity, and compassion; both imagine a future devoid of and even hostile to elements from the ecofeminist critique such as empathy, care, and relational arrangements.

A novel from the 1930s is anachronistic in a chapter on the most recent Spanish American environmental literature, but I conclude with it here because I believe it illustrates one of the central arguments in this and each of the preceding chapters. I propose that the environmentalist novels of the 1990s and 2000s represent a salient and distinct current in what has been somewhat feebly called the post-"Boom" of Spanish American literature; the efforts by Luis Sepúlveda, Arturo Arias, Gioconda Belli, Mempo Giardinelli, Homero Aridjis, Fernando Contreras Castro, Anacristina Rossi, and Fernando Raga attest to that. But the ideology of the suspicious and pessimistic vision of the future in *Barranquilla 2132* seems to prefigure many of the most of the critical positions in Raga's novels. Each finds it difficult to place a great deal of trust in the course of future events with regard to improvements in human life; the narratives from both authors are rather pessimistic in that regard. But what is remarkable about Osorio Lizarazo's novel is what I find remarkable about ecology and Spanish American literature as a whole. Raga's trilogy is a work in progress, yet *Barranquilla 2132* anticipated many of its most salient criticisms by nearly one hundred years. Likewise, the explicitly environmentalist novels by authors from Central and South America are phenomena of the last twenty years, yet the insights from these most recent texts have a far longer history in the literature, a history that includes titles from nineteenth century foundational fictions, the fanciful imagery of the *modernistas*, the perceptions of seas, mountains, plains, and jungles in the regional novelists, and the environmental ideologies of *indigenismo*.

NOTES

1. The use of terms First and Third World, North and South, or developed and undeveloped intends to make a distinction between wealthier and poorer nations or between economically and industrially developed countries that typically lie north of the 30th parallel in the northern hemisphere (Australia is a notable exception) and underdeveloped regions south of that latitude. The distinction between North and South, First and Third Worlds, and developed and underdeveloped nations speaks to unequal economic relations between the two sides. These terms have gone out of style somewhat, but I use them here because they appear as such in Belli's novel.

2. See my article, "Garbage Out: Space, Place, and Neo-imperial Anti-development in Gioconda Belli's *Wasala*," *Ecozon@* 1, no. 2 (2010): 38–50, for an extended discussion of these issues.

3. Gioconda Belli, *Waslala* (Barcelona: Emecé, 1998), 278.

4. Ibid., 114.

5. William Cronon, "The Trouble with Wilderness," in *Uncommon Ground: Rethinking the Human Place in Nature* (New York: Norton, 1995), 75.

6. Belli, *Waslala*, 65.

7. Ibid., 61.

8. Marisa Pereyra, "Paradise Lost: A Reading of *Waslala* from the Perspectives of Feminist Utopianism and Ecofeminism," in *The Natural World in Latin American Literatures: Ecocritical Essays on Twentieth Century Writings*, ed. Adrian Taylor Kaye (Jefferson, NC: McFarland, 2010), 148.

9. *The Encyclopedia of Philosophy*, 1st ed., s.v. "utopias and utopianism."

10. Ibid.

11. Belli, *Waslala*, 59.

12. DeVries, "Garbage Out," 45.

13. Donald Worster, "Nature and the Disorder of History," in *Reinventing Nature: Responses to Postmodern Deconstruction*, ed. Michael E. Soulé and Gary Lease (Washington, DC: Island, 1995), 67.

14. Fernando Raga, *Los hijos de Gaia* (Buenos Aires: Distal, 2005), 31.

15. Fernando Raga, *Revolución en Gaia* (Santiago de Chile: Editorial Forja, 2010), 23.

16. Raga, *Los hijos*, 120.

17. Ibid., 121.

18. Ibid., 148.

19. Ibid., 294.

20. Wendell Berry, "In Distrust of Movements," *Orion* (Summer 1999): 15–18; Peter Sauer, "Reinhabiting Environmentalism: Picking Up where Carson and Leopold Left Off," *Orion* (Summer 1999): 30-41.

21. The name was likely inspired by the appearance of the Buck Rodgers comic strip character in the 1920s: Osorio Lizarazo was a journalist and the character's syndication by the National Newspaper Service in 1929 suggests that the Colombian author was aware of the connotations of the use of the surname "Rogers" for a person who wakes up centuries in the future.

22. J. A. Osorio Lizarazo, *Barranquilla 2132* (Barranquilla: Tip. Delgado, 1932), 31, 111–12.

Conclusions

I conclude with a word on the texts that have been selected to construct this history: first, some thoughts on the way in which the ecological aspects of the texts that I trace to before Spanish America achieved its independence from Spain culminated in the publication of the series of environmentalist novels that I consider in chapters 8, 9, and 10. It is characterisitic of literary scholarship to put quotes around the term "Boom" as descriptive of Spanish American literature from the 1960s and early 1970s. So it is even more risky to write of a post-"Boom" or to attempt a description of the parameters of such a literary period. Nevertheless, the novels by Luis Sepúlveda, Arturo Arias, Mempo Giardinelli, Homero Aridjis, Gioconda Belli, and Fernando Raga make for a veritable current within recent fiction. This has been well-documented in the monographs by Laura Barbas-Rhoden, Adrian Kane, Carmen Flys Junquera, José Manuel Marrero Henriquez, Julia Barella Vigal, Beatriz Rivera-Barnes and Jerry Hoeg to which I make reference in the introduction. The literary works are characteristic of the ecological tendency in recent Spanish American literature; the scholarship demonstrates the way in which these titles contribute to environmentalist fiction as a current of post-"Boom" Spanish American literature. What I have attempted to document is the way in which the environmentalist discourse in these later novels represents a culmination from ecological moments in earlier texts. Political ecology, the contaminating tendencies of industrialized extractive practices, concerns about deforestation, desertification, and species extinction: the environmentalist novels of the post-"Boom" and those texts considered in the first 7 chapters share a similar ethical discourse with regard to these issues.

Next, I should reflect on my methodology of canon which may explain what does not appear here. The idea was to uncover poetry, essays, novels, and stories where ecological ideas were present; but for the texts that had disappeared from view, the ecocritical approach intends a rescue. Often, it took the form of the selection of titles because of current environmental concerns whose urgency I projected onto past literary efforts; but in several instances, this ecocritical posture failed. Take *La reina de Rapa Nui* (The queen of Easter Island) (1914) by Chilean Pedro Prado. The circumstances of the island of Rapa Nui (Easter Island in English) is one of the classic examples of human-caused ecological devastation and an ideal case study for archaeologists and anthropologists for its isolation—a small island in the middle of the South Pacific, more than 1,000 miles from the closest inhabited island. In fact, Jared Diamond's *Collapse: How Societies Choose to Fail or Succeed* makes the case for the island as an ominous warning to all societies of the present:

> The parallels between Easter Island and the whole modern world are chillingly obvious. Thanks to globalization, international trade, jet planes, and the Internet, all countries on earth today share resources and affect each other, just as did Easter's dozen clans. . . . People see the collapse of Easter Island society as a metaphor, a worst-case scenario, for what may lie ahead of us in our own future.[1]

So when I saw the title of Prado's book, I began to intently scan its pages for something that might predict in 1914 what Diamond concludes in 2005. But, other than a mention of the islanders' intense desire for water—"—¡Evai o'quimai! (venid agua!) gritaban" ["Come, water! they shouted"][2]—there was nothing about ecological apocalypse. Something similar happened in the course of my research with several other titles.

The Chilean mining novels *Subterra* (Underground) (1904) by Baldomero Lillo, *Llampo de sangre* (Bloody ore) (1950) by Óscar Castro, and *Hijo del salitre* (Child of nitrate) (1952) by Volodia Teitelboim, the whaling stories *Mónica Sanders* (1951) by Salvador Reyes and *El camino de la ballena* (The way of the whale) (1962) by Francisco Coloane (both Chileans), and the petro-novel *Oficina no. 1* (Office number 1) (1961) by Venezuelan Miguel Otero Silva are more illustrative examples of the way in which current ecological concerns influenced what I looked for in earlier literature. But these mining, whaling, and petro-texts do not have very much to say about the ecological aspects of such endeavors: the surprise for present readers is that it seems impossible to imagine such industries without an awareness of their potential for environmental harm. Neither *Subterra, Hijo del salitre, Mónica Sanders,* nor *Llampo de sangre* have anything of the contamination of local watersheds, landscape alteration, or the devastating toxic by-products of mining and the challenge of waste disposal that each industry implies. From among all of these novels, I could only locate one short passage in *Hijo de salitre* where a miner, Elias, and his mother discuss the deforestation of ancient Chilean forests:

> Elías pensó en un tiempo que el desierto se veía tan calvo e implacable porque había vivido millones de años desnudo bajo el sol. Su madre le explicó que aquellas no siempre fueron tierras eriales. "Allí donde descubres tamarugos y algarrobos muertos, existió en otros tiempos una gran región forestal, de tonalidad esmeralda. Toda la vegetación que aquí vivió, cayó fulminada por la arena y el desarrollo de las salinas. Así murieron los bosques antiguos."[3]
>
> [Elias once thought about the fact that the desert seemed so barren and implacable because it had lived for millions of years underneath the scorching sun. His mother had explained to him that these lands had not always been so fallow. "Where you see dying locust trees and tamarugos, there was once a great emerald forest. All the plants that were living here were struck down by the spread of sand and salt flats. That is how the ancient forests died."]

From current ecocritical perspectives, the issues raised in this passage seem absolutely essential to any literary exploration of the mining industry. Indeed, they are central to the Bolivian *indigenista* mining novels considered in chapter 7: *En las tierras de Potosí* (In the lands of Potosí) (1911) by Jaime Mendoza, *Socavones de angustia* (Shafts of anguish) (1947) by Fernando Ramírez Velarde, and *Altiplano* (Highlands) (1945) by Raúl Botelho Gosálvez. Yet the authors of the Chilean novels had something else in mind entirely: they seemed much more interested in exploring the ways in which the mining industry victimized its labor force with impunity. Similarly, the whaling novels by Coloane and Reyes, set in the waters off Southern Chile's *Tierra del fuego*, were published more than one hundred years after Atlantic whale populations began to decline precipitously in the nineteenth century due to over-exploitation; but, apart from brief commentaries by narrators or characters, there is nothing else about whale species endangerment. In *El camino de la ballena,* a harpooner is made to feel guilty for killing a nursing mother blue whale: "Ha muerto una ballena madre, recién parida. . . . ¡Eso no lo debe hacer nunca un ballenero que se precie!" ["A mother-whale who recently gave birth was killed. . . . This should never have happened, especially by the hand of an experienced and esteemed whaler!"][4] In *Mónica Sanders*, there is a poignant moment where a visiting *gringo* observer imagines an absurd scene of human triumph over nature:

> La humanidad enloquecida aullaba y alzaba los brazos saludando su victoria sobre las fuerzas supremas de la naturaleza y el aniquilamiento del animal cuyo último y más apocalíptico ejemplar era aquél; vió a los hombres danzando y cantando en torno al cuerpo de la ballena, celebrando el quedar ellos como los únicos seres vivos del planeta, como las únicas máquinas de carnes y sangre en un mundo de máquinas de hierro y acero construidas por ellos mismos ¡Victoria del hombre! ¡Ya no hay más bestias en las tierras ni en los mares! ¡Máquinas, sólo máquinas! Pronto tocará su turno al reino vegetal.[5]

> [The crazy human horde howled and raised their hands to salute their victory over nature's supreme forces and the killing of the very last of an animal species; the men were dancing and singing around the corpse of the whale, celebrating as if they were the last human beings on earth, the last troglodytes in a world of machines of iron and steel that they themselves had made. A human victory! But now, there were no longer any animals on land or in the sea! Only machines! Machines! Soon, the vegetable world will suffer a similar fate.]

Lines like the above demonstrate that an ecological sensibility was not absent from the literary moment when Reyes's novel was written, but these ideas are never developed and the cetacean carnage is merely a fictional backdrop for bizarre love triangles and the anxiety of an aging harpooner as his whaling venture is bought out by foreign interests.

Finally, Miguel Otero Silva seems to explore the cultural significance of the advent of the petroleum industry in Venezuela with regard to immigration, the influence of foreign economic interests, and the development of interpersonal relationships in the new social milieu of petro-cities. But a representation of the environmental impacts occasioned by the extractive practices of the petroleum industry is quite simply not explored in the novel. Instead, *Oficina no. 1* and these other texts take industries with historical propensities for environmental contamination or irreversible alterations as their primary inspiration, but representations of the ecological consequences of mining, drilling, and whaling just do not make anything more than brief and isolated appearances.

Then there is the case of the historical regional novels where it would be anachronistic to incorporate an environmentalist discourse in stories about conquistadors when that type of thinking did not even begin to be imagined until several hundred years after the Conquest. Among these are *El quijote de El Dorado* (The Quijote of the Golden City) (1964) by Ecuadoran Demetrio Aguilera-Malta and compatriot Leopoldo Benites Vinueza's *Los descubridores del Amazonas: La expedición de Orellana* (*Argonautas de la selva*) (The discoverers of the Amazon: Orellana's expedition (Explorers of the jungle)) (1945) on Francisco de Orellana's discovery of the Amazon; and two stories about the infamous Lope de Aguirre by Venezuelans Arturo Uslar-Pietri in *El camino de el Dorado* (The way to the City of Gold) (1947) and Miguel Otero Silva in *Lope de Aguirre: Príncipe de la libertad* (Lope de Aguirre: The prince of liberty) (1979). These narratives, and other historical fictions like them, seem to stay away from ecological questions because such concerns did not exist during the historical moments that they recount.

I should also explain why I largely refrained from analysis of poetry and essay after the *modernista* period when there has been a good deal of ecological elements in these literary forms throughout the twentieth century to the present. I chose to focus the analysis for the chapters on the regional novel, *indigenismo*, the "Boom" and the post-"Boom" on narrative fiction and elected to not include the poetry of authors that explore ecological themes such as Mexicans Homero Aridjis and Emilio Pacheco, Chileans Pablo Neruda and Nicanor Parra, and Nicaraguans Pablo Antonio Cuadra, Gioconda Belli, and Ernesto Cardenal. The decision was motivated by an effort to arrive without detour at Chilean Luis Sepúlveda's *Un viejo que leía novelas de amor* (*The Old Man who Read Love Stories*) (1989) which I consider the touchstone of Spanish American environmentalist literature. As I argue, particularly in chapter 4, there were clear precedents for Sepúlveda's work and those connections seem to come more from narrative fiction than from poetry. Also, several critics have recently done quite a bit of work on the poetry of the authors listed above; the environmentalist tendencies and/or ecological awareness in those texts and poets have been well-rehearsed.[6]

Finally, I propose that my analysis in the previous ten chapters leaves us with two tasks. First, I advocate for the inclusion of certain texts in what is considered the "canon" of Spanish American literature. It took the advent of a mature expression of ecocritical analysis such as has occurred over the last decade to make it possible to apply ecological perspectives to the literature of the region. Among the first were the recent works by authors like Luis Sepúlveda, Homero Aridjis, Gioconda Belli, and Mempo Giardinelli considered in chapters 8, 9, and 10; by now, these have been fairly well-covered from the ecocritical perspective. However, I would like to see a more prominent literary place for utterly forgotten works such as Nicaraguan Hernán Robleto's *Una mujer en la selva* (A woman in the jungle) (1936) with its recognition of the rights and identity of animals or Mexican Rafael Bernal's *Su nombre era muerte* (His name was death) (1947) for its undoing of the special status of humans as supreme over and separate from the natural world. These authors seem unusually prescient with regard to issues that have become central to the contemporary concerns of ecocriticism and critical animal studies. Ecocritical readings of Spanish American literature from its canon and corpus should reveal more and more such titles; I have done part of the work here, but there are those that I did not include (several other titles from the *indigenista* tradition might have been included in chapters 6 and 7 but would have simply repeated ideas from texts that were considered); others I may have simply overlooked; and others may have to be analyzed from ecocritical perspectives still to be articulated. The present tendency in literary studies toward opening the canon can only make this work easier.

Secondly, I find it imperative to undertake an ecocritical reevaluation of the whole of Spanish American literature, both canonical and non-canonical. In the commentary at the close of chapter 10, I make reference to the long history of ecology in the literature of the region. And this book attempts to account for it, but other scholars have begun to uncover this history in more depth, particularly with regard to the regional literature[7] and the more contemporary, more explicitly environmentalist works.[8] But in-depth analyses of singular authors or literary periods must now make their appearance. Writing this book has given me some ideas, but I would not at all be surprised to see them taken up by others: monographs on the ecology of the *Asociación de mayo* authors or *modernista* short stories or the voluminous oeuvre of Guatemalan Miguel Ángel Asturias. There are also specific aspects of ecocriticism that could be pursued in greater detail than I have done here: an analysis of political ecology as articulated in nineteenth century literature and politics or an examination of nonhuman creatures from the perspective of Critical Animal Studies. Several other ecocritical approaches suggest themselves as well: articulation of place and space, the representation of environmentalist advocacy, environmental justice, the meaning of and possible correlatives between the English term "wilderness" and Spanish terms like "*selva*," "*monte*," "*bosque*," or "*desierto*." In the interest of presenting a more global picture of ecology and Spanish American literature, I have made limited use of these approaches and have preferred the examination of a more extensive corpus of texts. But the next step is to focus such approaches to undertake the revaluation of canon and commit to sustained and detailed studies of the literature from the ecocritical perspectives that I suggest above. The game is afoot!

NOTES

1. Jared Diamond, *Collapse: How Societies Choose to Fail or Succeed* (New York: Viking, 2005), 119.
2. Pedro Prado, *La reina de Rapa Nui* (Santiago de Chile: Nascimiento, 1938), 158.
3. Volodia Teitelbaum, *Hijo del salitre* (Santiago de Chile: Editorial Austral, 1952), 103.
4. Francisco Coloane, *El camino de la ballena* (Santiago de Chile: Zig-Zag, 1962), 186.
5. Salvador Reyes, *Mónica Sanders* (Santiago de Chile: Zig-Zag, 1951), 102–3.
6. Analisa De Grave, "Ecoliterature and Dystopia: Gardens and Topos in Modern Latin American Poetry," *Confluencia* 22, no. 2 (2007): 89–104; Roberto Forns-Broggi, "Ecology and Latin American Poetry" in *Literature of Nature: An International Sourcebook*, ed. Patrick D. Murphy (Chicago: Fitzroy Dearborn, 1998), 374-84; Niall Binns, "Landscapes of Hope and Destruction: Ecological Poetry in Spanish America," *ISLE* 9, no. 1 (2002): 105–20; and Jeremy Rochelle, "Writing under the Shadow of the Chullachaqui: Amazonian Thought and Ecological Discourse in Recent Amazonian Poetry," *Review: Literature and Arts of the Americas* 45, no. 2 (2012): 198-206.
7. Jennifer L. French, *Nature, Neo-Colonialism, and the Spanish American Regional Writers* (Hanover, NH: Dartmouth University Press, 2005). Also, see the numerous articles by Jorge Marcone listed in the bibliography.
8. See Introduction, note 6.

Bibliography

Acevedo Díaz, Eduardo. *Grito de gloria.* Montevideo: Biblioteca Artigas, 1964.

———. *Ismaél.* Montevideo: Biblioteca Artigas, 1953.

———. *Nativa.* Montevideo: Biblioteca Artigas, 1964.

———. *Soledad.* Montevideo: Biblioteca Artigas, 1954.

Adamson, Joni, Mei Mei Evans, and Rachel Stein. *The Environmental Justice Reader: Politics, Poetics, Pedagogy.* Tucson: University of Arizona Press, 2002.

Adamson, Joni, and Scott Slovic. "The Shoulders We Stand On: An Introduction to Ethnicity and Ecocriticism." *Melus* 34, no. 2 (Summer 2009): 5–24.

Alegría, Ciro. *La serpiente de oro.* Madrid: Aguilar, 1963.

Alonso, Carlos J. *The Spanish American Regional Novel.* New York: Cambridge University Press, 1990.

Alonso, Diego. "José Enrique Rodó: una retórica para la democracia." *Revista canadiense de estudios hispánicos* 34, no. 2 (Winter 2001): 183–97.

Alstrum, James J. "*La Diana cazadora* de Clímaco Soto Borda: Un discurso disonante del fin de siglo." *Estudios de Literatura Colombiana* 5 (July-December 1999): 9–22.

Altamirano, Ignacio M. *Clemencia y La navidad en las montañas.* Mexico City: Porrúa, 1966.

———. *El zarco.* Habana: Casa de las Américas, 1976.

Amato, Mariana. "Force Majeure: Leopoldo Lugones toward a Vitalist Fascism." *Dissidences: Hispanic Journal of Theory and Criticism* 3, no. 6 (Spring 2010): 1–28.

Amaya-Amador, Ramón. *Prisión verde.* Buenos Aires: Agepe, 1957.

Anderson Imbert, E. *Historia de la literatura hispanoamericana.* 2 vols. Mexico City: Fondo de cultura económica, 1997.

Archibald, Priscilla. "Urban Transculturations." *Social Text* 25, no. 4 (Winter 2007): 91–113.

Area, Lelia. "Geografías imaginarias: El *Facundo* y la *Campaña en el ejército grande* de Domingo Faustino Sarmiento." *Revista Iberoamericana* 67, nos. 194–95 (January–June 2001): 91–103.

Aréstegui, Narciso. *El padre Horán.* 2 vols. Cuzco: Universidad Nacional del Cuzco, 1954.

Arguedas, Alcides. *Raza de bronce.* La Paz: González y Medina, 1919.

Arguedas, José María. *El zorro de arriba y el zorro de abajo.* Nanterre, France: ALLCA XX, 1990.

———. *Yawar fiesta.* Austin: University of Texas Press, 1985.

Arias, Arturo. *Los caminos de Paxil.* Guatemala City: Editorial cultura, 1990.

———. *The Rigoberta Menchú Controversy.* Minneapolis: University of Minnesota Press, 2001.

Aridjis, Homero. *¿En quién piensas cuando haces el amor?* Mexico City: Alfaguara, 1995.

———. *La leyenda de los soles.* Mexico City: Fondo de cultura, 1993.

Armbruster, Karla. "A Poststructuralist Approach to Ecofeminist Criticism." In *The Green Studies Reader: From Romanticism to Ecocriticism*, edited by Laurence Coupe, 198–203. New York: Routledge, 2000.

Asturias, Miguel Ángel. *Hombres de maíz*. Nanterre, France : ALLCA XX, 1992.

Balseca, Fernando. "En busca de nuevas regiones: la nación y la narrativa ecuatoriana." *Procesos: Revista ecuatoriana de historia* 8 (1995–1996): 141–55.

Barbas-Rhoden, Laura. *Ecological Imaginations in Latin American Fiction*. Gainesville: University Press of Florida, 2011.

Barrios, Valentín. *El mensú que triunfó en la selva*. Buenos Aires: Librería Perlado, 1951.

Basye Sander, Jennifer, and Peter Sander, with Anne Basye. *Green Christmas: How to Have a Joyous, Eco-Friendly Holiday Season*. Avon, MA: Adams Media, 2008.

Bate, Jonathan. "Culture and Environment: From Austen to Hardy." *New Literary History* 30, no. 3 (Summer 1999): 541–60.

———. "From 'Red' to 'Green.'" In *The Green Studies Reader: From Romanticism to Ecocriticism*, edited by Laurence Coupe, 167–72. New York: Routledge, 2000.

Behrndt, Stephen C. "Remapping the Landscape: The Romantic Literary Community." In *Comparative Romanticisms: Power, Gender, Subjectivity*, edited by Larry H. Peer and Diane Long Hoeveler, 11–32. Columbia, SC: Camden House, 1998.

Belli, Gioconda. *Waslala*. Barcelona: Emecé, 1998.

Bello, Andrés. *Poesías de Andrés Bello*. Barcelona: Casa editorial Maucci, 1909.

Bernal, Rafael. *Su nombre era muerte*. Mexico City: Editorial Jus, 1947.

Berry, Wendell. "In Distrust of Movements." *Orion* (Summer 1999): 15–18.

Best, Stephen, Anthony J. Nocella, II, Richard Kahn, Carol Gigliotti, and Lisa Kemmerer. "Introducing Critical Animal Studies." *Journal for Critical Animal Studies* 5, no. 1 (2007): 4–5.

Biersack, Aletta. "Reimagining Political Ecology: Culture/Power/History/Nature." In *Reimagining Political Ecology*, edited by Aletta Biersack and James B. Greenberg, 3–40. Durham, NC: Duke University Press, 2006.

Binns, Niall. "Landscapes of Hope and Destruction: Ecological Poetry in Spanish America." *ISLE* 9, no. 1 (2002): 105–20.

Blake, William. "Nature as Imagination." In *The Green Studies Reader: From Romanticism to Ecocriticism*, edited by Laurence Coupe, 16. New York: Routledge, 2000.

Blest Gana, Alberto. *La aritmética en el amor*. Santiago de Chile: Zig-Zag, 1950.

———. *Martín Rivas*. Madrid: Cátedra, 1983.

Boff, Leonardo. *Ecology & Liberation: A New Paradigm*. Maryknoll, NY: Orbis, 1995.

Borges, Jorge Luis. "Nathaniel Hawthorne." In *Prosa completa*. Barcelona: Editorial Bruguera, 1980.

Botelho Gosálvez, Raúl. *Altiplano*. Lima: Ediciones Nuevo Mundo, 1967.

Botero Restrepo, Jesús. *Andágueda*. Bogotá: Librería y Ediciones Teoría, n. d.

Bramwell, Anna. *Ecology in the 20th Century*. New Haven, CT: Yale University Press, 1989.

Branch, Michael P., and Scott Slovic, eds. *The ISLE Reader: Ecocriticism, 1993–2003*. Athens: University of Georgia Press, 2003.

Bryant, Raymond L., and Sinéad Bailey. *Third World Political Ecology*. New York: Routledge, 1997.

Buell, Lawrence. *The Environmental Imagination: Thoreau, Nature Writing, and the Formation of American Culture*. Cambridge, MA: Harvard University Press, 1995.

———. *The Future of Environmental Criticism: Environmental Crisis and Literary Imagination*. Malden, MA: Blackwell, 2005.

———. *Writing for an Endangered World*. Cambridge, MA: Harvard University Press, 2001.

Bula, Germán, and Ronald Bermúdez. *Alteridad y pertenencia*. Bogotá: Universidad de La Salle, 2009.

Burgos-Debray, Elizabeth. *Me llamo Rigoberta Menchú*. Habana: Casa de las Américas, 1983.

Camacho, David E., ed. "The Environmental Justice Movement: A Political Framework." In *Environmental Injustices, Political Struggles: Race, Class, and the Environment*, 11–30. Durham, NC: Duke University Press, 1998.

Cambaceres, Eugenio. *Sin rumbo*. Buenos Aires: Ediciones Jackson, 1944.

Cané, Miguel. *La familia de Sconner*. Buenos Aires: Imprenta de la universidad, 1930.

Caro, José Eusebio. *Poesías de Caro i Vargas Tejada*. Bogotá: Imprenta de Ortiz, 1857.

las Casas, Bartolomé de. *Brevísima relación de la destrucción de las Indias*. Madrid: Cátedra, 2001.

Casós, Fernando. *Los amigos de Elena*. 2 vols. Paris: La Librería Española de E. Denné Schmitz, 1874.

Castrillón Arboleda, Diego. *José Tombe*. Bogotá: Editorial Antena, 1942.

Castro, Baltazar. *Mi camarada padre*. Santiago de Chile: Zig-Zag, 1958, 1967.

Castro, Ernesto L. *Campo arado*. Buenos Aires: Losada, 1953.

Castro Herrera, Guillermo. "Naturaleza, sociedades y culturas en José Martí." *Cuadernos Americanos*, 2nd series, 51 (May-June 1995): 92–121.

Castro-Urioste, José. "Utopía y transgresión: El imaginario nacional en la obra de Rómulo Gallegos." *Hispanic Journal* 17, no. 2 (Fall 1996): 329–44.

Chang-Rodríguez, Eugenio. *La literatura política de González Prada, Mariátegui y Haya de la Torre*. Mexico City: Ediciones de Andrea, 1957.

Chang-Rodríguez, Raquel and Malva E. Miller, eds. *Voces de Hispanoamérica: Antología literaria*. Boston: Heinle & Heinle, 1996.

Clark, Brett, and John Bellamy Foster. "Guano: The Global Metabolic Rift and the Fertilizer Trade." In *Ecology and Power: Struggles over land and material resources in the past, present, and future*, edited by Alf Hornborg, Brett Clark, and Kenneth Hermele, 68–82. New York: Routledge, 2012.

Clark, Zoila. "Rasgos modernistas y naturalistas en *Sin rumbo* (1885) de Eugenio Cambaceres." *Hispanófila* 154 (2008): 47–58.

Coloane, Francisco. *El camino de la ballena*. Santiago de Chile: Zig-Zag, 1962.

Colón, Cristobal. *Textos y documentos completos*. Edited by Consuelo Varela (Madrid: Alianza Editorial, 1992

Contreras Castro, Fernando. *Única mirando el mar*. San José, Costa Rica: Ediciones Farben, 1994.

Cornejo-Polar, Antonio. "*Indigenismo* and Heterogenous Literatures: Their Double Sociocultural Statute." In *The Latin American Cultural Studies Reader*, edited by Ana del Sarto, Alicia Ríos, and Abril Trigo, 100–15. Durham, NC: Duke University Press, 2004.

Cortés, Manuel José. "A la naturaleza del oriente de Bolivia." *La semana*, August 13, 1859, 203–4.

Coupe, Laurence, ed. *The Green Studies Reader: From Romanticism to Ecocriticism*. New York: Routledge, 2000.

Crellis Secco, Susana. "La nación uruguaya en la literatura: Acevedo Díaz y la creación de una nación." *Alba de América: Revista Literaria* 20, nos. 37–38 (2001): 415–24.

Cronon, William, ed. *Uncommon Ground: Rethinking the Human Place in Nature*. New York: Norton, 1995.

———. "The Trouble with Wilderness." In *Uncommon Ground: Rethinking the Human Place in Nature*, edited by William Cronon, 69–90. New York: Norton, 1995.

Da Jandra, Leonardo. *Huatalqueños*. Mexico City: Editorial Joaquín Mortiz, 1991.

———. *La almadraba*. Mexico City: Planeta, 2008.

———. *Samahua*. Mexico City: Seix Barral, 1997.

Darío, Rubén. *Azul*. San Jose, Costa Rica: Editorial Universitaria Centroamericana, 1981.

———. *El modernismo y otros ensayos*. Madrid: Alianza, 1989.

———. *Obra Poética*. Madrid: Biblioteca Corona, 1914–1916.

———. *Prosas profanas*. Madrid: Editorial Mundo Latino, 1917.

De Grave, Analisa. "Ecoliterature and Dystopia: Gardens and Topos in Modern Latin American Poetry." *Confluencia* 22, no. 2 (2007): 89–104.

de la Rosa, Raúl. *La revolución ecológica*. Barcelona: Icaria, 2001.

de Vivanco, Lucero. "Modernidad y apocalipsis en *los zorros* de Arguedas." *Revista chilena de literatura* 78 (April 2011): 49–68.

del Solar, Alberto. *Huincahual*. Paris: Garnier hermanos, 1911.

DeVries, Scott. "Garbage Out: Space, Place, and Neo-imperial Anti-development in Gioconda Belli's *Wasala*." *Ecozon@* 1, no. 2 (2010): 38–50.

Diamond, Jared. *Collapse: How Societies Choose to Fail or Succeed.* New York: Viking, 2005.

Díaz Rodríguez, Manuel. *Narrativa y ensayo.* Caracas: Ayacucho, 1982.

Díaz Sánchez, Ramón. *Mene.* Buenos Aires: Editorial Universitaria de Buenos Aires, 1966.

Dickman, Max. *Madre América.* Buenos Aires: Santiago Rueda, 1960.

Di Chiro, Giovanna. "Nature as Community: The Convergence of Environment and Social Justice." In *Uncommon Ground: Rethinking the Human Place in Nature*, edited by William Cronon, 298–320. New York: Norton, 1995.

Diéguez Olaverri, Juan. "A Los Cuchumatanes," *El Jardín* 1, no. 21 (June 14, 1908): 168.

Dobson, Andrew. "Citizenship." In *Political Theory and the Ecological Challenge*, edited by Andrew Dobson and Robin Eckersley, 216–31. New York: Cambridge University Press, 2006.

Donoso, José. *Historia personal del "Boom."* Barcelona: Seix Barral, 1983.

Drengson, Alan, and Yuichi Inoue. *The Deep Ecology Movement: An Introductory Anthology.* Berkeley, CA: North Atlantic Books, 1995.

Echeverría, Esteban. *La cautiva. La Guitarra = Elvira.* Buenos Aires: La Cultura Argentina, 1916.

Escobar, Arturo. "Constructing Nature: Elements for a Poststructuralist Political Ecology." In *Liberation Ecologies: Environment, Development, Social Movements*, edited by Richard Peet and Michael Watts, 46–68. London: Routledge, 1996.

Fallas, Carlos Luis. *Mamita Yunai.* Mexico City: Fondo de Cultura Popular, 1957.

Ferguson, Moira. "Breaking in Englishness: *Black Beauty* and the Politics of Gender, Race, and Class." *Women: A Cultural Review* 5, no. 1 (1994): 34–52.

Fernández, Teodosio. Introduction to *Huasipungo* by Jorge Icaza. Madrid: Cátedra, 1997.

Fernández de Lizardi, José Joaquín. *El periquillo Sarniento.* Mexico City: Editores Mexicanos Unidos, 2002.

Fernández Meijide, Graciela. *Las cifras de la guerra sucia.* Buenos Aires: Asamblea Permanente por los Derechos Humanos, 1988.

Flys Junquera, Carmen, José Manuel Marrero Henríquez, and Julia Barella Vigal, eds. *Ecocríticas: Literatura y medio ambiente.* Orlando, FL: Iberoamericana Vervuert Publishing Company, 2010.

Forns-Broggi, Roberto. "Ecology and Latin American Poetry." In *Literature of Nature: An International Sourcebook*, edited by Patrick D. Murphy, 374–84. Chicago: Fitzroy Dearborn, 1998.

Franco, Jean. *An Introduction to Spanish American Literature.* New York: Cambridge University Press, 1994.

French, Jennifer L. *Nature, Neo-Colonialism, and the Spanish American Regional Writers.* Hanover, NH: Dartmouth University Press, 2005.

Freyre, Ricardo Jaimes. *Los sueños son vida. Anadiomena. Las víctimas.* Buenos Aires: "Buenos Aires" Sociedad Cooperativa Editorial Limitada, 1917.

Fuentes, Carlos. *Cristóbal Nonato.* Mexico City: Fondo de Cultura Económica, 1987.

———. *La nueva novela hispanoamericana.* Mexico City: Cuadernos Joaquín Mortiz, 1969.

Galeano, Eduardo. *Las venas abiertas de América Latina.* Mexico City: Siglo XXI, 1971.

———. *Úselo y tírelo.* Buenos Aires: Planeta, 1994.

Gallegos, Rómulo. *Canaima.* Madrid: Austral, 1982.

———. *Doña Bárbara.* Caracas: Monte Avila Editores, 1977.

———. *Sobre la misma tierra.* Buenos Aires: Espasa Calpe, 1961.

Galván, Manuel de Jesús. *Enriquillo.* Santo Domingo: Ediciones de Taller, 1993.

Galvis Yepes, Carmen. "Jose Eusebio Caro y su poética de la muerte del padre." *Estudios de Literatura Colombiana* 7 (July-December, 2000): 60–69.

Gamboa, Federico. *Santa.* Madrid: Cátedra, 2002.

García Piedrahita, Eduardo. "Dos novelas colombianas olvidadas." *Boletín Cultural y Bibliográfico* 17 (1980): 217–28.

Garganigo, John F., Rene de Costa, Ben A. Heller, Alessandra Luiselli, Georgina Sabat-Rivers, and Elzbieta Sklodowska, eds. *Huellas de las literaturas hispanoamericanas.* Upper Saddle River, NJ: Prentice Hall, 2002.

Garrard, Greg. *Ecocriticism.* New York: Routledge, 2004.

Gerassi-Navarro, Nina. *Pirate Novels: Fictions of Nation Building in Spanish America*. Durham, NC: Duke University Press, 1999.

Giardinelli, Mempo. *Imposible equilibrio*. Mexico City: Joaquín Mortiz, 1995.

Giménez, Teresa Vicente. *Justicia ecológica y protección del medio ambiente*. Madrid: Trotta, 2002.

Gómez Cardona, Fabio. "Silencios y ocultamientos en la novela *José Tombé*." *Poligramas* 33 (January–June 2010): 183–211.

Gómez de Avellaneda, Gertrudis. *Sab*. Habana: Instituto Cubano del Libro, 1973.

Gómez García, Juan Guillermo. "La significación continental de Manuel González Prada: sobre la génesis del anarquismo en Hispanoamérica." *Estudios Políticos* 37 (July–December 2010): 207–32.

González, Aníbal. *Abusos y admoniciones: Ética y escritura en la narrativa hispanoamericana moderna*. Mexico City: Siglo Veintiuno Editores, 2001.

González, Joaquín V. *Mis montañas*. Buenos Aires: William Jackson, n. d.

González, Natalicio. *La raíz errante*. Mexico City: Editorial Guarania, 1953.

González Martínez, Manuel. *Llanura, soledad y viento*. Buenos Aires: Sudamericana, 1965.

González Prada, Manuel. *Antologia poética*. Mexico City: Editorial Cultura, 1940.

———. *Exóticas*. Lima: Tipografía De "El Lucero," 1911.

González-Stefan, Beatriz. "Martí y la alta modernidad." In *Escribiendo la Independencia: Perspectivas postcoloniales sobre la literatura hispanoamericana del siglo XIX*, edited by Robert Folger and Stephan Leopold, 343–71. Madrid: Iberoamericana, 2010.

Graham-Yooll, Andrew. "Light at the End of a Tunnel: The New Writers of Latin America." *Antioch Review* 52, no. 4 (Fall 1994): 566–79.

Grinberg Pla, Valeria Mackenbach, and Werner Mackenbach. "*Banana novel revis(it)ed*: etnia, género y espacio en la novela bananera centroamericana. El caso de *Mamita Yunai*." *Iberoamericana,* 2nd series 6, no. 23 (September 2006): 161–76.

Grove, Richard. *Green Imperialism*. New York: Cambridge University Press, 1995.

Grugel, Jean, and Pía Riggirozzi, eds. *Governance after Neo-Liberalism in Latin America*. New York: Palgrave Macmillan, 2009.

Guerra, Rosa. *Lucía Miranda*. Buenos Aires: Imprenta Americana, 1860.

Guillén Pinto, Alfredo. *Lágrimas indias*. La Paz: Casa Editora Mundial, 1920.

Güiraldes, Ricardo. *Don Segundo Sombra*. Buenos Aires: Editorial Losada, 1974.

Gunnels, Bridgette W. "An Ecocritical Approach to Horacio Quiroga's 'Anaconda' and 'Regreso de Anaconda.'" *Mosaic* 39, no. 4 (December 2006): 93–110.

Gutiérrez, Juan María. *El capitán de patricios*. Buenos Aires: Imprenta de la Universidad, 1928.

Gutiérrez Najera, Manuel. *Cuentos y cuaresmas del duque Job*. Mexico City: Porrúa, 1966.

Hallum, Anne. "Ecotheology and Environmental Praxis in Guatemala." *Nova Religio: The Journal of Alternative and Emergent Religions* 7, no. 2 (2003): 55–70.

Harvey, David. *Justice, Nature, and the Geography of Difference*. Cambridge, MA: Blackwell, 1996.

Hazera, Lydia de León. *La novela de la selva hispanoamericana: Nacimiento, desarrollo y transformación*. Bogotá: Caro y Cuervo, 1971.

Heise, Ursula K. *Sense of Place and Sense of Planet: The Environmental Imagination of the Global*. New York: Oxford University Press, 2008.

Heredia, José María. "Niagara." In *Poetry of Niagara*, edited by Myron Thomas Pritchard, 34–38. Boston: Lothrop, 1901.

Hernández, Arturo D. *Bubinzana*. Lima: Luis E. Valcarcel, 1960.

———. *Sangama*. Lima: Juan Mejía Vaca, 1958.

———. *Selva trágica*. Lima: Juan Mejía Vaca & P.L. Villanueva, 1956.

Hernández, José. *El gaucho Martín Fierro*. Buenos Aires: Librería Martín Fierro, 1894.

Herrera y Reissig, Julio. *Las pascuas del tiempo (poesías y otros poemas)*. Montevideo: O. M. Bertani, 1913.

———. *Los éxtasis de la montaña, eglogánimas*. Buenos Aires: Ediciones Selectas América, 1920.

Hirsch, Eric. "Introduction. Landscape: Between Place and Space." In *The Anthropology of Landscape: Perspectives on Place and Space*, edited by Eric Hirsch and Michael O'Hanlon, 1–30. Oxford: Clarendon, 1995.

Hirschberg, S, G. Spiekerman, and R. Dones. *Comprehensive Assessment of Energy Systems: Severe Accidents in the Energy Sector*. Villigen, Switzerland: Paul Scherrer Institute, 1996.

Hochman, Jhan. "Green Cultural Studies." In *The Green Studies Reader: From Romanticism to Ecocriticism*, edited by Laurence Coupe, 187–92. New York: Routledge, 2000.

Hoeg, Jerry. "The Landscape of the Consumer Society: Fernando Contreras Castro's *Única mirando al mar*." In *Reading and Writing the Latin American Landscape*, edited by Beatriz Rivera-Barnes and Jerry Hoeg, 177–85. New York: Palgrave, 2009.

Hornborg, Alf. "Accumulation: Land as a medium of domination." In *Ecology and Power: Struggles over land and material resources in the past, present, and future*, edited by Alf Hornborg, Brett Clark, and Kenneth Hermele, 13–22. New York: Routledge, 2012.

Hornborg, Alf, Brett Clark, and Kenneth Hermele, eds. *Ecology and Power: Struggles over land and material resources in the past, present, and future*. New York: Routledge, 2012.

Hostos, Eugenio María de. *Obra literaria selecta*. Caracas: Biblioteca Ayacucho, 1988.

Howarth, William. "Ecocriticism in Context." In *The Green Studies Reader: From Romanticism to Ecocriticism*, edited by Laurence Coupe, 163–66. New York: Routledge, 2000.

Husband, Andrew. "Environmental Affects." *ISLE: Interdisciplinary Studies in Literature and Environment* 17, no. 4 (Autumn 2010): 783–85.

Ibáñez, Jaime. *Cada voz lleva su angustia*. Bogotá: Ediciones Gamma, 1961.

Icaza, Jorge. *Huasipungo*. Madrid: Cátedra, 1997.

Iglesia, Cristina. "Mejor se duerme en la pampa: Deseo y naturaleza en *Una excursión a los indios ranqueles* de Lucio V. Mansilla." *Revista iberoamericana* 63, nos. 178–79 (January–June 1997): 185–92.

Isaacs, Jorge. *María*. Buenos Aires: Editorial Kapelusz, 1962.

Jaramillo Arango, Rafael. *Barrancabermeja: novela de proxenetas, rufianes, obreros y petroleros*. Bogotá: Editorial E.S.B., 1934.

Jrade, Cathy L. *Modernismo, Modernity, and the Development of Spanish American Literature*. Austin: University of Texas Press, 1998.

Kane, Adrian Taylor, ed. *The Natural World in Latin American Literatures: Ecocritical Essays on Twentieth Century Writings*. Jefferson, NC: McFarland, 2010.

Kearns, Sofia. Introduction to *La loca de Gandoca/The Madwoman of Gandoca*, by Anacristina Rossi. Translated by Terry J. Martin. Lewiston, NY: Edwin Mellen, 2006.

Kern, Robert. "Ecocriticism: What is it Good for?" In *The ISLE Reader: Ecocriticism, 1993–2003*, edited by Michael P. Branch and Scott Slovic, 258–81. Athens: University of Georgia Press, 2003.

Kogl, Alexandra. *Strange Places: The Political Potentials and Perils of Everyday Spaces*. Lanham, MD: Lexington Books, 2008.

Laguerre, Enrique. *Cauce sin río*. Río Piedras, Puerto Rico: Editorial Cultural, 1991.

Lara, Jesús. *Yanakuna: Novela quechua*. La Paz: Librería y Editorial Juventud, 1958.

Latorre, Manuel. *Zurzulita*. Santiago de Chile: Editorial Chilena, 1920.

Lomboy, Reinaldo. *Ranquil: Novela de la tierra*. Santiago de Chile: Editorial Orbe, 1966.

López, Mariano. "Stress, psicoanálisis y sentimiento de culpa en *Beba* de Carlos Reyles." *Texto Crítico* 12 (1986): 179–88.

López, Vicente F. *La novia del hereje*. Buenos Aires: La Cultura Argentina, 1917.

López-Lozano, Miguel. *Utopian Dreams, Apocalyptic Nightmares: Globalization in Recent Mexican and Chicano Narrative*. West Lafayette, IN: Purdue University Press, 2008.

Lowy, Michael, and Robert Sayre. *Romanticism Against the Tide of Modernity*. Durham, NC: Duke University Press, 2001.

Lugones, Leopoldo. *El espejo negro y otros cuentos*. Buenos Aires: Editorial Abril, 1988.

———. *El libro de paisajes*. Buenos Aires: Otero y García, 1917.

———. *Montañas de oro*. Montevideo: Editorial Rioplatense, 1919.

Lynch, Benito. *El inglés de los güesos*. Mexico City: El Libro Popular, 1955.

Magariños Cervantes, Alejandro. *Caramurú*. Buenos Aires: Imprenta del Orden, 1865.

Magnarelli, Sharon. "La mujer y la naturaleza en *La vorágine*: A imagen y semejanza del Hombre." In *La vorágine: Textos críticos*, edited by Montserrate Ordóñez, 333–52. Bogotá: Alianza, 1987.

Malamud, Randy. "Poetic Animals and Animal Souls." *Society and Animals* 6, no. 3 (1998): 263–77.

Mandrillo, Cósimo. "Antiimperialismo, tecnología y petróleo en la novela venezolana." *Enl@ce: Revista Venezolana de Información, Tecnología y Conocimiento* 9, no. 1 (January–April 2012): 51–59.

Mansilla, Lucio. *Una visita a los indios ranqueles*. Buenos Aires: Espasa-Calpe, 1943.

Marcone, Jorge. "Cultural Criticism and Sustainable Development in Amazonia: A Reading from the Spanish American Romance of the Jungle." *Hispanic Journal* 19, no. 2 (Fall 1998): 281–94.

———. "Del retorno a lo natural: *La serpiente de oro*, la 'novela de la selva' y la crítica ecológica." *Hispania: A Journal Devoted to the Teaching of Spanish and Portuguese* 81, no. 2 (May 1998): 299–308.

———. "*El hablador* de Mario Vargas Llosa y la imagen de la Amazonía en el Perú contemporáneo." In *LA CHISPA '93: Selected Proceedings*, edited by Gilbert Paolini, 134–40. New Orleans, LA: Tulane University, 1993.

———. "Jungle Fever: Primitivism in Environmentalism: Rómulo Gallegos's *Canaima* and the Romance of the Jungle." In *Primitivism and Identity in Latin America: Essays on Art, Literature, and Culture*, edited by Erik Camayd-Freixas and José Eduardo González, 157–72. Tucson: University of Arizona Press, 2000.

———. "A Painful Pastoral: Migration and Ecology in Chicana/o Literature." *Pacific Coast Philology* 46, no. 2 (2011): 194–209.

Marcone, Jorge, and Priscilla Solis Ibarra. "Mexican and Chicana/o Environmental Writing: Unearthing and Inhabiting." In *Teaching North American Environmental Literature*, edited by Laird Christensen, Mark C. Long, and Fred Waage, 93–111. New York: Modern Language Association of America, 2008.

Marín, Juan. *Paralelo 53 sur*. Santiago de Chile: Editorial Nascimiento, 1955.

Marrero Henríquez, José Manuel. "Ecocrítica y hispanismo." In *Ecocríticas: Literatura y medio ambiente*, edited by Carmen Flys Junquera, José Manuel Marrero Henríquez, and Julia Barella Vigal, 193–218. Orlando, FL: Iberoamericana Vervuert Publishing Company, 2010.

Marroquín, José Manuel. *El moro*. Bogotá: Editorial Oveja Negra, 1985.

Martí, José. *Ismaelillo. Versos sencillos. Versos libres*. Habana: Impr. De Rambla, Bouza, 1913.

———. *Nuestra América*. Caracas: Biblioteca Ayacucho, 1977.

———. *Versos de José Martí*. Buenos Aires: Ediciones Mínimas, 1919.

Martín, Eugenio Alonso et al. *Siglo XIX. Vol 2. Manual de literatura hispanoamericana*, edited by Felipe B. Pedraza Jiménez. Berriozar, Navarra: Cenlit Ediciones, 1991.

Martínez, Luis. *A la costa*. Madrid: Ediciones de Cultura Hispánica, 1992.

Mata Ordóñez, G. Humberto. *Sumag Allpa*. Cuenca, Ecuador: Biblioteca Cenit, 1940.

Mateos. Juan A. *El cerro de las campanas*. Mexico City: Editorial Porrúa, 1971.

Meléndez, Concha. *La novela indianista en Hispanoamérica*. Río Piedras, PR: Ediciones de la Universidad de Puerto Rico, 1961.

Mendoza, Daniel. *El llanero*. Madrid: Editorial América, 1918.

Mendoza, Jaime. *En las tierras de Potosí*. Sucre: Talleres de la imprenta universitaria, 1976.

Mera, Juan León. *Cumandá*. Boston: DC Heath, 1932.

———. *La virgen del sol. Melodías indígenas*. Barcelona: Escudillers, 1887.

Merchant, Carolyn. "Reinventing Eden: Western Culture as a Recovery Narrative." In *Uncommon Ground: Rethinking the Human Place in Nature*, edited by William Cronon, 132–59. New York: Norton, 1995.

Milanés, José Jacinto. *Obras de Don José Milanés*. New York: Juan F. Trow, 1865.

Miller, Shawn William. *An Environmental History of Latin America*. New York: Cambridge University Press, 2007.

Milton, Kay. *Environmentalism and Cultural Theory: Exploring the Role of Anthropology in Environmental Discourse*. New York: Routledge, 1996.

———. *Loving Nature*. New York: Routledge, 2002.

Mires, Fernando. *El discurso de la naturaleza: Ecología y política en América Latina*. San Jose, Costa Rica: DEI, 1990.

Monteforte Toledo, Mario. *Entre la piedra y la cruz*. Guatemala City: Editorial "El Libro de Guatemala," 1948.

Mora, Gabriela. *El cuento modernista hispanoamericano*. Lima: Latinoamericana Editores, 1996.

Moraña, Mabel. "Indigenismo y globalización." In *Indigenismo hacia el fin del milenio: Homenaje a Antonio Cornejo-Polar*, edited by Mabel Moraña, 243–52. Pittsburgh: Instituto Internacional de Literatura Iberoamericana, 1998.

Muñoz Arbelaez, Santiago. Review of *Banana Cultures: Agriculture, Consumption, and Environmental Change in Honduras and the United States* by John Soluri. *Historia Crítica* 38 (May-August 2009): 208–14.

Murphy, Patrick D. *Ecocritical Explorations in Literary and Cultural Studies: Fences, Boundaries, and Fields*. Lanham, MD: Lexington Books, 2009.

Nervo, Amado. "Éxtasis." In *Antología poética hispano-americana; con notas biográficas y críticas*, edited by Calixto Oyuela. Buenos Aires: A. Estrada y cía., 1919.

———. "La nieve misteriosa de la montaña." In *Los cien mejores poemas de Amado Nervo*, edited by Enrique González Martínez. Mexico City: Tip. Murguia, 1920.

Operé, Fernando. "*La cautiva* de Echeverría: El trágico señuelo de la frontera." *Bulletin of Spanish Studies* 53, no. 5 (2003): 545–54.

Ordóñez, Montserrat. "Introduction." In *La Vorágine*, by José Eustasio Rivera. Madrid: Cátedra, 1995.

———, ed. *La vorágine: Textos críticos*. Bogotá: Alianza, 1987.

Osorio Lizarazo, J. A. *Barranquilla 2132*. Barranquilla: Tip. Delgado, 1932.

Othón, Manuel José. *Poemas escogidos; selección de Agustín Loera y Chávez*. Mexico City: Impr. Victoria, 1917.

Palma, Clemente. *Cuentos malévolos*. Barcelona: Imp. Salvat y Compañía, 1904.

Palma, Ricardo. *Tradiciones peruanas*. Madrid: Cátedra, 1994.

Paredes, Jorge, and Benjamin McLean. "Hacia una tipología de la literatura ecológica." *Ixquic: Revista Hispánica Internacional de Análisis y Creación* 2 (August 2000): 1–37.

Paulson, Susan. "Land-claims: Racialized environmental struggles in Latin America." In *Ecology and Power: Struggles over land and material resources in the past, present, and future*, edited by Alf Hornborg, Brett Clark, and Kenneth Hermele, 262–73. New York: Routledge, 2012.

Peet, Richard, and Michael Watts, eds. *Liberation Ecologies: Environment, Development, Social Movements*. London: Routledge, 1996.

———. "Liberation Ecology: Development, Sustainability, and Environment in an Age of Market Triumphalism." In *Liberation Ecologies: Environment, Development, Social Movements*, edited by Richard Peet and Michael Watts, 1–45. London: Routledge, 1996.

Pérez, Rolando. "Irony, Love, and Political Economy in José Asunción Silva's *De sobremesa*." *Hispanófila* 150 (May 2007): 87–102.

Pereyra, Marisa. "Paradise Lost: A Reading of *Waslala* from the Perspectives of Feminist Utopianism and Ecofeminism." In *The Natural World in Latin American Literatures: Ecocritical Essays on Twentieth Century Writings*, edited by Adrian Taylor Kane, 136–53. Jefferson, NC: McFarland, 2010.

———. *Toá*. Manizales, Colombia: Arturo Zapata, 1933.

Portillo y Rojas, José López. *La parcela*. Mexico City: Imprenta de El Tiempo, 1904.

Poss, J. E. "The Meanings of Tuberculosis for Mexican Migrant Farmworkers in the United States." *Social Science Medicine* 47, no. 2 (July 1998): 195–202.

Prado, Pedro. *La reina de Rapa Nui*. Santiago de Chile: Nascimiento, 1938.

Quiroga, Horacio. *Nuevos cuentos de la selva*. 3 vols. Buenos Aires: Solaris, 1997.

Raga, Fernando. *Los hijos de Gaia*. Buenos Aires: Distal, 2005.

———. *Revolución en Gaia*. Santiago de Chile: Editorial Forja, 2010.

Ramírez Velarde, Fernando. *Socavones de angustia.* La Paz: Biblioteca Paceña, 1953.

Read, Kay Almere. *Time and Sacrifice in the Aztec Cosmos*. Bloomington: Indiana University Press, 1998.

Renfrew, Daniel. "In the margins of contamination: lead poisoning and the production of neoliberal nature in Uruguay." *Journal of Political Ecology* 16 (2009): 87–103.

Reyes, Salvador. *Mónica Sanders*. Santiago de Chile: Zig-Zag, 1951.

Reyles, Carlos. *Beba.* Montevideo: Biblioteca Artigas, 1965.

———. *El terruño.* Montevideo: Agencia General de Librería y Publicaciones, 1927.

Reyna, Ernesto. *El Amauta Atusparia.* Lima : Ediciones de Frente, 1932.

Riofrío, Miguel. *La emancipada.* Guayaquil: Editorial de la Universidad de Guayaquil, 1988.

Ríos, Enrique. "El grito de rebeldía de Hostos *Bayoán*: Perfiles del 'peregrino de la libertad' ante el 98." *Historia y Comunicación Social* 3 (1998): 163–76.

Riva Palacio, Vicente. *Calvario y Tabor*. Mexico City: Editorial Porrúa, 1985.

Rivera, José Eustasio. *La vorágine*. Buenos Aires: Losada, 1952.

Rivera-Barnes, Beatriz. "Yuyos Are Not Weeds: An Ecocritical Approach to Horacio Quiroga." *ISLE: Interdisciplinary Studies in Literature and Environment* 16, no. 1 (Winter 2009): 35–52.

Rivera-Barnes, Beatriz, and Jerry Hoeg, eds. *Reading and Writing the Latin American Landscape*. New York: Palgrave, 2009.

Roa-Bastos, Augusto. *Hijo de hombre*. Madrid: Espasa Calpe, 1993.

Robleto, Hernán. *Una mujer en la selva*. Santiago de Chile: Editorial Ercilla, 1936.

Rochelle, Jeremy. "Writing under the Shadow of the Chullachaqui: Amazonian Thought and Ecological Discourse in Recent Amazonian Poetry." *Review: Literature and Arts of the Americas* 45, no. 2 (2012): 198–206.

Rodó, José Enrique. *Ariel*. Buenos Aires: Kapelusz, 1966.

Rodríguez, Alberto. *Donde haya Dios*. Buenos Aires: Lautaro Bolsillo, 1958.

Rodríguez, Ana Patricia. *Dividing the Isthmus: Central American Transnational Histories, Literatures, & Cultures.* Austin: University of Texas Press, 2009.

Rodríguez, Fermín Adrián. "Nacionalidad forzada. Leopoldo Lugones, entre la tradición del espíritu y la herencia del monstruo." *Revista Iberoamericana* 75, no. 227 (April–June 2009): 375–97.

Rojas Pérez, Walter. *Costa Rica violada. El caso de "Mamita Yunai."* San Jose, Costa Rica: Editorial Porvenir, 2006.

Rosa, Rodrigo Rey. *Lo que soñó Sebastián*. Barcelona: Seix Barral, 1994.

Rossi, Anacristina. "Author's note." *ISLE* 15, no. 1 (Winter 2008): 195–215.

———. *La loca de Gandoca*. San Jose, Costa Rica: Educa, 1996.

Rotker, Susana. *Captive Women: Oblivion and Memory in Argentina*, translated by Jennifer French. Minneapolis: University of Minnesota Press, 2003.

———. *Ensayistas de nuestra América*. 2 vols. Buenos Aires: Losada, 1994.

Roza, Jesús S. *El último rei de los muiscas*. Bogotá: Echeverría Hermanos, 1864.

Rozelle, Lee. *Ecosublime: Environmental Awe and Terror from New World to Oddworld.* Tuscaloosa: University of Alabama Press, 2006.

Ruiz, Bladimir. "Incorporaciones y exclusiones al proyecto de nación: El negro y el indio en la narrativa decimonónica venezolana." *Revista Iberoamericana* 49, no. 205 (2003): 883–94.

Ruskin, John. "Landscape, Mimesis and Morality." In *The Green Studies Reader: From Romanticism to Ecocriticism*, edited by Laurence Coupe, 26–31. New York: Routledge, 2000.

Sacoto, Antonio. Introduction to *La emancipada*, by Miguel Riofrío, 7–19. Guayaquil: Editorial de la Universidad de Guayaquil, 1988.

Sarduy, Severo. *Colibrí*. Barcelona: Argos Vergara, 1984.

Sarmiento, Domingo Faustino. *Facundo*. Madrid: Ediciones escolares, 2001.

Sauer, Peter. "Reinhabiting Environmentalism: Picking Up where Carson and Leopold Left Off." *Orion* (Summer 1999): 30–41.

Schlickers, Sabine. "La novelística naturalista-criollista uruguaya: *Beba* de Carlos Reyles y *Gaucha, Campo* y *Guri* de Javier de Viana." *Arrabal* 4 (2002): 177–90.

Schwartz, Kessel. *A New History of Spanish American Fiction*. 2 vols. Coral Gables, FL: University of Miami Press, 1971.

Senkman, Leonardo. "Jewish Latin American Writers and Collective Memory." In *Tradition and Innovation: Reflections on Latin American Jewish Writers*, edited by Robert E. DiAntonio and Nora Glickman, 33–43. Albany: State University of New York Press, 1993.

Sepúlveda, Luis. *Historia de una gaviota y del gato que le enseñó a volar*. Barcelona: Tusquets, 1996.

———. *Mundo del fin del mundo*. Barcelona: Tusquets, 1994.

———. *Patagonia Express*. Barcelona: Tusquets, 1995.

———. *Un viejo que leía novelas de amor*. Barcelona: Tusquets, 1993.

Shaw, Walter. "El neoindigenismo y *Hombres de maíz*." In *Cien años de magia: Ensayos críticos sobre la obra de Miguel Ángel* Asturias, edited by Oralia Preblei-Niemi, 97–119. Guatemala City: F & G Editores, 2006.

Silva, José Asunción. *Obras completas*. Caracas: Ediciones Ayacucho, 1977.

———. *Los mejores poemas de José Asunción Silva*. Mexico City: Cultura, 1917.

Sinardet, Emanuelle. "*A la costa* de Luis A. Martínez: ¿La defensa de un proyecto liberal para Ecuador?" *Bulletin de Institut francais des etudes andines* 27, no. 2 (1998): 285–307.

Sklodowska, Elzbieta. *La parodia en la nueva novela hispanoamericana (1960–1985)*. Philadelphia: John Benjamins, 1991.

Skurski, Julie. "The Ambiguities of Authenticity in Latin America: *Doña Bárbara* and the Construction of National Identity." *Poetics Today* 15, no. 4 (Winter 1994): 605–42.

Slater, Candace. "Amazonia as Edenic Narrative." In *Uncommon Ground: Rethinking the Human Place in Nature*, edited by William Cronon, 114–31. New York: Norton, 1995.

Slovic, Scott. "Ecocriticism: Containing Multitudes, Practising Doctrine." In *The Green Studies Reader: From Romanticism to Ecocriticism*, edited by Laurence Coupe, 160–62. New York: Routledge, 2000.

———. *Going Away to Think*. Reno: University of Nevada Press, 2008.

Snider, Dixie E., et al. "Tuberculosis and Migrant Farm Workers." *JAMA* 265, no. 13 (April 3, 1991): 1732.

Sommer, Doris. *Foundational Fictions: The National Romances of Latin America*. Berkeley: University of California Press, 1991.

Soto Borda, Clímaco. *Diana cazadora*. Bogotá: Editorial ABC, 1942.

Spektorowski, Alberto. "The making of an Argentine fascist. Leopoldo Lugones: from revolutionary left to radical nationalism." *History of Political Thought* 17, no. 1 (1996): 79–108.

Stoll, David. *Rigoberta Menchú and the Story of All Poor Guatemalans*. Phildelphia: Westview, 2008.

Tallmadge, John. "Toward a Natural History of Reading." In *The ISLE Reader: Ecocriticism, 1993–2003*, edited by Michael P. Branch and Scott Slovic, 282–95. Athens: University of Georgia Press, 2003.

Teitelbaum, Volodia. *Hijo del salitre*. Santiago de Chile: Editorial Austral, 1952.

Thayer, Robert L., Jr. *Gray World, Green Heart: Technology, Nature, and the Sustainable Landscape*. New York: John Wiley & Sons, 1994.

Tinker Salas, Miguel. *The Enduring Legacy: Oil, Culture, and Society in Venezuela*. Durham, NC: Duke University Press, 2009.

Tittler, Jonathan. "Una relectura del canon *criollista*." *Tabula Rasa* 7 (July–December 2007): 197–210.

Thoreau, Henry David. *Walden*. Boston: Houghton, Mifflin and Company, 1897.

Toro, Fermín. "La sibila de los Andes." In *Biblioteca de escritores venezolanas*, edited by José María Rojas, 454–59. Caracas: Rojas Hermanos Editores, 1875.

Toyosato, Mayumi. "Grounding Self and Action: Land, Community, and Survival in *I, Rigoberta Menchú, No Telephone to Heaven*, and *So Far from God*." *Hispanic Journal* 19, no. 2 (Fall 1998): 295–311.

Uliánova, Olga. "Levantamiento campesino de Lonquimay y la Internacional Comunista." *Estudios Públicos* 89 (Summer 2003): 173–223.

Urbina, Luis G. *Puestas de sol*. Paris: Librería de la Vda de Ch. Bouret, 1910.

Uribe Piedrahita, César. *Mancha de aceite*. Bogotá: Oveja Negra, 1985.

Vargas Llosa, Mario. *El hablador*. Barcelona: Seix Barral, 1987.

Vitale, Luis. *Hacia una historia del ambiente en América Latina*. Mexico City: Editorial Nueva Imagen, 1983.
Warren, Karen J. *Ecofeminist Philosophy: A Western Perspective on What it is and Why it Matters*. Lanham, MD: Rowman & Littlefield Publishers, 2000.
Weart, Spencer R. *The Discovery of Global Warming*. Cambridge, MA: Harvard University Press, 2003.
Wilde, Eduardo. *Aguas abajo*. Buenos Aires: Fondo nacional de las artes, 1999.
Williams, Raymond. "The Green Language." In *The Green Studies Reader: From Romanticism to Ecocriticism*, edited by Laurence Coupe, 50–58. New York: Routledge, 2000.
———. *Keywords: A Vocabulary of Culture and Society*. New York: Oxford University Press, 1983.
———. *Marxism and Literature*. New York: Oxford University Press, 1977.
Williams, Raymond L. *The Twentieth Century Spanish American Novel*. Austin: University of Texas Press, 2003.
Worster, Donald. "Nature and the Disorder of History." In *Reinventing Nature: Responses to Postmodern Deconstruction*, edited by Michael E. Soulé and Gary Lease, 65–85. Washington, DC: Island, 1995.
Wyld Ospina, Carlos. *Los lares apagados*. Guatemala: Edit. Universitaria, 1958.
Wylie, Lesley. "Frontier Fictions: The Place of Amazonia in César Uribe Piedrahita's *Toá*." *Bulletin of Spanish Studies* 87, no. 7 (2010): 959–74.
Yankas, Lautaro. *El vado de la noche: Novela de la raza*. Santiago de Chile: Zig-Zag, 1963.
Yepes, José Ramón. *Novelas y estudios literarios de José Ramón Yepes*. Maracaibo: Imprenta Americana, 1882.
Zaniboni Filho, Evoy, and Uwe H. Schulz. "Migratory Fishes of the Uruguay River." In *Migratory Fishes of South America: Biology, Fisheries, and Conservation Status*, edited by Joachim Carolsfield, Brian Harvey, Carmen Ross, and Anton Baer, 157–94. Ottawa: World Bank and International Development Research Center, 2003.
Zeno Gandía, Manuel. *La charca*. Río Piedras, Puerto Rico: Ediciones Huracán, 1999.
Zevallos Aguilar, Ulises Juan. *Las provincias contraatacan: Regionalismo y anticentralismo en la literatura peruana del siglo XX*. Lima: Vicerrectorado Académico UNMSM, 2009.
Zorilla de San Martín, Juan. *Tabaré*. Montevideo: Biblioteca de autores uruguayos, 1888.

Index

About the Author

Scott DeVries is associate professor of Spanish at Bethel College, Indiana. He is the author of a variety of articles on Spanish American literature and film whose work has appeared in *Hispania, Ecozon@: European Journal of Literature, Culture, and Environment*, and the *Christian Scholar's Review*.